I0072581

Framework of Digital Signal Processing

Framework of
Digital Signal Processing

Edited by **George Pilato**

NY RESEARCH
P R E S S

New York

Published by NY Research Press,
23 West, 55th Street, Suite 816,
New York, NY 10019, USA
www.nyresearchpress.com

Framework of Digital Signal Processing
Edited by George Pilato

© 2015 NY Research Press

International Standard Book Number: 978-1-63238-205-4 (Hardback)

This book contains information obtained from authentic and highly regarded sources. Copyright for all individual chapters remain with the respective authors as indicated. A wide variety of references are listed. Permission and sources are indicated; for detailed attributions, please refer to the permissions page. Reasonable efforts have been made to publish reliable data and information, but the authors, editors and publisher cannot assume any responsibility for the validity of all materials or the consequences of their use.

The publisher's policy is to use permanent paper from mills that operate a sustainable forestry policy. Furthermore, the publisher ensures that the text paper and cover boards used have met acceptable environmental accreditation standards.

Trademark Notice: Registered trademark of products or corporate names are used only for explanation and identification without intent to infringe.

Printed in the United States of America.

Contents

Preface

This book aims to highlight the current researches and provides a platform to further the scope of innovations in this area. This book is a product of the combined efforts of many researchers and scientists, after going through thorough studies and analysis from different parts of the world. The objective of this book is to provide the readers with the latest information of the field.

The framework of digital signal processing is explained in this book with the help of descriptive information. Digital signal processing (DSP) comprises of a broad variety of functions in which the execution of high-performance systems to meet severe requirements and performance constraints is fast becoming a topic of interest, both in the practical and educational contexts. Assumed to be accessible to a significant number of people, the aim of this book is to provide students, researchers, engineers and the industrial society with a reference to the most recent developments in rising issues in the design and execution of DSP systems for purpose-oriented circuits and programmable tools. The book deals with various topics including real-time audio functions, optical signal processing, image and video processing and superior architectures and executions. It will allow students and experts to deal with the significant gap in information in the changeover from algorithm requirement to the design of architectures for VLSI achievements.

I would like to express my sincere thanks to the authors for their dedicated efforts in the completion of this book. I acknowledge the efforts of the publisher for providing constant support. Lastly, I would like to thank my family for their support in all academic endeavors.

Editor

Real-Time Audio Applications

Low Computational Robust F_0 Estimation of Speech Based on TV-CAR Analysis

Keiichi Funaki and Takehito Higa

Additional information is available at the end of the chapter

1. Introduction

The F_0 estimation determines a performance of speech processing such as speech coding, tonal speech recognition, speaker recognition, and speech enhancement. F_0 estimation named "YIN" has been proposed [1] and it is being prevalently used around the world due to its high performance and open-source policy. Speech processing is commonly applied in realistic noisy environments; hence, the performance is degraded seriously. It is well known that YIN does not perform well for noisy speech although it does perform best for clean speech. Accordingly, more robust F_0 estimation algorithm is desired and the robust F_0 estimation is long lasting problem in speech processing. We have already proposed robust F_0 estimation algorithm based on time-varying complex speech analysis for analytic speech signal [2][3]. Analytic signal is a complex-valued signal in which its real part is speech signal and its imaginary part is Hilbert transform of the real part. Since the analytic signal provides the spectrum only on positive frequencies, the signals can be decimated by a factor of two with no degradation. As a result, the complex analysis offers attractive features, for example, more accurate spectral estimation in low frequencies. In [2] and [3], complex LPC residual is used to calculate the criterion of weighted autocorrelation function (AUTOC) with a reciprocal of Average Magnitude Difference Function (AMDF) [6]. The complex residual is calculated from analytic speech signal by means of time-varying complex AR (TV-CAR) speech analysis method [4][5]. In [2], MMSE-based TV-CAR speech analysis [4] is introduced and in [3], ELS-based TV-CAR speech analysis [5] is introduced to calculate complex LPC residual signal. It has been reported in [2] that the method can estimate more accurate F_0 for IRS (Intermediate Reference System) filtered speech corrupted by white Gauss noise. Moreover, it has been reported in [3] that the ELS-based complex speech analysis can perform better even for additive pink noise. Furthermore, in order to investigate the effectiveness of the time-varying analysis, the performance was compared for the frame with respect to degree of voiced nature [7]. The experiments using IRS filtered speech corrupted by white Gauss noise or pink noise demonstrate that ELS-based robust time-varying complex speech analysis can perform better for stationary voiced speech and ELS-based time-invariant speech analysis can perform better for ordinary voiced frame. However the computational cost turns to be larger by introducing time-varying analysis. In this paper, in order to reduce the computational cost, pre-selection is introduced. The pre-selection is performed by peak picking of speech spectrum based on the TV-CAR analysis [8]. The evaluation is carried out using Keele Pitch Database [9]. The reminder of the chapter is organized as follows. In Section 2, TV-CAR speech analysis is explained. Analytic signal and Time-Varying Complex AR (TV-CAR) model are explained. Two kinds of the TV-CAR parameter estimation algorithms from an analytic signal, viz., MMSE and ELS methods are explained. In Section 3, F_0 estimation algorithm is explained in detail. Sample-based pre-selection is explained and frame-based final–selection is explained. In Section 4, experimental results are explained and these confirm the effectiveness of the proposed method.

2. TV-CAR speech analysis

In this section, ELS-based robust TV-CAR speech analysis method is explained. Before the explanation, analytic signal and TV-CAR model is explained, in which analytic signal is output of the TV-CAR model. In 2.6, the benefit of the robust TV-CAR analysis is explained by showing the estimated sprctra from natural speech.

2.1. Analytic speech signal

Target signal of the time-varying complex AR (TV-CAR) method is an analytic signal that is complex-valued signal defined by an all-pole model as follows.

$$y^c(t) = \frac{y(2t) + j \cdot y_H(2t)}{\sqrt{2}} \tag{1}$$

where $y^c(t)$,$y(t)$ and $y_H(t)$ denote an analytic signal at time t, an observed signal at time t, and a Hilbert transformed signal for the observed signal, respectively. Notice that superscript c denotes complex value in this paper. Since analytic signals provide the spectra only over the range of $(0, \pi)$ analytic signals can be decimated by a factor of two. $2t$ means the decimation. The term of $1/\sqrt{2}$ is multiplied in order to adjust the power of an analytic signal with that of the observed one.

2.2. Time-varying complex AR (TV-CAR) model

Conventional LPC model is defined by

$$Y_{LPC}\left(z^{-1}\right)=\frac{1}{1+\displaystyle\sum_{i=1}^{1}a_i z^{-1}} \tag{2}$$

where a_i and I are i-th order LPC coefficient and LPC order, respectively. Since the conventional LPC model cannot express the time-varying spectrum, LPC analysis cannot extract the time-varying spectral features from speech signal. In order to represent the time-varying features, the TV-CAR model employs a complex basis expansion shown as

$$a_i^c\left(t\right)=\sum_{l=0}^{L-1}g_{i,l}^c f_l^c\left(t\right) \tag{3}$$

where $a_i^c(t)$,I,L, $g_{i,l}^c$,l and $f_i^c(t)$ are taken to be i-th complex AR coefficient at time t, AR order, finite order of complex basis expansion, complex parameter, and a complex-valued basis function, respectively. By substituting Eq.(3) into Eq.(2), one can obtain the following transfer function. Eq.(4) means the TV-CAR model.

$$Y_{TVCAR}\left(z^{-1}\right)=\frac{1}{1+\displaystyle\sum_{i=1}^{I}\sum_{l=1}^{L}g_{i,l}^c f_l^c\left(t\right)z^{-i}} \tag{4}$$

The input-output relation is defined as

$$y^c\left(t\right)=-\sum_{i=1}^{I}\sum_{l=0}^{L-1}g_{i,l}^c f_l^c\left(t\right)y^c\left(t-i\right)+u^c\left(t\right) \tag{5}$$

where $u^c(t)$ and $y^c(t)$ are taken to be complex-valued input and analytic speech signal shown in Eq.(1), respectively. In the TV-CAR model, the complex AR coefficient is modeled by a finite number of arbitrary complex basis functions such as Fourier basis, wavelet basis or so on. Note that Eq.(3) parameterizes the AR coefficient trajectories that continuously change as a function of time so that the time-varying analysis is feasible to estimate continuous time-varying speech spectrum. In addition, as mentioned above, the complex-valued analysis facilitates accurate spectral estimation in the low frequencies, as a result, this feature allows for more accurate F_0 estimation if formant structure is removed by the inverse filtering. Eq.(5) can be represented by vector-matrix notation as

$$\bar{d} = i, \, l^{\,f}$$

$$\theta^T = \left[\bar{g}_0^T, \, \bar{g}_1^T, \, \cdots, \, \bar{g}_l^T, \, \cdots, \, \bar{g}_{L-1}^T \right]$$

$$\bar{g}_l^T = \left[g_{1,l}^c, \, g_{2,l}^c, \, \cdots, \, g_{i,l}^c, \, \cdots, \, g_{I,l}^c \right]$$

$$\bar{y}_f^T = \left[y^c(I), \, y^c(I+1), \, y^c(I+2), \, \cdots y^c(N-1) \right]$$

$$\bar{u}_f^T = \left[u^c(I), \, u^c(I+1), \, u^c(I+2), \, \cdots u^c(N-1) \right] \qquad (6)$$

$$\bar{\Phi}_f = \left[\bar{D}_0^f, \, \bar{D}_1^f, \, \cdots \bar{D}_l^f, \, \cdots, \, \bar{D}_{L-1}^f \right]$$

$$\bar{D}_l^f = \left[\bar{d}_{1,l}^f, \, \cdots \bar{d}_{i,l}^f, \, \cdots, \, \bar{d}_{I,l}^f \right]$$

$$\bar{d}_{i,l}^f = \begin{bmatrix} y^c(I-i)f_i^c(I), \ y^c(I+1-i)f_i^c(I+1) \\ \cdots, \ y^c(N-1-i)f_i^c(N-1)^T \end{bmatrix}$$

where N is analysis interval, $\bar{y}_f$ is $(N - I, 1)$ column vector whose elements are analytic speech signal, $\bar{\theta}$ is $(L \cdot I, 1)$ column vector whose elements are complex parameters, Φ_f is $(N - I, L \cdot I)$ matrix whose elements are weighted analytic speech signal by the complex basis. Superscript T denotes transposition.

2.3. MMSE-based algorithm [4]

There are several algorithms that estimate the TV-CAR model parameter from complex-valued signal such as MMSE, WLS(Weighted Least Square), M-estimation, GLS(Generalized Least Square), and ELS(Extended Least Square). The MMSE-algorithm is basic algorithm and used for initial estimation of the ELS. Before explaining the ELS, the MMSE algorithm is explained.

MSE criterion is defined by

$$\bar{r}_f = \left[r^c(I), r^c(I+1), \cdots, r^c(N-1) \right]^T = \bar{y}_f + \bar{\Phi}_f \hat{\theta} \qquad (7)$$

$$r^c(t) = y^c(t) + \sum_{I=1}^{I} \sum_{l=0}^{L-1} \hat{g}_{i,l}^c f_l^c(t) y^c(t-i) \qquad (8)$$

$$E = \bar{r}_f^H \bar{r}_f = \left(\bar{y}_f + \bar{\Phi}\hat{\theta} \right)^H \left(\bar{y}_f + \bar{\Phi}\hat{\theta} \right) \qquad (9)$$

Where $\hat{g}_{i,l}^c$ is the estimated complex parameter, $r^c(t)$ is an equation error, or complex AR residual and E is Mean Squared Error (MSE) for the equation error. To obtain optimal complex

AR coefficients, we minimize the MSE criterion. Minimizing the MSE criterion of Eq.(9) with respect to the complex parameter leads to the following MMSE algorithm.

$$\left(\overline{\Phi}_f^H \overline{\Phi}_f\right)\hat{\theta} = -\overline{\Phi}_f^H \overline{y}_f \tag{10}$$

Superscript H denotes Hermitian transposition. After solving the linear equation of Eq.(10), we can get the complex AR parameter ($a_i^c(t)$) at time t by calculating the Eq.(3) with the estimated complex parameter $\hat{g}_{i,l}^c$.

2.4. ELS-based algorithm [5]

Figure 1 shows block diagram of ELS estimation. If the equation error shown as in Eq.(8) is white Gaussian, the MMSE estimation is optimal, however, it is rare case. As a result, MMSE estimation suffers from biased estimation. In the ELS method, an AR filter is adopted to whiten the equation error as follows (Figure 1(2)).

$$r^c(t) = -\sum_{k=1}^{K} b_k^c r^c(t-k) + e^c(t) \tag{11}$$

where b_k^c is k-th parameter of the AR filter whose order is K and $e^c(t)$ is 0-mean white Gaussian of equation error at time t. The inverse filter of Eq.(11) is called a whiten filter. The TV-CAR model can be represented using Eq.(5) and Eq.(11) as follows.

$$y^c(t) = -\sum_{i=1}^{I}\sum_{l=0}^{L-1} g_{i,l}^c f_l^c(t) y^c(t-i) - \sum_{k=1}^{K} b_k^c r^c(t-k) + e^c(t) \tag{12}$$

Eq.(12) is the ELS model shown as in Figure 1(3). The parameter is estimated so as minimize the MSE for the whitened equation error in the ELS algorithm whereas the parameter is estimated so as minimize the MSE for the equation error in the MMSE algorithm shown as in Figure 1(1).

Eq.(12) can be expressed by the following vector-matrix notation.

$$\overline{y}_f = -\overline{\Phi}_f \overline{\theta} - \overline{R}_f \overline{b} + \overline{e}_f = -\left(\overline{\Phi}_f \overline{R}_f\right)\binom{\overline{\theta}}{\overline{b}} + \overline{e}_f \tag{13}$$

Where

$$\overline{R}_f = \begin{pmatrix} r^c(I-1) & r^c(I-2) & \cdots & r^c(I-K) \\ r^c(I) & r^c(I-1) & \cdots & r^c(I+1-K) \\ \vdots & \vdots & \ddots & \vdots \\ r^c(t) & r^c(t-1) & \cdots & r^c(t-K) \\ \vdots & \vdots & \ddots & \vdots \\ r^c(N-2) & r^c(N-3) & \cdots & r^c(N-1-K) \end{pmatrix} \tag{14}$$

$$\overline{b} = \left[b_1^c, b_2^c, \cdots, b_K^c \right]^T$$

$$\overline{e}_f = \left[e^c(I), e^c(I+1), e^c(I+2), \cdots, e^c(N-1) \right]^T$$

By minimizing the MSE for Eq.(13), one can get the following equation.

$$\begin{pmatrix} \overline{\Phi}_f^H \overline{\Phi}_f & \overline{\Phi}_f^H \overline{R}_f \\ \overline{R}_f^H \overline{\Phi}_f & \overline{R}_f^H \overline{R}_f \end{pmatrix} \begin{pmatrix} \hat{\theta} \\ \hat{b} \end{pmatrix} = - \begin{pmatrix} \overline{\Phi}_f^H \overline{y}_f \\ \overline{R}_f^H \overline{y}_f \end{pmatrix} \tag{15}$$

By applying the well-known inversion Matrix lemma to Eq.(15), one can obtain the following equation.

$$\left(\overline{\Phi}_f^H \overline{\Phi}_f \right) \hat{\theta}_{bias} = \overline{\Phi}_f^H \overline{R}_f \hat{b} \tag{16}$$

$$\hat{\theta} = \hat{\theta}_0 - \hat{\theta}_{bias} \tag{17}$$

The MMSE estimated parameter $\hat{\theta}_0$ contains the biased element $\hat{\theta}_{bias}$. The unbiased estimation of $\hat{\theta}$ is calculated by $\hat{\theta}_0 - \hat{\theta}_{\,bias}$. The ELS algorithm is equivalent to the GLS (Generalized Least Square) algorithm and more sophisticated algorithm. Since the equation error $r^c(t)$ cannot be observed, the iteration algorithm is required by estimating the A(z) and B(z). The iteration procedure is shown as follows.

1. Initial $\hat{\theta}_0$ is estimated by MMSE (Eq.(10)).

2. The equation error is calculated by Eq.(8).
3. $\hat{b}$ is estimated so as to minimize Eq.(18) using $r^c(t)$.

4. The bias parameter $\hat{b}$ is calculated by Eq.(16).

5. The unbiased parameter $\hat{\theta}$ is calculated by Eq.(17).

6. Go to 2.

$$\frac{1}{2\pi j}\oint_{|z|=1}\left|R\left(z\right)B\left(z\right)\right|^2\frac{dz}{z}=0 \tag{18}$$

In Eq.(18), R(z) is z-transform of $r^c(t)$ and B(z) is the transfer function of the whiten filter. The procedures from 2 to 5 are iterated with the pre-determined number. The ELS algorithm estimates two kinds of AR filters, A(z) and B(z), iteratively. Since the ELS algorithm can estimate unbiased and less effected speech spectrum against additive noise, more accurate F_0 and formants frequencies can be estimated. Thus, more accurate F_0 trajectories can be estimated than the MMSE estimation.

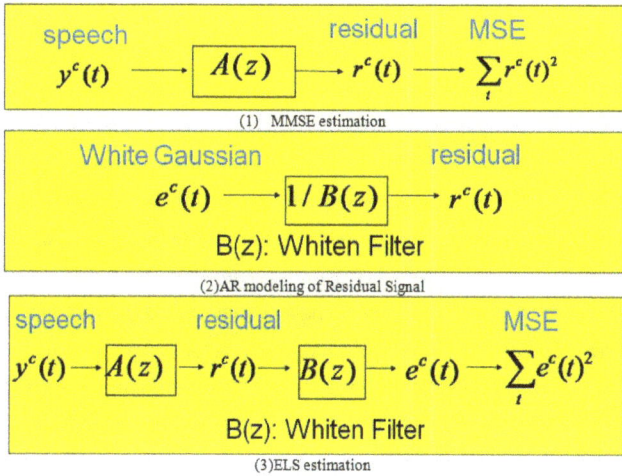

Figure 1. Block diagrams of MMSE and ELS estimation.

2.5. Benefit of robust TV-CAR speech analysis

In this paragraph, we explain the benefit of robust TV-CAR speech analysis by showing the estimated speech spectrum and explain its effectiveness on F_0 estimation of speech. Figure 2 shows example of the estimated speech spectra of natural Japanese vowel /o/ for analytic signal and conventional LPC analysis for speech signal.

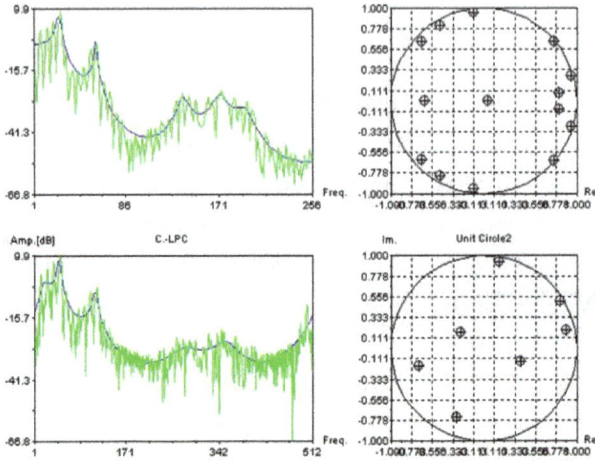

Figure 2. Estimated Spectra of vowel /o/ with complex and conventional LPC analysis.

In Figure 2, left side denote the estimated spectra. Upper is for real-valued LPC analysis. Lower is for complex-valued LPC analysis. Blue line means estimated spectrum by LPC analysis and green line means estimated DFT spectrum. Right side means estimated poles from the estimated AR filter. Figure 3 shows the estimated running spectrum for clean natural speech /arayu/ and for the speech corrupted by white Gaussian (10[dB]). In Figure 3, (1) means speech waveform, (2),(3),(4),(5) and (6) mean the estimated spectrum by MMSE-based time-invariant real-valued AR speech analysis, by MMSE-based time-invariant complex-valued AR speech analysis (L=1), by MMSE-based time-varying complex AR (TV-CAR) speech analysis (L=2), by ELS-based time-invariant complex-valued AR speech analysis (L=1), and by ELS-based time-varying complex AR (TV-CAR) speech analysis (L=2), respectively. Analysis order I is 14 for real analysis and 7 for complex analysis. Basis function is 1st order polynomial function (1,t). One can observe that the complex analysis can estimate more accurate spectrum in low frequencies whereas the estimation accuracy is down in high frequencies. Since speech spectrum provides much energy in low frequencies, it is expected that the high spectral estimation accuracy in low frequencies makes it possible to improve the performance on F_0 estimation. Furthermore, the ELS analysis can estimate more accurate spectrum than MMSE, so that the ELS analysis makes it possible to estimate more accurate F_0. Time-varying analysis can estimate tive-varying spectrum from speech. It is expected that the time-varying analysis enables to estimate more accurate F_0 since F_0 is varying in the analysis interval.

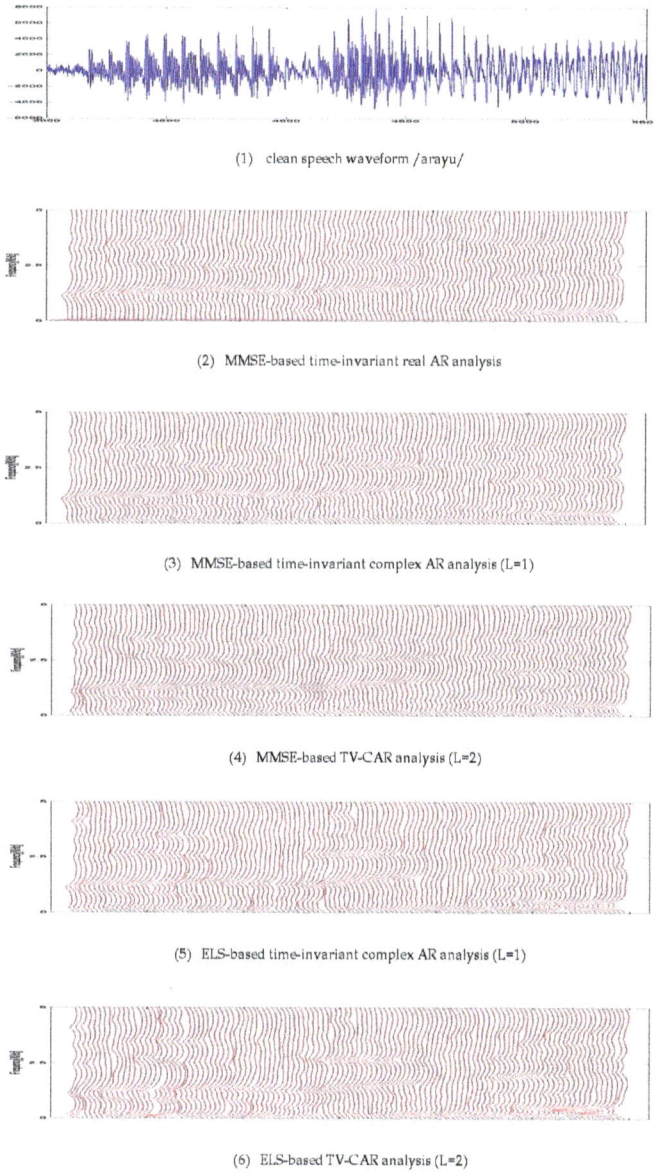

(1) clean speech waveform /arayu/

(2) MMSE-based time-invariant real AR analysis

(3) MMSE-based time-invariant complex AR analysis (L=1)

(4) MMSE-based TV-CAR analysis (L=2)

(5) ELS-based time-invariant complex AR analysis (L=1)

(6) ELS-based TV-CAR analysis (L=2)

Figure 3. Estimated spectrum for noise corrupted speech /arayu/ (10[dB]).

3. F_0 Estimation method

Proposed method employs two-stage search of F_0. In first stage, pre-selection, F_0 and F_1 are estimated by using sample-based F_0 contour estimation [8]. In second stage, final-selection, F_0 is estimated by using frame-based F_0 estimation [3] within limited range based on the pre-estimated F_0 and F_1. The two-stage estimation makes it possible to reduce the computation with less degradation In 3.1, pre-selection algorithm is explained. In 3.2, final-selection algorithm is explained.

3.1. Sample-based pre-selection

F_0 and F_1 are estimated as the lowest two peak frequency, viz., glottal and first formant frequencies by peak-picking for the estimated time-varying speech spectrum. The procedure of F_0 and F_1 contour estimation is shown as in Figure 4

1. The set of complex-valued parameter $\hat{g}_{i,1}^c$ is estimated by the ELS algorithm for each analysis frame.

2. By using Eq.(3) and Eq.(4) with the estimated parameter $\hat{g}_{i,1}^c$, the speech power spectrum for each sample t is calculated, and the two peaks of the estimated spectrum are searched by the peak-picking.

The peak-pinking is carried out from low frequency to high frequency shown as in Figure 5. The estimated two peaks correspond to glottal formant (F_0) and first formant (F_1). The formant frequencies are estimated by solving the equation of the reciprocal of Eq.(4).

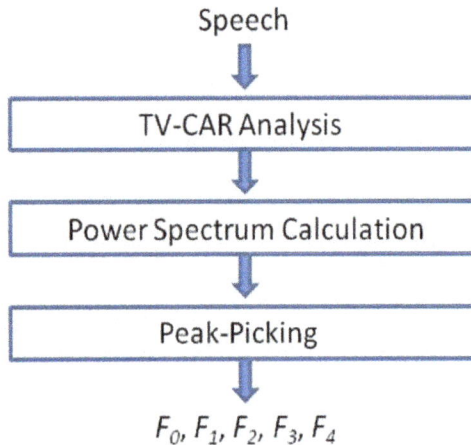

Speech

⬇

TV-CAR Analysis

⬇

Power Spectrum Calculation

⬇

Peak-Picking

⬇

F_0, F_1, F_2, F_3, F_4

Figure 4. Flow of F_0 and F_1 contour estimation

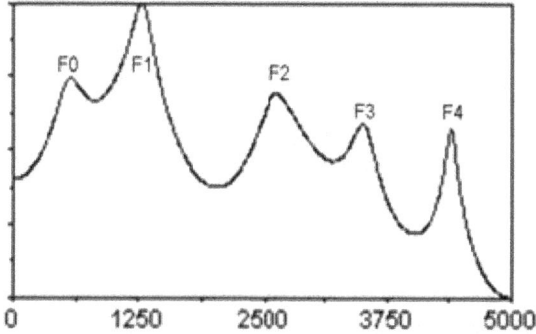

Figure 5. Peak Picking

3.2. Frame-based final-selection

In frame-based F_0 estimation, autocorrelation or AMDF is commonly used. In this paragraph, autocorrelation and AMDF are explained and then adopted weighted autocorrelation is explained.

Autocorrelation function (AUTOC) is defined by

$$f(\tau) = \frac{1}{N} \sum_{t=0}^{N-1} x(t) x(t+\tau) \tag{19}$$

where x(t) is target signal such as speech signal, LPC residual or so on, N is frame length and τ means delay. F_0 is selected as peak frequency for Eq.(19) within certain range of F_0.

AMDF is defined as follows.

$$p(\tau) = \frac{1}{N} \sum_{t=0}^{N-1} |x(t) - x(t+\tau)| \tag{20}$$

F_0 is selected as notch frequency for Eq.(20) within certain range of F_0. In Shimamura method [6], the AUTOC is weighted by a reciprocal of the AMDF shown as Eq.(21). Since the weighting makes it possible to suppress other peaks, the method can estimate more accurate F_0 than AUTOC or AMDF. The value of m is set to be 1 in order to avoid the value of 0 at the denominator.

$$G(\tau) = \frac{f(\tau)}{p(\tau) + m} \tag{21}$$

where $f(\tau)$ and $p(\tau)$ are AUTOC shown as in Eq.(19) and AMDF shown as in Eq.(20), respectively. In the frame-based method, Shimamura criterion shown as Eq.(21) is applied to complex AR residual extracted by the ELS-based TV-CAR speech analysis. The time-varying complex parameter is estimated and complex AR residual is calculated with the estimated complex parameter with Eq.(17). Note that pre-emphasis is operated for speech analysis such as real-valued AR or TV-CAR speech analysis, and inverse filtering is applied for the non pre-emphasized speech signal so as not to eliminate F_0 spectrum on the residual signal. Real part of AUTOC is used to calculate the AUTOC for complex-valued signal. F_0 is estimated within the range corresponding to 50-400[Hz]. In order to reduce the computational amount, the range is shortened by setting the upper value as follows.

$$\min\left(F_0^S + \left(F_1^S - F_0^S\right)\delta / 100, 400\right) \tag{22}$$

where F_0^S and F_1^S are estimated F_0 and F_1 by the sample-based pre-selection. Setting upper bound below F_1 can not only reduce the computational cost but also can reduce the estimation error.

4. Experiments

Speech signals used in the experiment are 5 long sentences uttered by 5 male speaker and 5 long sentences uttered by 5 female speaker of Keele pitch database [9]. Speech signals are filtered by an IRS filter [10]. The IRS filter is band pass FIR filter whose frequency response corresponds to that for analog part of the transmitter of telephone equipment. The frequency response is shown in Figure 6. In order to evaluate the proposed method for the speech data processed by speech coding, the IRS filter has to be introduced shown as in [2]. The experimental conditions are summarized in Table 1. Frame length is 25.6[msec] and frame shift length is 10[msec]. Analysis orders are 14 and 7 for real-valued analysis and complex-valued analysis, respectively. The basis expansion order L is set to be 1(time-invariant) or 2(time-varying) in the experiments. First order polynomial function is adopted as a basis function. White Gauss noise or pink noise [11] is adopted for additive noise and the levels are 30, 20, 10, 5, 0, and -5 [dB]. In order to extract more accurate F_0, 3-point Lagrange's interpolation is adopted. Commonly used criterion for F_0 estimation, Gross Pitch Error (GPE), is adopted for objective evaluation. F_0 estimation error is defined as

$$e_p(n) = F_e(n) - F_t(n) \qquad (23)$$

where $F_t(n)$ is true F_0 value and $F_e(n)$ is the estimated one. The true values are derived by pitch file in Keele database. In Eq.(14), if $|e_{p(n)}| \geq F_t(n) \times THR / 100$ then the estimation error is regarded as ERROR and GPE is the probability of the error frames. Otherwise, the estimation is regarded as SUCCESS and FPE is standard deviation of the error. Figures 7,8,9 and 10 show the experimental results setting the THR as 10[%]. Figure 7 and 9 means the results for male speech. Figure 8 and 10 means the results for female speech. In Figures, (1) shows the results of GPEs or FPEs for additive white Gauss noise. (2) shows the results of GPEs or FPEs for additive pink noise. PROPOSED means the GPEs or FPEs for the proposed method with δ being 25. SP means the Shimamura method [6], viz., Shimamura criterion for speech signal. Other lines mean the GPEs or FPEs for the analysis method shown in Table 2. In all figures, X-axis means noise level of 30, 20, 10, 5, 0,−5[dB]. Y-axis means GPE[%] or FPE[Hz].

Figures 7 and 8 demonstrate that the proposed method can perform slightly better than the full-search method(TVC_E) for male speech while it can perform equivalently to the full search method(TVC_E) for female speech. Figures 9 and 10 show that the proposed one does not perform well in terms of FPE although the Shimamura method performs better in terms of FPE.

Speech data	Keele Pitch database [9]
	Male 5 long sentences
	Female 5 long sentences
IRS filter	64-th FIR [10]
Target signal	complex AR residual
	Sampling 10kHz/16bit
Analysis window	Window Length: 25.6[ms]
	Shift Length: 10.0[ms]
F_0 **search range**	50 to Eq.Eq.(22)
Complex-valued AR	I=7, L=2 (time-varying)
	Pre-emphasis $1 - z{-1}$
Criterion	AUTOC/AMDF [6]
Noise	(1)white Gauss noise
	(2)pink noise [11]
	Noise Level 30,20,10,5,0,-5[dB]
Interpolation	3 point Lagrange's

Table 1. Experimental Conditions

	Line	Real or Complex	Non or TV	MMSE or ELS
LPC	Red Dotted	Real	Non	MMSE
TVR	Blue Dotted	Real	TV	MMSE
LPC_E	Magenta Dotted	Real	Non	ELS
TVR_E	Green Dotted	Real	TV	ELS
CLPC	Red Solid	Complex	Non	MMSE
TVC	Blue Solid	Complex	TV	MMSE
CLPC_E	Magenta Solid	Complex	Non	ELS
TVC_E	Green Solid	Complex	TV	ELS

Table 2. Analysis methods

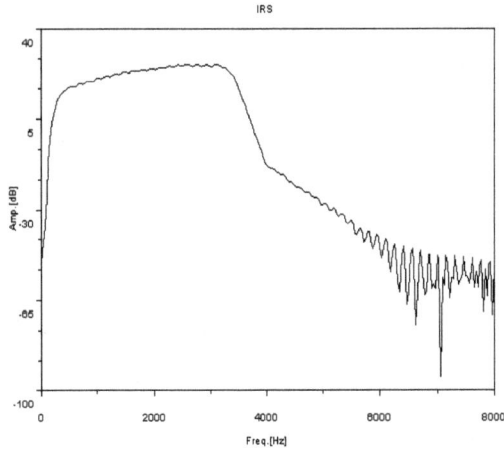

Figure 6. Frequency response of IRS filter

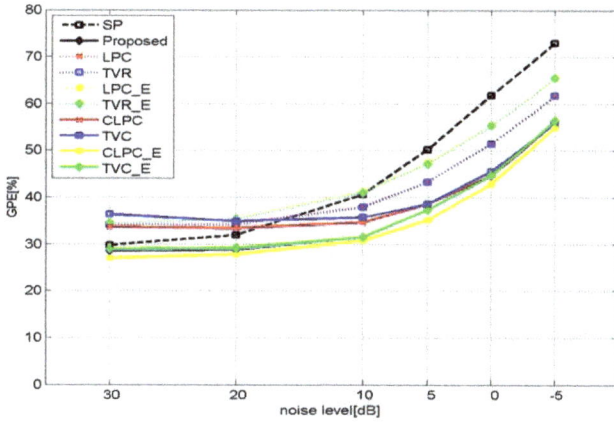

(1) GPEs for additive white Gauss noise

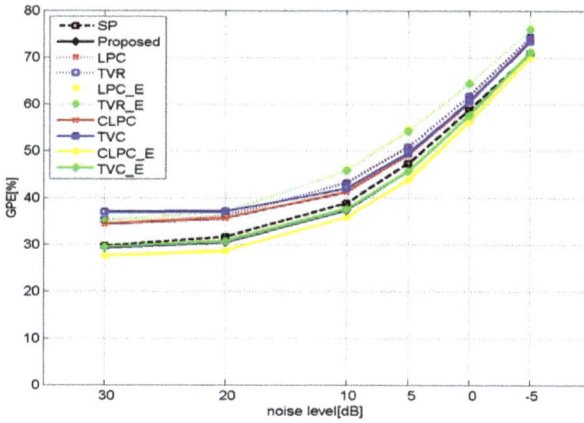

(2)GPEs for additive pink noise

Figure 7. Experimental Results for Male speech

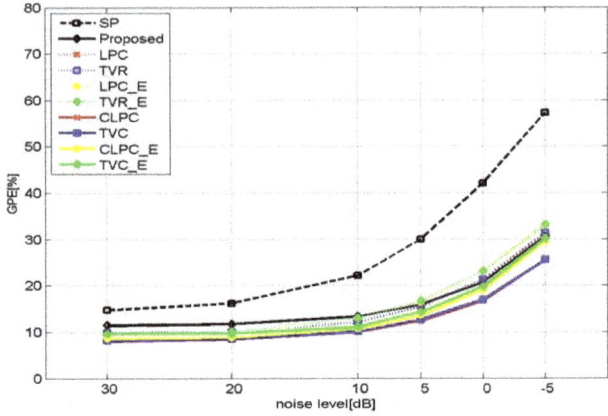

(1) GPEs for additive white Gauss noise

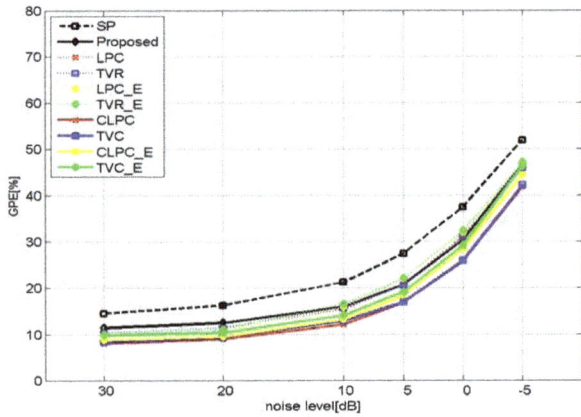

(2)GPEs for additive pink noise

Figure 8. Experimental Results for Female speech

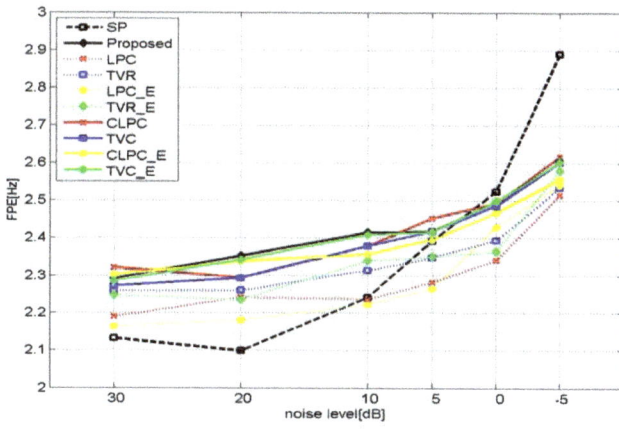

(1) FPEs for additive white Gauss noise

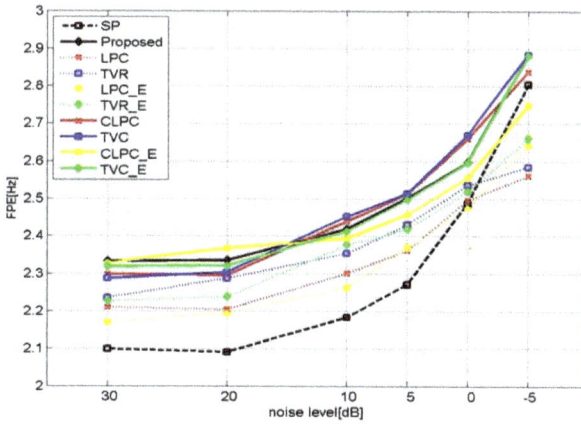

(2) FPEs for additive pink noise

Figure 9. Experimental Results for Male speech

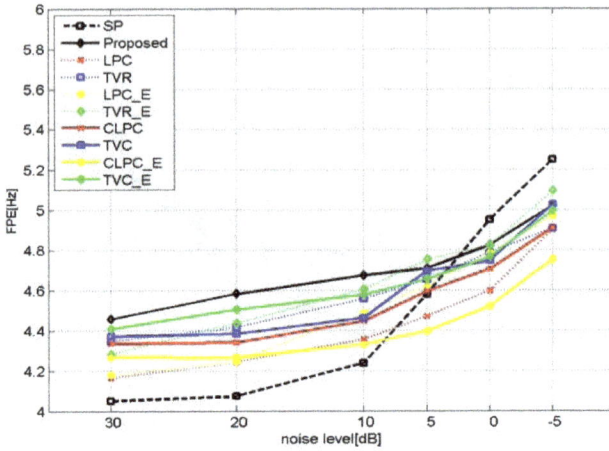

(1) FPEs for additive white Gauss noise

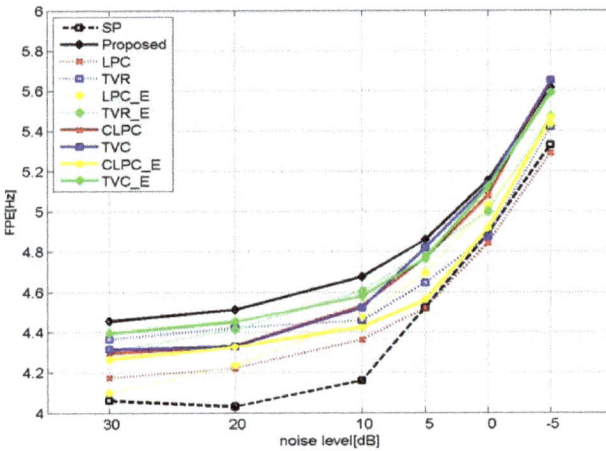

(2)FPEs for additive pink noise

Figure 10. Experimental Results for Female speech

5. Conclusions

This paper proposed fast robust fundamental frequency estimation algorithm based on robust TV-CAR speech analysis. The method provides two stage of search procedure, pre-selection and final-selection. In the pre-selection, F_0 and F_1 are estimated by using time-varying F_0 contour estimation. In the final-selection, F_0 is estimated for only the shorten range based on the pre-selected F_0 and F_1. The proposed method can perform better for male speech in terms of GPE with reduced computation.

Acknowledgements

This work was supported by Grand-in-Aid for Scientific Research (C), Research Project Number:20500158.

Author details

Keiichi Funaki[1*] and Takehito Higa[2]

*Address all correspondence to: funaki@cc.u-ryukyu.ac.jp

1 Computing & Networking Center, University of the Ryukyus, Okinawa, Japan

2 Graduate School of Engineering and Science, University of the Ryukyus, Okinawa, Japan

References

[1] Alan de Cheveigne and H.Kawahra, YIN (2002). A fundamental frequency estimator for speech and music. Journal of the Acoustical Society of America , 111(4), 1917-1930.

[2] K., Funaki, et al. (2007). Robust F0 Estimation Based on Complex LPC Analysis for IRS Filtered Noisy Speech. IEICE Trans. on Fundamentals Aug , E90-A(8)

[3] K., Funaki. (2008). F_0 estimation based on robust ELS complex speech analysis. Proc. EUSIPCO-2008, Lausanne, Switzerland Aug

[4] K., Funaki, Y., Miyanaga, & K., Tochinai. (1998). On a time-varying complex speech analysis. Proc. EUSIPCO-98,Rodes,Greece Sep

[5] K., Funaki. (2001). A time-varying complex AR speech analysis based on GLS and ELS method. Proc.EUROSPEECH2001, Aalborg Denmark Sep

[6] T., Shimamura, & H., Kobayashi. (2001). Weighted Autocorrelation for Pitch Extraction of Noisy Speech. IEEE Trans. Speech and Audio Processing , 9(7), 727-730.

[7] K., Funaki. (2010). On Evaluation of the F_0 Estimation Based on Time-Varying Complex Speech Analysis. Makuhari, Japan Sep

[8] K., Funaki. (2011). F_0 Contour Estimation Using ELS-based Robust Time-Varying Complex Speech Analysis. IEEE DSP/SPE workshop, Sedona, AZ, USA Jan

[9] Keele Pitch Database University of Liverpool http://www.liv.ac.uk/Psychology/hmp/projects/pitch.html

[10] ITU-T Recommendation G.191. (2000). Software tools for speech and audio coding standardization. Nov.

[11] NOISE-X92,. http://spib.rice.edu/spib/selectnoise.html.

Chapter 2

Dynamic Reconfigurable on the Lifting Steps Wavelet Packet Processor with Frame-Based Psychoacoustic Optimized Time-Frequency Tiling for Real-Time Audio Applications

Alexey Petrovsky, Maxim Rodionov and
Alexander Petrovsky

Additional information is available at the end of the chapter

1. Introduction

The discrete wavelet packet transform (DWPT) as a generalization of the standard wavelet transform provides a more flexible choices for time–frequency (time-scale) representation of signals [1] in many applications, such as the design of cost-effective real-time multimedia systems and high quality audio transmission and storage. In parallel to the definition of the ISO/MPEG standards, several audio coding algorithms have been proposed that use the DWPT, in particular, adaptive wavelet packet transform, as the tool to decompose the signal [2],[3]. In practice, DWPT are often implemented using a tree-structured filter bank [2], [3],[4]. The DWPT is a set of transformations that admits any type of tree-structured filter bank, that provides a different time–frequency tiling map. Many architectures have been proposed for computing the discrete wavlet transform in the past. However, it is not the case for the DWPT. There are very few papers regarding the development of specific architectures for the DWPT. In [5] is designed a programmable DWPT processor using two-buffer memory system and a single multiplier–accumulator (MAC) to calculate different subbands. Method [6] exploits the in-place nature of the DWPT algorithm and uses a single processing element consisting of multipliers in parallel and adders for each low-pass and high-pass filters – wavelet butterflies (is the number of filter taps) to increase the throughput. In [7] is also proposed a folded pipelined architecture to speed up the throughput. It consists of MACs communicated by memory banks to compute each level of the total decomposition levels.

Applying the lifting scheme [8] for the construction of wavelets filter bank allows significantly reduce the number of arithmetic operations that are necessary to compute the transform. A folded parallel architecture for lifting-based DWPT was presented in [9]. It consists of a group of MACs operating in parallel on the data prestored in a memory bank. In [10] is proposed an architecture using a direct implementation of a lifting-based wavelet filter to perform one level of DWPT at a time. The main drawback of these existing architectures is that they all use memory to store the intermediate coefficients and involve intense memory access during the computation. A recursive pyramid algorithm based folded architecture for computing lifting-based multilevel DWPT is presented in [11]. However, the scheduling and control complexity is high, which also introduce large numbers of switches, multiplexers and control signals. The architecture is not regular and need to be modified for different number of level of DWPT computation. A folded architecture for lifting-based wavelet filters is proposed in [12] to compute the wavelet butterflies in different groups simultaneously at each decomposition level. According to the comparison results, the proposed architecture is more efficient than the previous proposed architectures in terms of memory access, hardware regularity and simplicity, and throughput. It is necessary to notice, that the architecture of the given processor is effective only for calculation of full tree DWPT. Here there is no technique of management for DWPT with best tree searching.

Algorithm transformation techniques have been employed in high-speed DSP system design is presented in [13]. All of the above mentioned techniques are applied during the processor design phase and their implementation is time invariant. Therefore, this class of signal processing techniques is referred as static techniques. Recently, dynamic techniques both of the circuit level and algorithmic level have been proposed [14]. These techniques are based on the principles that the input signal is usually non-stationary, and hence, it is better (from a coding perspective) to adapt the algorithm and architecture to the input signal. Such systems are referred to as reconfigurable signal processing systems [15],[16]. The key goal of these techniques is to improve the algorithm performance by exploiting variability in the data and channel.

Our approach is to design of dynamic algorithm transform (DAT) for design of application-specific reconfigurable lifting-based DWPT pipeline processor, in particular, for audio signal processing in real-time. The principle behind DAT techniques is to define parameter of input audio signals (subband entropy) and output encoded sequences (subband rate) for the given embedded processor architecture. Adaptive wavelet analysis for audio signal processing purposes is particularly interesting if the psychoacoustic information is considered in the DWPT decomposition scale. Due to the lack of selectivity of wavelet filter banks, the psychoacoustic information is computed in the wavelet domain.

2. Flexible tree structured signal expansion based on DWPT

DWPT algorithm is a generalization of the discrete wavelet transform that can be represented as a filter bank with a tree structure [3] (see figure 1). Within a given node number n of the tree

Dynamic Reconfigurable on the Lifting Steps Wavelet Packet Processor with Frame-Based
Psychoacoustic Optimized Time-Frequency Tiling for Real-Time Audio Applications

25

at any level l ($n = 0 .. .2^{l-1}, l \in Z$) input $x_{l, n,k}$, ($k-$ signal samples) is separated by low-frequency (LF) $x_{l+1,2 \cdot n,k}$ and high frequency (HF) $x_{l+1,2 \cdot n+1,k}$ components using a pair of wavelet filters $h(z)$ and $g(z)$ with finite impulse response (FIR), after which each subband signal down-sampling by factor of two. Function block that implements this separation of the input signal is called a dual-channel filter bank analysis.

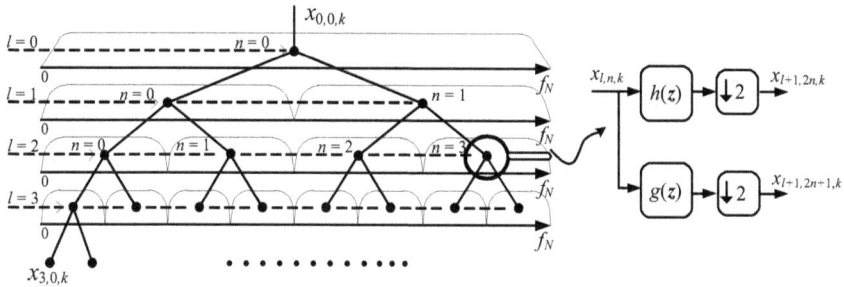

Figure 1. DWPT tree structure (left) and dual-channel filter bank – wavlet baterfily (right)

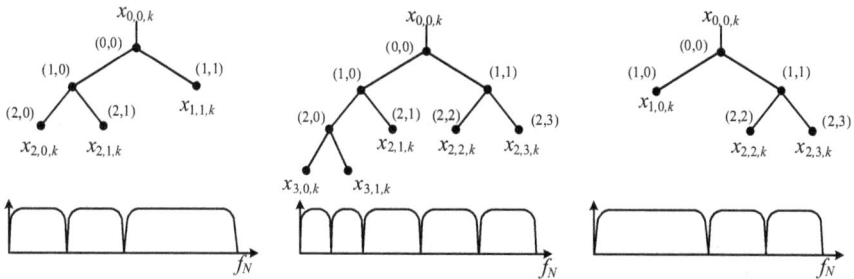

Figure 2. DWPT tree structure examples and corresponding magnitude response of then filter bank

Thus, a specific node (l, n) corresponds to the frequency range $(n \cdot 2^{-l}, (n + 1) \cdot 2^{-l})$, normalized to the Nyquist frequency (f_N). At each level of decomposition frequency resolution increases twice, but twice the resolving power decreases over time. DWPT is a complete decomposition of the signal in the low and high frequencies. Variation of the resolution in frequency and time domains allows for a more detailed decomposition, for example, in the lower frequencies, which leads to an increase in the frequency resolution and reduced over time. This feature has adapted DWPT. The advantage is the ability to DWPT sufficiently flexible selection of tree decomposition (see figure 2), based on the nature of the signal. The choice of tree structure can be performed based on pre-known features of the signal, and executed dynamically, "arranged" for the current frame processing [2].

3. Dynamic transformation of DWPT decomposition

We present adaptive DWPT tree derived via DAT's. The principle behind DAT is to define parameter of input signals (subband entropy) and output sequences (subband rate) for the given embedded processor architecture. In other hands, DAT techniques is to construct a minimum cost subband decomposition of DWPT by maximizing the minimum masking threshold (which is limited by the perceptual entropy (PE)) in every subband for the given embedded processor architecture and temporal resolution. Achieving this purpose, we suppose that the tree structure of DWPT decomposition is adapted, as closely as possible, to the critical bands $(CB - WPD : (l, n) \in E_{CB})$ as shown in [14]. For the DWPT tree structure E_i the information density H belong to tree E_i is estimated as

$$H_{E_i} = \sum_{\forall (l,n) \in E_i} \sum_k w_{E_i}(k) \cdot ln(w_{E_i}(k)),$$

(1)

where

$$w_{E_i}(k) = \frac{|x_{l,n,k}|}{\sum_{\forall (l,n) \in E_i} |x_{l,n,k}|},$$

(2)

here $x_{l,n,k}$ are wavelet coefficients, l is a decomposition level, n is the node number of decomposition level, k is the index of the current wavelet coefficient of the node (l, n). H_{E_i} is estimates based on the wavelet coefficients of terminated nodes (nodes is a grey area in a figure 3).

The growing decision for DWPT tree based on the given H is being taken in terms of allowing the further decomposition of the WP tree can be expressed as:

$$H_{E_i} < H_{E_{i-1}}.$$

(3)

If (3) is true we continue the subband splitting process in DWPT tree, otherwise the suboptimal decomposition for the given frame of signal is founded.

The subband splitting process is managed based on the estimated values of PE in parent and child nodes of current DWPT tree structure. PE estimation is described in [17],[18],[19] and expressed as

$$PE_{l,n} = \sum_{k=0}^{K_{l,n}-1} log_2(2[n \text{ int}(SMR_{l,n,k})] + 1),$$

(4)

Dynamic Reconfigurable on the Lifting Steps Wavelet Packet Processor with Frame-Based
Psychoacoustic Optimized Time-Frequency Tiling for Real-Time Audio Applications

27

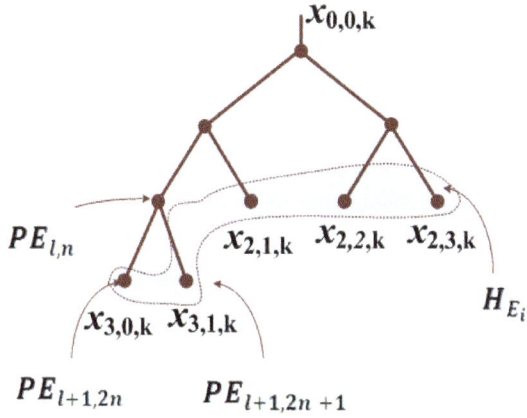

Figure 3. DWPT tree growing process

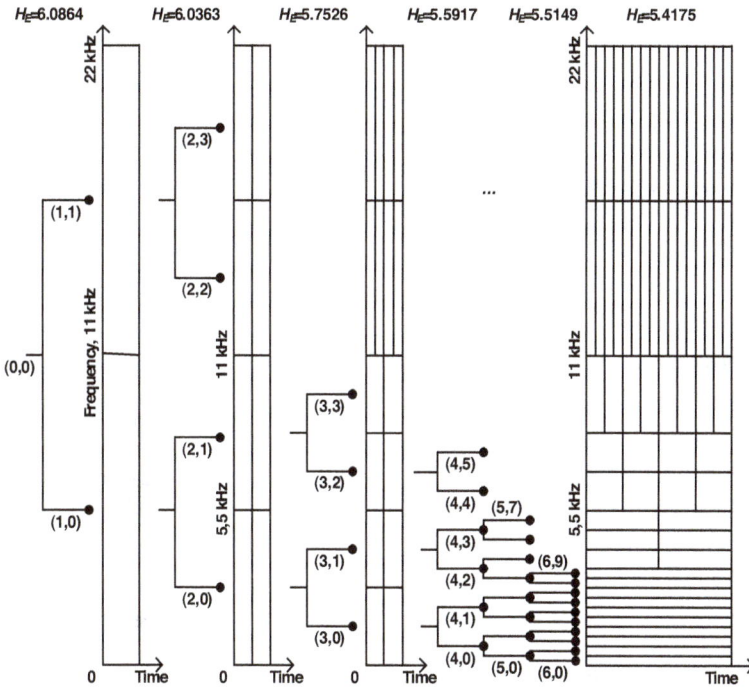

Figure 4. DWPT tree structure creation and corresponding time-frequency tilling map

where $SMR_{l,n,k}$ is a ration between the absolute value of the wavelet coefficients $x_{l,n,k}$ in a subbabnd of tree E_i (node (l, n)), and the corresponding masking threshold $T_{l,n}$, which is linearly spread among the $K_{l,n}$ coefficients $x_{l,n,k}$, $k = \{0, K_{l,n}\}$ of node (l, n). The large magnitude of $SMR_{l,n,k}$ determines node (l, n) significance for PE formation.

Each allowed parent node (l, n) is split on two child nodes $(l+1,2n)$ and $(l+1,2n+1)$, if and only if the sum of $PE_{l+1,2n}$ and $PE_{l+1,2n+1}$ in the child nodes less than in the current node $PE_{l,n}$, that can written as.

$$PE_{l,n} > PE_{l+1,2n} + PE_{l+1,2n+1}. \qquad (5)$$

Schematically the DWPT tree growing process is shown in the figure 3. The example of dynamic DWPT tree structure growing level by level based on H and corresponding time-frequency tilling map are demonstrated in the figure 4.

Applying the information density H, the perception entropy PE, the limited WP tree structure CB - WPD and the maximum allowed computation resource together in DWPT growing procedure allows us to found suboptimal solution for input signal analysis on the given hardware architecture.

4. DWPT implementation based on lifting scheme

4.1. Factoring wavelet filters in to lifting steps

In the tree-based scheme of the DWPT, each node of the tree consists of a two-channel filter bank. Each node can be broken down into a finite sequence of simple filtering steps, which are called lifting steps or ladder structures. In [20] for two-channel filter bank proposed a method of transition from the implementation on the basis of FIR filters to architecture at the based on lifting scheme. The decomposition is essentially a factorization of the polyphase matrix of the wavelet filters into elementary matrices. As discussed in [20], the lifting steps scheme consists of three phases: the first step splits the data into two subsets: even and odd; the second step recalculates the coefficients (high-pass) as the failure to predict the odd set based on the even; finally the third step updates the even set using the wavelet coefficients to compute the scaling function coefficients (low-pass). This method allows to reduce on halve the number of multiplications and summations. In terms of the z-transform transition to the implementation of the filter bank based on the lifting scheme can be viewed in two steps.

The first step is to move towards the implementation of polyphase filtering algorithm [20]. The process of calculating the LF and HF components of the signal $x_{l,n,k}$ in any node of the tree can be written as the following expression:

$$[X_{l+1,2n}(z) \quad X_{l+1,2n+1}(z)] = [X_{l,n}^e(z) \quad z^{-1}X_{l,n}^o(z)] \cdot \tilde{\mathbf{P}}, \qquad (6)$$

Dynamic Reconfigurable on the Lifting Steps Wavelet Packet Processor with Frame-Based
Psychoacoustic Optimized Time-Frequency Tiling for Real-Time Audio Applications

29

where $X_{l+1,2\,n}(z)$ and $X_{l+1,2\,n+1}(z)$ are z-representation in the field of low and high frequency components, $X_{l,\,n}^{e}(z)$, and $X_{l,n}^{o}(z)$ are representation of sequences, respectively, consisting of the even and odd samples the input sequence $x_{l,n,k}$, $\tilde{\mathbf{P}}$ is a polyphase matrix that can be written as

$$\tilde{\mathbf{P}} = \begin{bmatrix} \tilde{h}_e(z) & \tilde{g}_e(z) \\ \tilde{h}_o(z) & \tilde{g}_o(z) \end{bmatrix}. \tag{7}$$

The elements $\tilde{h}_e(z)$, $\tilde{h}_o(z)$ and $\tilde{g}_e(z)$, $\tilde{g}_o(z)$ of polyphase matrix from (7) are in the following dependence to the original coefficients of low-pass $\tilde{h}(z)$ and high-pass $\tilde{g}(z)$ wavelet filters correspondingly:

$$\tilde{h}(z) = \tilde{h}_e(z^2) + z^{-1}\tilde{h}_o(z^2), \tag{8}$$

$$\tilde{g}(z) = \tilde{g}_e(z^2) + z^{-1}\tilde{g}_o(z^2). \tag{9}$$

This approach does not give the gain to the computational cost, but the hardware implementation allows us to reduce the operation frequency in double in compare with input data rate due to parallel computation (see figure 5).

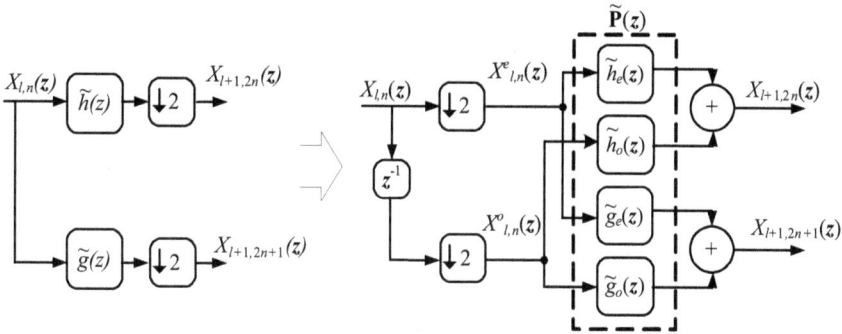

Figure 5. The transition to the polyphase implementation of analysis filter bank

	$s_1(z)$	$s_2(z)$	$s_3(z)$	$t_1(z)$	$t_2(z)$
b_0	3.1029	-5.1995	0.3141	0.0763	-3.1769
b_1	0	1.6625	0	-0.2920	-0.0379
u	0	-1	-3	1	3

Table 1. The parameters of lifting scheme for db4 (8 taps)

The second step is a factorization of the polyphase matrix into simpler triangular matrices. The result is, in a general, the original matrix $\tilde{\mathbf{P}}$ that can be expressed as

$$\tilde{\mathbf{P}}=\prod_{i=1}^{I/2}\left(\begin{bmatrix} 1 & \check{s}_i(z) \\ 0 & 1 \end{bmatrix}\begin{bmatrix} 1 & 0 \\ \check{t}_i(z) & 1 \end{bmatrix}\right)\begin{bmatrix} c_1 & 0 \\ 0 & c_2 \end{bmatrix}, \tag{10}$$

where I is the number of elementary triangular matrices derived from the factorization of polyphase matrices; $\check{s}_i(z)$ and $\check{t}_i(z)$ are low-order polynomials; c_1, c_2 are real coefficients. In general, the polynomials $\check{s}_i(z)$ and $\check{t}_i(z)$ can represented as $(b_0+b_1 z^{-1})z^u$, where b_0, b_1 are constants, u is the integer exponent. For example, the b_0, b_1 and u parameters of lifting scheme for db4 wavelet mother function are presented in table 1. K_1 and K_2 are equal -0.1202 and -8.3192 correspondingly. For fixed point DWPT implementation an arithmetic with an arbitrary number of integer and fractional bits is used as proposed in [21],[22]. The advantage of this number representation is the fact that it can be realized using conventional integer arithmetic resource.

The scaling $X_{l+1,2n}(z)$ and wavelet $X_{l+1,2n+1}(z)$ coefficients relative to the input signal $X_{l,n}(z)$ in z domain are two-channel analysis filter bank results in according to (6) and (10) can be written as follows:

$$[X_{l+1,2n}(z) \quad X_{l+1,2n+1}(z)]=[X_{l,n,e}(z) \quad X_{l,n,o}(z)]\times\prod_{i=1}^{I/2}\left(\begin{bmatrix} 1 & \check{s}_i(z) \\ 0 & 1 \end{bmatrix}\begin{bmatrix} 1 & 0 \\ \check{t}_i(z) & 1 \end{bmatrix}\right)\begin{bmatrix} c_1 & 0 \\ 0 & c_2 \end{bmatrix}, \tag{11}$$

where $X_{l,n,e}(z)$ and $X_{l,n,o}(z)$ are z-representation of two sequences consisting of even and odd samples of the input signal $x_{l,n,k}$. The block diagram for the direct implementation of two-channel analysis filter bank based on lifting scheme is shown on figure 6.

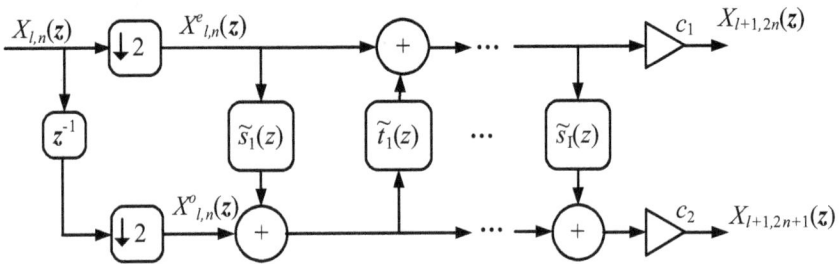

Figure 6. Block diagram of two-channel analysis filter bank based on the lifting scheme

The inverse decomposition of a two-channel filter bank in the same terms can be expressed as

$$\left[\hat{X}^e_{l,n}(z) \quad z^{-1}\hat{X}^o_{l,n}(z)\right] = \left[X_{l+1,2n}(z) \quad X_{l+1,2n+1}(z)\right] \begin{bmatrix} 1/c_1 & 0 \\ 0 & 1/c_2 \end{bmatrix} \prod_{i=l/2}^{1} \left(\begin{bmatrix} 1 & 0 \\ -\check{t}_i(z) & 1 \end{bmatrix} \begin{bmatrix} 1 & -\check{s}_i(z) \\ 0 & 1 \end{bmatrix} \right), \quad (12)$$

and corresponding block diagram of two-channel synthesis filter bank based on the lifting scheme shown in a figure 7.

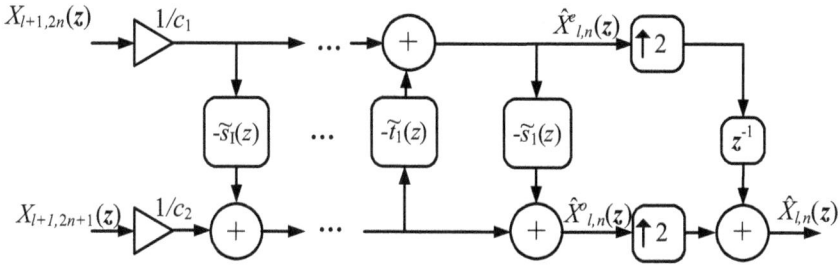

Figure 7. The block diagram of two-channel synthesis filter bank based on lifting scheme

According to the block diagram (see figure 7), the synthesis procedure is implemented as follows: at first, the input coefficients $X_{l+1,2n}(z)$ and $X_{l+1,\,2n+1}(z)$ of each channel are multiplied on the coefficients $1/c_1$, $1/c_2$ correspondingly, at second, two-channel synthesis filter bank implements the inverse operations of algorithm analysis. This implementation uses the same polynomial $\check{s}_i(z)$ and $\check{t}_i(z)$ from (10) with opposite signs. Reconstructed signal $X_{l,n}(z)$ is obtained from the calculating sequences $X^e_{l,n}(z)$, $X^o_{l,n}(z)$.

Together the analysis and synthesis filter bank implementation based on lifting scheme is required the same number of operations (summation and multiplication) in compare with analysis only implementation based on regular FIR filter implementation.

4.2. The algorithm implementation based on fixed-point variable format arithmetic

A number of the target application requirements (work in real time, greater throughput, and other) make it necessary to use fixed-point arithmetic to perform the specified computation. With the implementation of two-channel filter bank based on lifting structures using integer arithmetic, the number of difficulties arise, related to the fact that values of the coefficients of the polynomials $s_i(z)$ and $t_i(z)$ can take both fractional and great integer values (that is well-known negative effect of polyphase matrix factorization). This feature causes to an increase of the arithmetic units and the word length of internal registers. Therefore, in this paper for the implementation of the algorithm DWPT on fixed-point arithmetic the approach based on [21], [22] is used, according to which the format of the numbers involved in the intermediate computation, is variable. This method assumes that the number of bits to be allocated under the integer and fractional parts of numbers in the different nodes of the algorithm is different.

In accordance with this approach, any number represented in two's complement fixed-point format, is given in the form of expression:

$$a = ma \cdot 2^{ex\,p_a}, \quad \text{where } ma = (-1)^s + \sum_{i=0}^{wl-2} a_i \cdot 2^{i-wl+1}. \tag{13}$$

Here, ma– value of the number presented in two's complement code that is interpreted as a fraction in the range $[-1,1)$; $ex\,p_a$ – the order of the scaling factor $2^{ex\,p_a}$; a_i - value of the i-th bit of the number equal to 0 or 1; s – the sign bit; wl – the word length. Thus for intermediate data in different nodes of the algorithm its value $ex\,p_a$ is determined. So, depending on the $ex\,p_a$ value the redistribution of bits for fractional and integer part of number at different sites of the algorithm is produced (see figure 8).

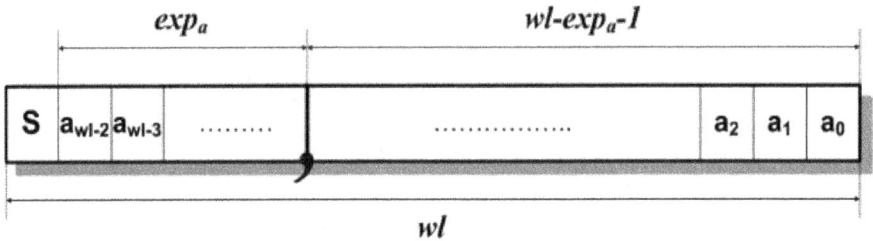

Figure 8. Data format in fixed-point arithmetic with variable word length

For a given in (13) format, the operations of addition and multiplication of a and b $(ex\,p_b \geq exp\_a)$ are defined as

$$c = a + b = mc \cdot 2^{ex\,p_c} = \left(ma \cdot 2^{ex\,p_a - ex\,p_b} + mb\right) \cdot 2^{ex\,p_b}, \tag{14}$$

$$c = a \cdot b = mc \cdot 2^{ex\,p_c} = ma \cdot mb \cdot 2^{ex\,p_a + ex\,p_b}. \tag{15}$$

The figure 9 schematically illustrates the process of performing operations described above in (14) and (15).

In this paper a generic set of processing elements is proposed to implement of analysis and synthesis banks on the lifting structures using variable arithmetic format (see table 2). In this table $x_e[k]$, $x_o[k]$ – input values, and $y_e[k]$, $y_o[k]$ – output values, respectively, in the upper and lower channels filter bank analysis (synthesis) in the k -th time value. The parameters $s0$, $s1$, $s2$ define the arithmetic shift values. These parameters are computed according to (14) and (15) for each node of the algorithm.

Dynamic Reconfigurable on the Lifting Steps Wavelet Packet Processor with Frame-Based
Psychoacoustic Optimized Time-Frequency Tiling for Real-Time Audio Applications

33

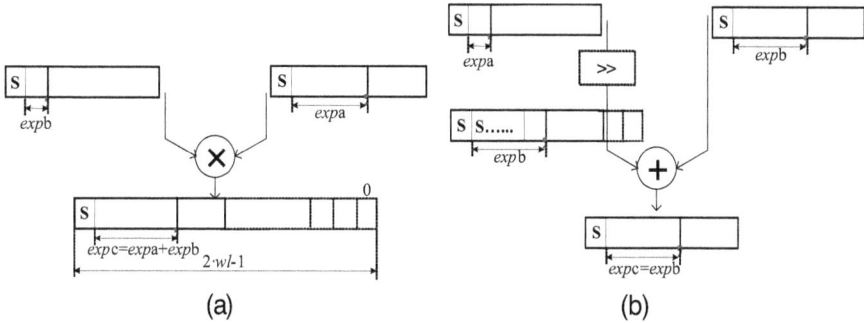

(a) **(b)**

Figure 9. Performing operations of a) addition and b) multiplication

Analysis bank		Synthesis bank	
Symbol	Computational operations	Symbol	Computational operations
$S1$	$y_e[k]=x_e[k];$	$S1^{-1}$	$y_e[k]=x_e[k];$
	$y_o[k]=x_o[k]\gg s0+(x_e[k]\cdot mb^0)\gg s1$		$y_o[k]=(x_o[k]-(x_e[k]\cdot mb^0)\gg s1)\ll s0$
$S2$	$y_e[k]=x_e[k];$	$S2^{-1}$	$y_e[k]=x_e[k];$
	$y_o[k]=x_o[k]\gg s0+(x_e[k]\cdot mb^0)\gg s1+$		$y_o[k]=(x_o[k]-(x_e[k]\cdot mb^0)\gg s1-$
	$+(x_e[k-1]\cdot mb^1)\gg s2;$		$-(x_e[k-1]\cdot mb^1)\gg s2))\ll s0$
$T1$	$y_e[k]=x_e[k]\gg s0+(x_o[k]\cdot mb^0)\gg s1$	$T1^{-1}$	$y_e[k]=(x_e[k]-(x_e[k]\cdot mb^0)\gg s1)\ll s0$
	$y_o[k]=x_o[k];$		$y_o[k]=x_o[k];$
$T2$	$y_e[k]=x_e[k]\gg s0+(x_o[k]\cdot mb^0)\gg s1+$	$T2^{-1}$	$y_e[k]=(x_e[k]-(x_e[k]\cdot mb^0)\gg s1-$
	$+(x_o[k-1]\cdot mb^1)\gg s2;$		$-(x_o[k-1]\cdot mb^1)\gg s2)\ll s0$
	$y_o[k]=x_o[k];$		$y_o[k]=x_o[k];$

Table 2. A set of processing elements for the based on the lifting structures algorithm using arithmetic variable format

For an explanation of the practical aspects related to the use of the proposed arithmetic, below an example of the two-channel analysis bank realization on the mother wavelet function db4 [23] is considered. In result of polyphase matrix factorization for a given wavelet basis in the MATLAB environment the following expression was obtained:

$$\tilde{P}=\begin{bmatrix}1 & b_1^0\\0 & 1\end{bmatrix}\begin{bmatrix}1 & 0\\(b_2^0+b_2^1z^{-1})z & 1\end{bmatrix}\begin{bmatrix}1 & (b_3^0+b_3^1z^{-1})z^{-1}\\0 & 1\end{bmatrix}\cdot\begin{bmatrix}1 & 0\\(b_4^0+b_4^1z^{-1})z^3 & 1\end{bmatrix}\begin{bmatrix}1 & b_5^0z^{-3}\\0 & 1\end{bmatrix}\begin{bmatrix}c_1 & 0\\0 & c_2\end{bmatrix}. \tag{16}$$

The coefficients b_m^n, and also their parameters mb and $expb$, calculated in accordance with (13), are presented in table 3.

Lifting step number, i	Type	u	b_i^0			b_i^1		
			Value	mb	expb	Value	mb	expb
1	$s_1(z)$	0	-3,1029	-0,7757	2	0	0	-
2	$t_2(z)$	1	-0,0763	-0,6104	-3	0,2920	0,5840	-1
3	$s_3(z)$	-1	5,1995	0,6499	3	-1,6625	-0,8313	1
4	$t_4(z)$	3	3,1769	0,7942	2	0,0379	0,6064	-4
5	$s_5(z)$	-3	0,3141	0,6282	-1	0	0	-

Table 3. The lifting structures parameters calculated for the wavelet filters db4 (8 taps)

In the figure 10 shown a block diagram of the implementation of two-channel filter bank analysis for this example. In this scheme, apart from computing elements $S1$, $S2$, $T2$ (see table 2) in the upper channel of the bank to satisfy the condition of causality delay registers are inserted (elements z^{-l}, $l \in \mathbb{Z}$).

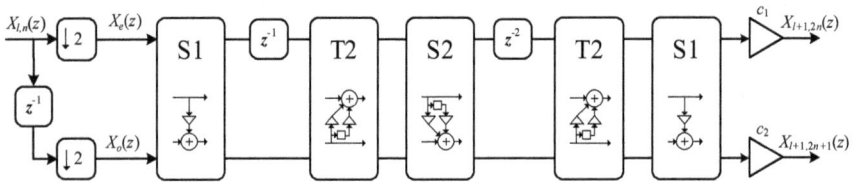

Figure 10. Block diagram of two-channel filter bank based on the lifting structures for db4 (see table 2)

In the figure 11a in more detail the first step realization of the analysis bank is considered and in the figure 11b the last step realization of synthesis bank is shown.

As can be seen from the figure 10 and figure 11, the computing units of analysis and synthesis procedures in terms of implementation differ only in the signs of constant multiplying coefficients and the arithmetic shifts positions and directions.

Based on the materials described above, a concrete realization of two-channel bank can be represented as a vector of parameters containing a set of multiplier constants, shift parameters and some additional information regarding the delay elements in the intermediate nodes of the algorithm.

4.3. Accuracy analysis of the algorithm for fixed-point variable format

To analyse of the proposed approach in MATLAB function library was written, which simulates the process of fixed-point calculating in filter bank with specified structure of the tree. In the figure 12 is shown the estimation error variance signal recovery, depending on the

Dynamic Reconfigurable on the Lifting Steps Wavelet Packet Processor with Frame-Based
Psychoacoustic Optimized Time-Frequency Tiling for Real-Time Audio Applications

35

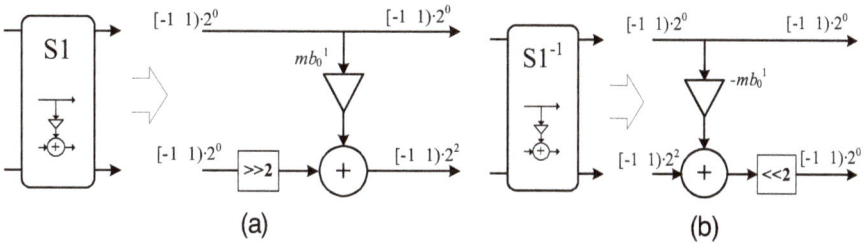

Figure 11. First step of the analysis (a) and the last step of the synthesis (b) bank implementation

choice of bit internal registers as a result of passing through the two-channel filter bank
analysis / synthesis (in the example used wavelet filters db8). This figure also shows the results
of an experiment using FIR filters, the underlying of the algorithm DWPT. It can be noted that
FIR filter implementation gives better results while using the same registers word length, but
requires twice as many calculations. So in order to achieve the level of error variance in the -70
dB for based on lifting structures implementation requires 16-bit, which are approximately
two bits more than the realization based on FIR. But this drawback is compensated by a
significant reduction of arithmetic operations compared to the direct implementation. Thus,
we conclude that the proposed approach is more efficient in hardware implementation
compared with the bank on the basis of FIR filters.

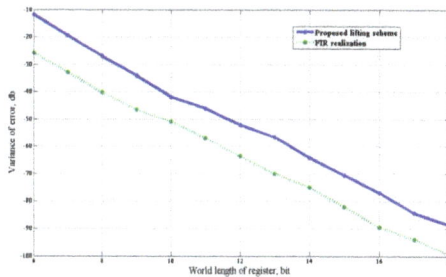

Figure 12. The variance of the reconstruction signal error dependence on the registers bit capacity in analysis/synthe-
sis two-channel filter bank systems (wavelet filters are db8): the solid line indicates the results for the proposed algorithm on the lifting structures, dotted - integer implementation of the same system using a FIR filter

Below we consider another experiment for demonstration of the energy localization properties
by using our fixed-point DWPT algorithm realization. For this a polyharmonic signal was
generated and passed through a five-level decomposition tree fast wavelet transform (division
of the tree is carried out only in the low-frequency components). As an example, the wavelet
functions db2 family was chosen. Thus all range of amplitudes of the wavelet coefficients was
divided by 40 thresholds (these values are plotted on the X-axis of figure 13). Each threshold

has been mapped to a vector of the obtained analysis wavelet coefficients on condition that these coefficients are greater than this threshold. Otherwise, the values of the coefficients were replaced by zeros (i.e. in each vector were discarded unimportant relative to a given threshold values). For all vectors was performed reconstruction procedure by synthesis filter bank. In figure 13a, figure 13b for the floating-and fixed-point implementations, respectively, are shown: solid line – the reconstructed to original signal energy relation (in percentage) depending on the threshold values; dotted line – the percentage of "discarded" wavelet coefficients depending on the chosen threshold.

Based on these results, we note almost complete compliance of floating-point model with the proposed fixed-point approach. Thus, the fixed-point variable format algorithm DWPT implementation preserves the energy localization inherent to the wavelet packets.

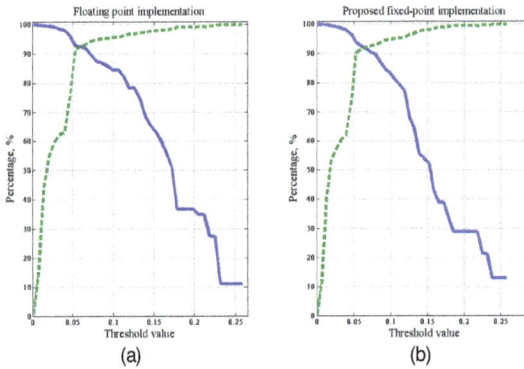

Figure 13. Energy estimates of signal reconstruction, depending on the threshold of significant wavelet coefficients for the based on a floating-point (a) and proposed fixed-point (b) DWPT algorithm implementations

5. DWPT pipeline processor with dynamic reconfigurable architecture

5.1. DAT based reconfigurable signal processing system

The structure of reconfigurable DSP system for signal analysis based on DAT approach consists of the specific microprocessor oriented on the signal processing (DSP microprocessor) and DWPT processor itself with the reconfigurable architecture. The DSP microprocessor perform several task, such as: processing wavelet coefficients $X_{l,n,k}$ in subbands (l, n) that corresponds to the current DWPT tree structure E_i; estimate H_{E_i} and $PE_{l,n}$; obtain the reconfiguration vector for DWPT processor $r_{l,n}$, $(l, n) \in E_i$. DWPT processor is realized on pipeline architecture with

Dynamic Reconfigurable on the Lifting Steps Wavelet Packet Processor with Frame-Based
Psychoacoustic Optimized Time-Frequency Tiling for Real-Time Audio Applications

37

dynamic reconfiguration for implementing adaptive DWPT. The length of the pipeline is obtained from the limited DWPT tree structure (CB - WPD). A great dependence of the process on the DWPT structure grows leads to the necessity of introducing an easily reconfigurable parallel-pipeline structure with computation resource C. Thus, the DSP system for audio processing based on DAT-approach consists as shown on figure 14.

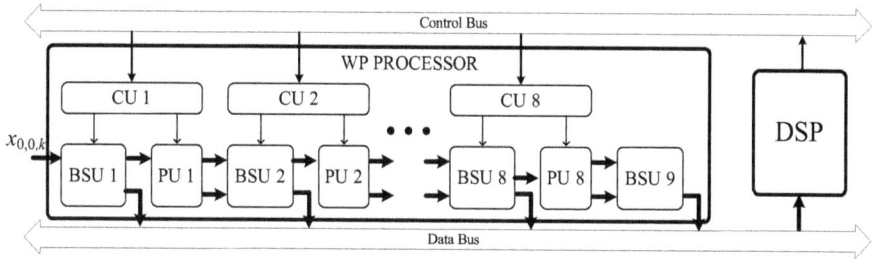

Figure 14. DAT-based reconfigurable signal processing system

The pipeline architecture is applied for effective implementation of the DWPT algorithm. We suggest the pipeline architecture for constructing the DWPT lifting based processor. This architecture integrates the sequential connection of the homogeneous block (buffer/switch unit (BSU) and processing unit (PU)) that implement a two-channel filter bank that allows parallel calculating DWPT with an arbitrary tree structure. The maximum number of decomposition level that can be realize is 8, it is associate with the depth of CB - WPD. The basic decomposition of DWPT expressed as PU which acts a two-channel filter bank based on the lifting scheme. The reconfiguration vector $r_{l,n}$ decoding, memory address generation, PU enabling, data exchange controlling and the pipe-line synchronizing are performed by the control units (CU). All this functions in CU at each DWPT processor stages is carried out in parallel. The pipe-line DWPT processor stages are synchronized according to the DAT's techniques.

5.2. DWPT lifting based pipeline Processing Unit (PU)

The block diagram of PU is shown in the figure 15. The input sequence $x_{l,n,k}$ is split into even X_e and odd X_o samples in PU before the processing is started according to the lifting scheme. The structure of PU has the following abbreviations (see figure 15): wl is a bit capacity; I is a number of elementary steps of the lifting scheme; V_{PE} is a vectors, each element of it is a set of the parameters of the same elementary step of the lifting scheme; V_{BUF} is a vector, each element of it specify the number of delays, respectively in the upper and lower channels after the same elementary step. The present elements corresponds to FIFO registers that, on the one hand are delay elements z^{-1} in the algorithm and, on the other hand, makes possible a pipelined realization in architecture for throughput performance increasing. The coefficients c_1 and c_2 applies to the result of lifting scheme as it described in (11). The estimated hardware resources required for PU implementation are shown in a table 4.

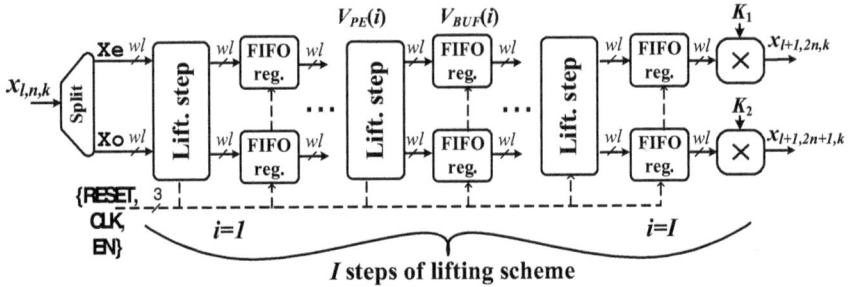

Figure 15. Block diagram of the PU

Resource type	Utilized
Multipliers wl×wl	$N+2$
Adders(wl – bit capacity)	N
Registers(wl – bit capacity)	$N+1$
Multiplexers 2-in-1 (wl – bit capacity)	1

Table 4. Estimation of hardware resources for PU implementation (N – is the number of filter taps)

5.3. Buffer/Switch Unit (BSU)

The BSU realizes double buffering scheme known as "ping-pong" for providing parallel access to the data for storing results and getting source data from/for PU. The additional channel is for outputting the result data. The two output streams of samples $x_{l+1,2n,k}$ and $x_{l+1,2n+1,k}$ from l-th PU are stored in BSU and simultaneously $l+1$-th PU can get the samples for the next processing stage. Unified block diagram of BSU is represented in the figure 16. Each BSU in parallel-pipeline architecture has addressed a different memory size that depends on the DWPT decomposition level.

The momory amount M_V (taking into account the requirement of double buffering) and the number of processing units of L , can be expressed as

$$M_V = 2 \cdot \sum_{j=1}^{J} \frac{K}{2^{l_j}}, \; L = \max_{j=1..J} l_j \qquad (17)$$

where J is amount of all nodes CB-WPD, l_j is a decomposition level, K is a initial frame length of input signal.

Dynamic Reconfigurable on the Lifting Steps Wavelet Packet Processor with Frame-Based
Psychoacoustic Optimized Time-Frequency Tiling for Real-Time Audio Applications

39

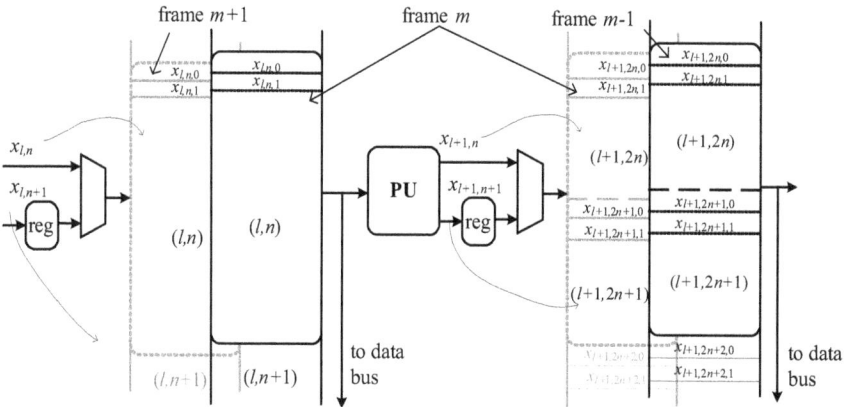

Figure 16. Unified block diagram of the BSU

In the figure 17 is shown an example of the tree decomposition, given by the set of nodes {(0,0), (1,0), (1,1), (2,0), (2,1), (2,2), (2.3), (3,0), (3,1)}. It also schematically illustrates the principle of distribution of blocks of memory for the structure of the tree.

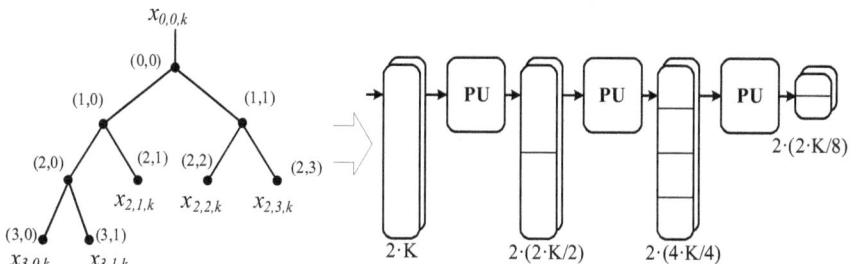

Figure 17. The parallel pipeline architecture for three-level DWPT tree structure

5.4. Rapid prototyping algorithm of pipeline DWPT processor

The prototype of the DWPT processor can be specified as parameters describing the structure of two-channel filter bank and the vector that defines the limit tree decomposition.

The method of rapid prototyping can be described by the following sequence of actions.

1. Calculating the lifting structure of the dual filter bank based on the original wavelet basis functions.

2. Translating the mathematical model for fixed-point arithmetic with the requirements of accuracy, and limitation of hardware resources (registers and bit computing units).

3. Forming a parameters vector for configure a DWPT processor prototype.

4. Estimating the cost of hardware prototype implementation.

5. Estimating computation characteristics of the DWPT procressor prototype.

6. Generating the output files of the synthesized VHDL-description of the DWPT processor.

5.5. FPGA based hardware implementation of the pipeline DWPT processor

For estimation of performance and resource utilization the present architecture has been implemented on Xilinx FPGA XC3s2000. The realized pipeline DWPT processor has following features. The number of decomposition levels is limited by eight. The mother wavelet function Db8 (16 taps) transformed into nine lifting steps is used. The input and output data has the 16 bits word length, the capacity of internal computing is 18 bits. The present implementation doesn't have FIFO stages in PU that allows minimizing hardware resources. The processed frame size can be selected in a range from 128 to 1024 samples. Each BSU contains the pair of two 1024×16 bits block RAMs that is used for realizing double buffering scheme. The PU hardware resources utilization are shown in a table 5 and complete processor implementation resources are presented in table 6.

Resource type	Utilized
4 input look-up tables	788
flip-flops	226
MULT18x18s	18

Table 5. Estimations of hardware resource for FPGA-based PU implementation.

Resource type	Utilized, pcs.	Percentage wise, %
4 input look-up tables	31356	76
flip-flops	3037	7
RAMB16	16	40
MULT18x18s	40	100

Table 6. Hardware resource estimations for WP processor implementation on XC3s2000

In the figure 18 the protopyte board of dynamic reconfigurable pipeline DWPT processor is shown.

The implemented design performance is 8 MSPS. So, if the sample rate of input audio signal is 44100 Hz then the time cost for computation of wavelet coefficients is 0.6% from all time resource. For example, the 512 samples frame size (~11.6 ms) processing take approximately 0.064 ms on presented DWPT lifting based pipeline processor. The rest time is distributed between the dynamic DWPT tree decomposition algorithms, wavelet coefficients post-processing and transfer operation.

Figure 18. Protopyte board of dynamic reconfigurable pipeline DWPT processor

5.6. DAT based dynamic reconfigurable architecture algorithm

Suppose that for some audio input frame is the space of trees structures E, which is processing a stream-flow or parallel reconfigurable processor $(m, r_{l,n})$, where m is a number of processor stages, $r_{l,n}$ is a processor reconfigurable parameters vector of the structure corresponding to the decomposition of tree DWPT $(l, n) \in E_i$. The limit corresponds to a tree $CB - WPD$: $(l, n) \in E_{CB}$. Next, on the basis of the growing algorithm, described in section 3, the DWPT tree structures are formed, for example, E_1, E_2, E_3 for which restrictions are checked E_{CB}, as well as the calculated information density H_{E_i}. Based on that, if it turns out $H_{E_3} < H_{E_2} < H_{E_1}$, the structure of the E_3 to the required frequency-time resolution processing of the frame. Reconfigurable processor DWPT determined by the current vector of reconfigurable parameters:

$$r_{l,n} = [(\alpha_1, \ \beta_0, \ \beta_1), \ (\alpha_2, \ \beta_0, \ \beta_1, \ \beta_2, \ \beta_3), \ \dots \ (\alpha_2, \ \beta_0, \ \dots, \ \beta_n)] \tag{18}$$

where α_l and β_n takes the values 0 or 1.

Parameters α_l determine the transition to a new level of large-scale tree DWPT l, i.e. include signal processing in the next processor step m:

$$\alpha_l = \begin{cases} 1, & if \ H_{E_i} < H_{E_{i-1}} \ and \ E_i \notin E_{CB} \\ 0, & otherwise \end{cases} . \tag{19}$$

In turn, a group of parameters β_n includes n nodes at the level l:

$$\beta_n = \begin{cases} 1, & ifP\,E_{l,n} > P\,E_{l+1,2n} + P\,E_{l+1,2n+1} \\ 0, & otherwise \end{cases}. \tag{20}$$

Thus, the transition of signal processing according to the DWPT tree structure E_i on the architecture of the processor $E_{m,i}$ for processing on the architecture of $E_{m+1,i+1}$, in accordance with the tree structure E_{i+1} is the vector according to the reconfigurable parameters $r_{l,n}$, $(l, n) \in E_{i+1}$:

$$E_{m+1,i+1} = r_{l,n} \cdot E_{m,i}. \tag{21}$$

From basic principles of psychoacoustics follows that human perception of acoustic information is quite inert, from 5 ms to 300 ms. Masking forward and backward is approximately 20 ms. With a input audio signal frame length of 5 ms. and processing delay determined by a single stage parallel pipeline processor, we can assume that the delay in processing the input signal $(l - 2)$ th levels of the processor (the maximum value of $l = 8$ for CB - WPD) much smaller than the temporal instability signal perceived by man. This allows you to organize multi-frame processing on the basis of parallel pipeline processors, a reconfiguration of the structure DWPT processor to determine the variability of the current signal frame - a frame for which to calculate the cost function H_{E_i}.

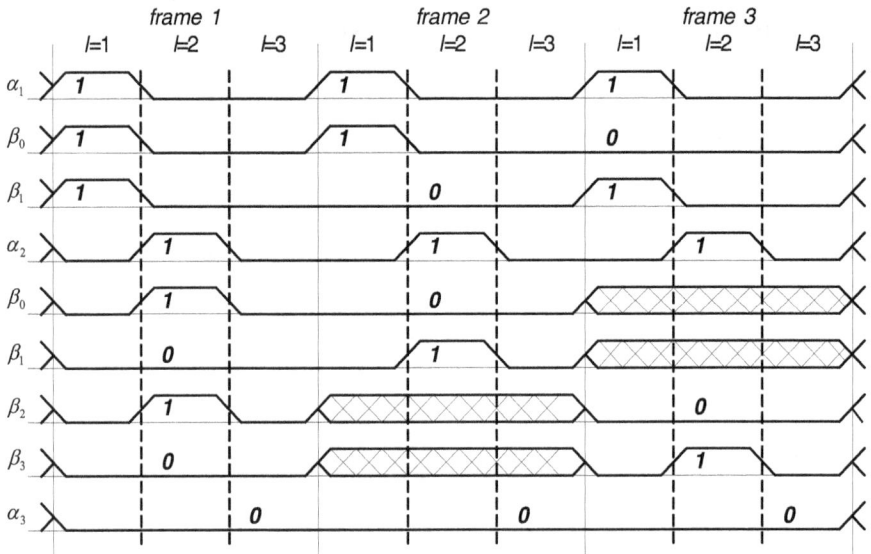

Figure 19. The timing diagram of control signals change for three consecutive frames of the audio

Dynamic Reconfigurable on the Lifting Steps Wavelet Packet Processor with Frame-Based
Psychoacoustic Optimized Time-Frequency Tiling for Real-Time Audio Applications

43

The profile of the time parameters α_l and β_n, the transformation vector processor $r_{l,n}$, in accordance with the tree structure $(l, n) \in E_i$ for three consecutive frames of the audio signal is shown in figure 19. DWPT tree structures that you see dotted line in figure 20 for the respective frames, determine options for their future growth in accordance with the obtained values of the perceptual entropy $PE_{l,n}$ at each node of the tree, but, for example, a value that indicates the informative density $H_{E_{m,i}}$ of the resulting decomposition tree DWPT shows the ineffectiveness of further growth of the tree structure.

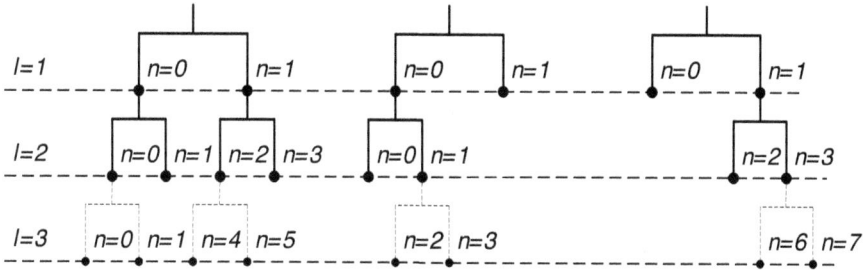

Figure 20. DWPT tree structures for figure 19

Thus, the DWPT trees structure E_i, described by the nodes (l, n), as well as the corresponding reconfiguration vector DWPT processor are obtained according to the algorithm of dynamic reconfiguration shown in the figure 21 and can be written as following:

for 1st frame: $E_1 = \{[(1,0);(1,1)],\ [(2,0);(2,1);(2,2)(2,3)]\}$ and vector is $r_1 = [(1,1,\ 1),\ (1,1,\ 0,1,\ 0),\ (0)]$;

for 2nd frame: $E_2 = \{[(1,0);(1,1)],\ [(2,0);(2,1)]\}$, and vector is $r_2 = [(1,1,\ 0),\ (1,0,\ 1,\ \times,\ \times),\ (0)]$;

for 3rd frame $E_3 = \{[(1,0);(1,1)],\ [(2,2);(2,3)]\}$, and vector is $r_3 = [(1,0,\ 1),\ (\times,\ \times,\ 0,1),\ (0)]$.

This algorithm of dynamic reconfiguration allows obtaining a suboptimal solution for DWPT analysis. The advantages of the above algorithm can be summarized as: pruning method is a top-down method, DWPT pruning can be viewed as a split process, i.e. we have the temporal construction DWPT tree for each signal frame that is ideal decision for real time processing implemented in a reconfigurable hardware.

The processing of the first nine frames in the pipeline DWPT processor is shown in the figure 22 where j is a number of the frame which was loaded to DWPT processor in the current time for processing according to the DWPT tree structure $E_{m,i}$ of the current frame i. The computation process at each stage of pipeline DWPT processor schematically shows with cubes, where cube mean a frame processing on corresponding DWPT processor stage. The cube "Master" means that on this stage a current frame is used for actual DWPT tree structure creation. The cube "Slave" means that on this stage a current frame is processed according to the actual DWPT tree structure. The cube "Master (suboptimal decomposition)" means that

Suppose, a computation resource $(m, r_{l,n})$ is limited by C, the limiting DWPT tree structure is $CB - WPD: (l, n) \in E_{CB}$, the required computation resource of i-th DWPT tree $(l, n) \in E_i$ is defined as value c_i, the split decision of dividing the parent node (l, n) on the two children $(l + 1, 2n)$ and $(l + 1, 2n + 1)$ is referred as a $split(l, n)$ where l is a transformation scale level, n is n-th node of the level l, m is a stage of DWPT processor and number of the input frame of the audio signal is j.

STEP 1.   Let $j = 1$, $m = 1$, $l = 0$ and $split(l, n) = YES$, $r_{l,n} = YES$, DWPT tree root node is $(0,0)$ for the first frame of the input audio signal with perceptual entropy $PE_{0,0}$, information density $H_{E_{0,0}}$ and reconfiguration process is allowed.

STEP 2. $i = j$, the first frame of the input audio signal defines growing process of DWPT tree structure. Making the signal decomposition is based on 1st PU.

STEP 3.   Estimate the perceptual entropy in each node $PE_{l,n}$ and information density $H_{E_{1,1}}$ of the DWPT tree structure.

STEP 4.   Check the DWPT tree structure information density $H_{E_{1,1}}$ in comparison with the DWPT tree $E_{0,0}$

　　　　　　　IF $H_{E_{0,0}} < H_{E_{1,1}}$,

　　　　　　　　　　THEN that is not an audio signal, the coefficients are not processed and GOTO STEP 1.

STEP 5. $l = l + 1$.

　　　　　　　IF $l - 1 >$ maximum of the scale level of the limiting DWPT tree structure $CB - WPD$,

　　　　　　　　　　THEN STOP – the growing process for the DWPT tree structure $E_{i,m}$ is finished.

STEP 6.   Check the DWPT tree structure $E_{i,m}$ nodes belonging to $CB - WPD: E_{CB}$:

　　　　　　　IF $(l, n) \in E_{m,i} = (l, n) \in E_{CB}$,

　　　　　　　　　　THEN $r_{l,n} = NO$.

STEP 7. $m = m + 1$. DWPT tree structure $E_{i,m}$ growing is performed as follows

　　　　　　　FOR each node n of the level l:

　　　　　　　　　　- estimate and check the adequacy of the DWPT tree computation resource $c_{i,m}$:

　　　　　　　IF $c_{i,m} > C$,

　　　　　　　　　　THEN $r_{l,n} = NO$ and $STOP$ – the growing process for the DWPT tree structure $E_{i,m}$ is finished.

　　　　　　　　　　　　　　- perform the decomposition of the parent node (l, n).

　　　　　　　　　　　　　　- calculate the perceptual entropy into the child nodes $PE_{l+1,2n}$ and $PE_{l+1,2n+1}$:

　　　　　　　IF $PE_{l,n} \geq PE_{l+1,2n} + PE_{l+1,2n+1}$,

　　　　　　　　　　THEN $split(l, n) = YES$, $r_{l,n} = YES$

　　　　　　　　　　　　　　ELSE $split(l, n) = NO$, $r_{l,n} = NO$.

STEP 8. $j = j + 1$. Read the next frame of the input audio signal. It will be processed according to DWPT tree structure $E_{i,m}$.

STEP 9.   Estimate the information density $H_{E_{i,m}}$ of the DWPT tree structure $E_{i,m}$:

　　　　　　　IF $H_{E_{i,m}} > H_{E_{i,m-1}}$,

　　　　　　　　　　THEN $r_{l,n} = NO$ and $STOP$ – the growing process for the DWPT tree structure $E_{i,m}$ is finished.

STEP 10.   GOTO STEP 5.

STOP.   Suboptimal DWPT tree structure for the ith input frame of the audio signal is $E_{i,m-1}$, $m = m - 2$, $l = l - 2$, $i = i + 1$. GOTO STEP 5.

Figure 21. Algorithm of dynamic reconfiguration

on this stage the suboptimal decomposition is fund for the current frame. The cube "Slave (suboptimal decomposition)" means that on this stage the suboptimal decomposition is fund for the current frame but it is slave. The last cube "Master (no optimal decomposition)" means that on this stage no optimal decomposition is detected. Here, the current frame i, for example $i = 0$, as it show in the figure 22, sets a master and begin from the stage $m = 0$ consequentially processed on each stage in DWPT processor, involving new frames $i = 1, 2, 3 \ldots$ as slave in processing while no optimal decomposition for the master will find at stage $m = 2$. The

Dynamic Reconfigurable on the Lifting Steps Wavelet Packet Processor with Frame-Based
Psychoacoustic Optimized Time-Frequency Tiling for Real-Time Audio Applications

45

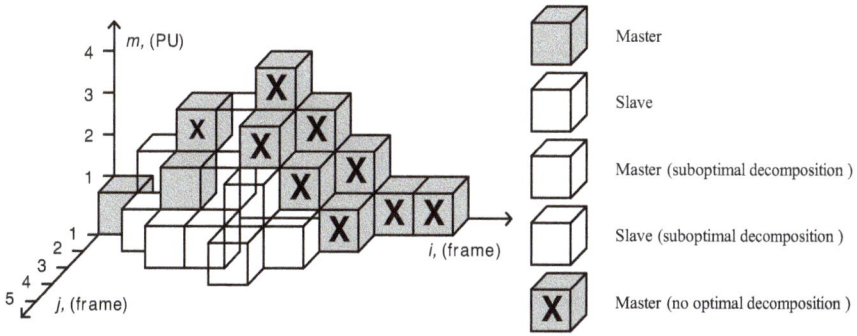

Figure 22. Diagram of the dynamic reconfiguration DWPT tree structures and multi-frame processing in the parallel-pipeline DWPT processor

suboptimal decomposition for master was founded on stage $m=1$. The result DWPT coefficients of the frame $i=0$ are removed from DWPT processor and next frame $i=1$ sets as a master and process is repeated while no optimal decomposition for the master will find at stage $m=3$. The suboptimal decomposition for the current frame $i=1$ was at stage $m=2$ then the result DWPT coefficients removed from DWPT processor and next frame $i=2$ sets as a master. As we can see, the decomposition is not optimal for the current frame $i=2$ then the suboptimal decomposition is at stage $m=1$ and next frame $i=3$ becomes a master. In the figure 22 is shown how DWPT tree is growing and involving a frames to computation process and how it rolling back with removing frames from the process. The reconfiguration in DWPT processor is based on the formation of transformation vectors $r_{l,n}$ according to the algorithm of dynamic transformation DWPT tree structure mentioned on figure 21 which is formed in DSP processor.

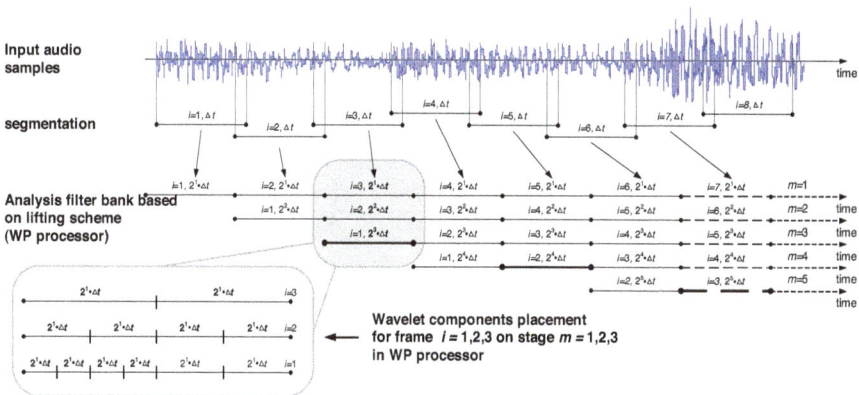

Figure 23. The input signal analysis in DWPT processor

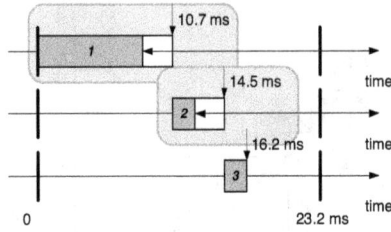

Figure 24. Time schedule operation in the DSP processor

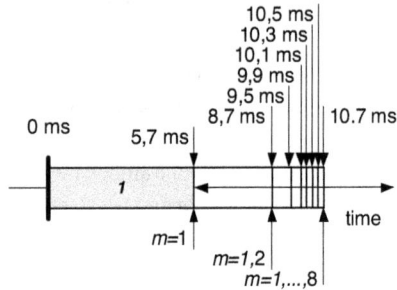

Figure 25. Run-time of the procedure 1, depending on a number of stages *m*

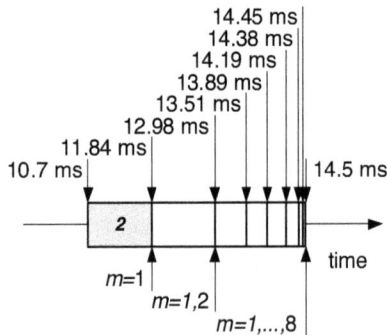

Figure 26. Run-time of the procedure 2, depending on the number of stages *m*

The complete input signal analysis in DWPT processor is demonstrated in the figure 23. The input signal is segmented on a frame with minimal overlapping and analysed. The frame length in a time is equal 22.3 ms. At the same time for frames to be processed under the current structure of the tree $E_{m,i}$ on the steps m of DWPT processor, a DSP processor shall monitor the implementation of the procedures such as: the masking threshold calculation algorithm as it described in appendix A [24] (procedure 1), perceptual entropy $PE_{l,n}$ assessment based on (4)-(5) (procedure 2) and the entropy of the DWPT tree structure H_E estimation according to (1)-(3) (procedure 3). The time schedule for DSP processor (250 MHz, 32 bit floating point DSP microprocessor) based on the listened above procedures are shown in the figure 24. The run time of procedure 1 and procedure 2 are showed in the figure 25 and figure 26 correspondingly as it is mentioned the computational time is not a constant, it depends on the number of stages m in DWPT processor involving in the input frames processing.

The output signal synthesis in DWPT processor is demonstrated in the Figure 27. The monitoring system loads the input frame i to the appropriate level m of DWPT processor. Move the frame i to the next stage of the processor is executed when the monitoring system takes the next frame i + 1, which will need to get involved is to step DWPT processor. To coordinate the work performed at each stage of the processor, it is necessary to introduce a delay, multiple processing time of one frame at one stage, the most rhythmic work will be provided by the parallel pipeline structure of DWPT processor.

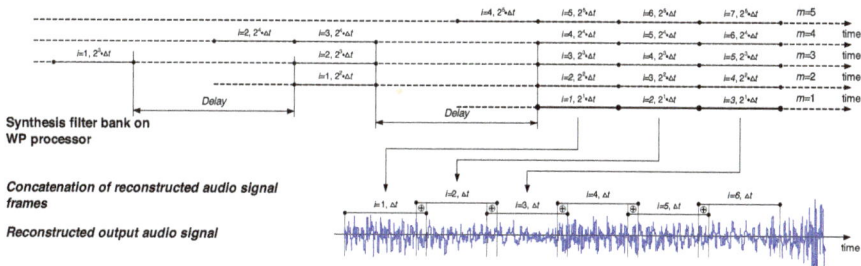

Figure 27. The output signal synthesis in DWPT processor

6. Conclusion

In the given paper the dynamic reconfigurable lifting based adaptive DWPT processor was presented. The lifting scheme allows to reduce on halve the number of multiplications and summations and increase the processing speed. Appling DAT-based approach as the design techniques for time-varying DWPT decomposition allows us to construct dynamically adapted to input signal DWPT analysis. The reconfigurable system offers several advantages over competing alternatives: faster and smaller than general purpose hardware solutions; lower development cost than dedicated hardware solutions; dynamic reconfigurable supports multiple algorithms within a single application; multi-purpose architecture generates volume demand for a single hardware design. The proposed techniques optimize system performance and, in addition, provide a convenient framework within which on-going research in the areas of non-uniform filter bank applied to speech/audio coding algorithms and reconfigurable architectures can be synergistically combined to enable the design of reconfigurable high-performance DSP systems.

Thus, the proposed dynamic reconfigurable DWPT processor with frame-based psychoacoustic optimized time-frequency tilling is successfully applicable for several application such as monophonic full-duplex audio coding system [18] and scalable audio coding based on hybrid signal decomposition where the transient part of the signal is modelled on psychoacoustic motivated frame based adaptive DWPT in marching pursuit algorithm [24]. The advantages of this DWPT processor is better viewed by considering the DWPT growing as a splitting process, i.e. the temporal construction DWPT tree created for each signal frame presents an ideal decision for real time processing implemented in a reconfigurable hardware.

Acknowledgements

This work was supported in part by Belarusian republican fund for fundamental research under the grants T04-217 and T08MC-040.

Author details

Alexey Petrovsky*, Maxim Rodionov and Alexander Petrovsky*

*Address all correspondence to: {petrovsky,post-rodmax,palex}@bsuir.by

Department of Computer Engineering, Belarusian State University of Informatics and Radioelectronics, Minsk, Belarus

References

[1] Mallat, S. A Theory of Multiresolution signal decomposition: the wavelet representation. IEEE transactions on pattern analysis and machine intelligence, july (1989). doi:, 11(7), 674-693.

[2] Coifman, R, & Wickerhauser, M. Entropy based algorithms for best basis selection. IEEE transaction of information theory, March (1992). doi:, 38, 712-718.

[3] Wickerhauser, M. V. Adapted wavelet analysis from theory to software. A Peters Wellesley, Massachusetts, (1994). p.

[4] Cohen, I, Raz, S, & Malah, D. Orthonormal shift-invariant adaptive wavelet packet decomposition and representation. Signal processing, 57 (3), march (1997). doi:S0165-1684(97)00007-8., 251-270.

[5] Wu, X, Li, Y, & Chen, H. Programmable wavelet packet transform processor. IEE electronics letters, (1999). doi:el:19990330., 35(6), 449-450.

[6] Trenas, M. A, Lopez, J, Sanchez, M, Arguello, F, & Zapata, E. L. Architecture for wavelet packet transform with best tree searching. Proceedings of IEEE international conference on Application-Specific Systems, Architectures and Processors, ASAP'00, (2000). doi:ASAP.2000.862399., 289-298.

[7] Trenas, M. A, Lopez, J, & Zapata, E. L. A configurable architecture for the wavelet packet transform. Journal of Signal Processing Systems, (2002). doi:A:1020221003822., 32(3), 255-273.

[8] Sweldens, W. The lifting scheme: A construction of second generation wavelets. SIAM: SIAM Journal on Mathematical Analysis, 29 (2) ((1997). , 511-546.

[9] Arguello, F, Lopez, J, Trenas, M. A, & Zapata, E. L. Architecture for wavelet packet transform based on lifting steps. Parallel Computting, (2002). doi:S0167-8191(02)00101-1., 28(7-8), 1023-1037.

[10] Aroutchelvame, S. M, & Raahemifar, K. Architecture of wavelet packet transform for 1-D signal. Proceedings of IEEE Canadian Conference on Electical and Computer Engineering, CCECE'05, may (2005). doi:CCECE.2005.1557216., 1304-1307.

[11] Paya, G, Peiro, M. M, Ballester, F, Herrero, V, & Mora, F. Lifting folded pipelined discrete wavelet packet transform architecture. VLSI Circuits and Systems. Edited by Lopez, Jose Fco.; Montiel-Nelson, Juan A.; Pavlidis, Dimitris. Proceedings of the SPIE, (2003). doi:, 5117, 321-328.

[12] Wang, C, & Gan, W. S. Efficient VLSI Architecture for Lifting-Based Discrete Wavelet Packet Transform. IEEE transactions on circuits and system- II: Express briefs, may (2007). doi:TCSII.2007.892410., 54(5), 422-426.

[13] Parhi, K. K. Algorithm transformation techniques for concurrent processors. Proceedings of the IEEE, dec. (1989). doi:, 77(12), 1879-1895.

[14] Petrovsky Al Petrovsky A. Dynamic algorithm transforms for reconfigurable real-time audio coding processor. Proceedings of the International Conference on Parallel Computing in Electrical Engineering, PARELEC'02, Warsaw, Poland, sep. 22-25, 2002, IEEE Computer Society Press, Los Alamitos, California, (2002). doi:PCEE. 2002.1115317., 422-424.

[15] Ackenhusen, J. G. Real-time signal processing: design and implementation of signal processing systems. Printice Hall, NJ, (1999). p.

[16] Villasenor, J, & Hutchings, B. The flexibility of configurable computing. IEEE Signal Processing Magazine, sep. (1998). doi:, 15(5), 67-84.

[17] Johnston, J. D. Transform coding of audio signals using perceptual noise criteria. IEEE Journal on Selected Areas in Communications, feb. (1988). doi:, 6(2), 314-323.

[18] Petrovsky Al Krahe D., Petrovsky A.A., Real-time wavelet packet-based low bit rate audio coding on a dynamic reconfigurable system. 114th AES Convention preprint 5778, March (2003). Amsterdam, Netherlands, 22 p.

[19] Petrovsky Al A multiresolution auditory model using adaptive WP excitation scalo-grams. ELEKTRONIKA, PAN, Warsaw, (2008). , 49(4), 65-70.

[20] Daubechies, I, & Sweldens, W. Factoring wavelet transforms into lifting steps. Journal of Fourier Analysis and Applications, (1998). , 4(3), 247-269.

[21] Coors, M, Keding, H, Luethje, O, & Meyr, H. Design and DSP implementation of fixed-point systems. EURASIP journal on advances in signal processing, (2002). doi:S1110865702205065., 908-925.

[22] Menard, D, Chillet, D, & Sentieys, O. Floating-to-fixed-point conversion for digital signal processors, EURASIP journal on applied signal processing, article ID 96421, doi:ASP/2006/96421., 2006, 1-19.

[23] Daubechies, I. Orthonormal bases of compactly supported wavelets II. Variations on theme. Communications on pure and applied mathematics, (1988). doi:, 41, 909-996.

[24] Petrovsky Al Azarov E., Petrovsky A. Hybrid signal decomposition based on instan-taneous harmonic parameters and perceptually motivated wavelet packets for scala-ble audio coding. Special Issue: "Fourier Related Transforms for Non-Stationary Signals", Signal Processing, june (2011). doi:j.sigpro.2010.09.005., 91(6), 1489-1504.

Optical Signal Processing

Chapter 3

Optical Signal Processing: Data Exchange

Jian Wang and Alan E. Willner

Additional information is available at the end of the chapter

1. Introduction

Optical signal processing is considered to be an attractive technique to enable fast signal ma-nipulation in the optical domain which can avoid cumbersome optical-electrical-optical (OEO) conversions [1]. Driven by the rapid increase of traffic rates, network capacity and complexity, advanced optical networks raise the significance of data traffic grooming and require different optical signal processing functions at network nodes to achieve enhanced network efficiency and flexibility. Typical optical signal processing operations include wavelength conversion, logic gate, format conversion, delay for buffer, regeneration, add/drop, (de)multiplexing, multicasting, etc [2-14]. One may note that most of these functions work in a similar fashion of unidirectional information transfer. For example, wavelength conversion copies information from one wavelength and transfers it onto another wave-length [2]. To achieve superior network performance, bidirectional information swapping, named data exchange, would be expected to provide enhanced flexibility of optical signal processing compared to unidirectional information transfer [15].

Generally speaking, as an important concept for efficiently utilizing network resources and im-proving network performance, data exchange refers to the information swapping between dif-ferent wavelengths/time-slots/polarizations or other degrees of freedom. In the wavelength domain (e.g., wavelength-division multiplexed (WDM) network), data exchange, which is al-so known as wavelength exchange or wavelength interchange, would require the swapping of data from one wavelength with the data from another wavelength. Extensions of data ex-change would expect the data swapping between different time-slots in the time domain (e.g., optical time-division multiplexed (OTDM) network), different polarization states in the polar-ization domain (e.g., polarization-multiplexed (pol-muxed) network), and different "twisted" light beams carrying different orbital angular momentum (OAM) values in the phase front do-main (e.g., OAM-multiplexed network). Moreover, the recently increasing interest of advanced modulation formats [16, 17] would require the data exchange to be available for different modulation formats, such as on-off keying (OOK), differential phase-shift keying (DPSK), differential quadrature phase-shift keying (DQPSK), pol-muxed, etc.

The emergence of nonlinear optics has triggered increased interest and paved a potential way to develop optical signal processing in high-speed optical networks [18, 19]. Optical nonlinearities (e.g. $\chi^{(2)}$ and $\chi^{(3)}$), including difference-frequency generation (DFG) [20, 21], cascaded sum- and difference-frequency generation (cSFG/DFG) [22-26], degenerate/non-degenerate four-wave mixing (FWM) [27-47], and Kerr-induced nonlinear polarization rotation [48-50], are potentially suitable candidates to enable data exchange. In some cases, simple linear optics may also provide an alternative approach to facilitating data exchange [51, 52]. To fulfill the rapid development of high-speed large-capacity optical communications with emerging multiplexing/demultiplexing techniques and advanced modulation formats, as shown in Fig. 1, a laudable goal would be to achieve robust data exchange in different degrees of freedom (wavelength, time, polarization, phase front), for different modulation formats (OOK, DPSK, DQPSK, pol-muxed), and at different granularities (entire data, groups of bits, tributary channels).

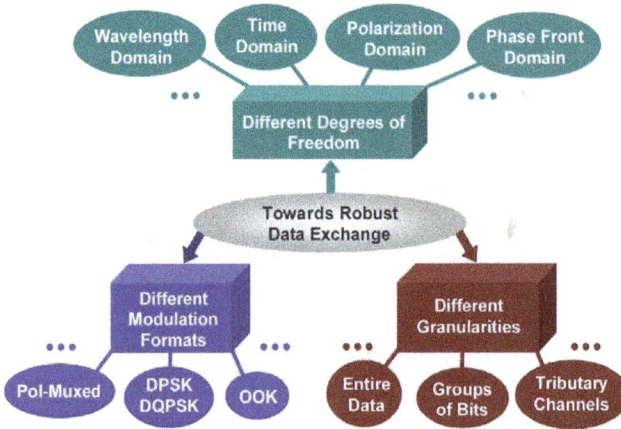

Figure 1. Schematic illustration of robust data exchange.

In this chapter, we tend to provide a comprehensive review of research works towards robust data exchange using various optical nonlinearities [22-50, 53] and simple linear optics [51, 52]. Several kinds of optical nonlinearities are employed: (1) cSFG/DFG in a periodically poled lithium niobate (PPLN) waveguide; (2) non-degenerate FWM in a highly nonlinear fiber (HNLF); (3) bidirectional degenerate FWM in an HNLF; (4) Kerr-induced nonlinear polarization rotation in an HNLF; (5) Conversion-dispersion-based tunable delays. We start with a conceptual description of data exchange followed by state-of-the-art results.

2. Concept of data exchange

Robust data exchange in the wavelength, time, polarization and phase front domains might be valuable for superior network performance. As an example, a desirable goal of data exchange would be to efficiently utilize nonlinearities in the wavelength domain, such that the data between two different wavelengths can be "exchanged", i.e., swapped, using nonlinear processes in a single device [54]. Figure 2(a) illustrates the basic concept of data exchange in the wavelength domain (wavelength exchange/interchange), which is a wavelength-domain data manipulation enabling the swapping of data between two different wavelengths. One straightforward way, as shown in Fig. 2 (b), is to use two separate wavelength converters (WCs) with one performing the wavelength conversion from signal A (Sig. A) to signal B (Sig. B), and the other from signal B to signal A. Towards single-device operation, one simple way of data exchange in the wavelength domain is to explore the combined signal depletion and wavelength conversion effects in a nonlinear device including a piece of HNLF or a PPLN waveguide [55-58]. Non-degenerate FWM ($\chi^{(3)}$) in an HNLF [29-45] and cascaded second-order nonlinearities ($\chi^{(2)} : \chi^{(2)}$) in a PPLN waveguide [22-26] are potential choices to realize such data exchange. As shown in Fig. 2 (c), due to the signal depletion and wavelength conversion effects, the data carried by signal A is consumed and converted to the wavelength of signal B and vice versa. This enables single-device-based data exchange in the wavelength domain. Similar concepts of data exchange in the time, polarization and phase front domains are also available enabled by various optical nonlinearities or linear optics.

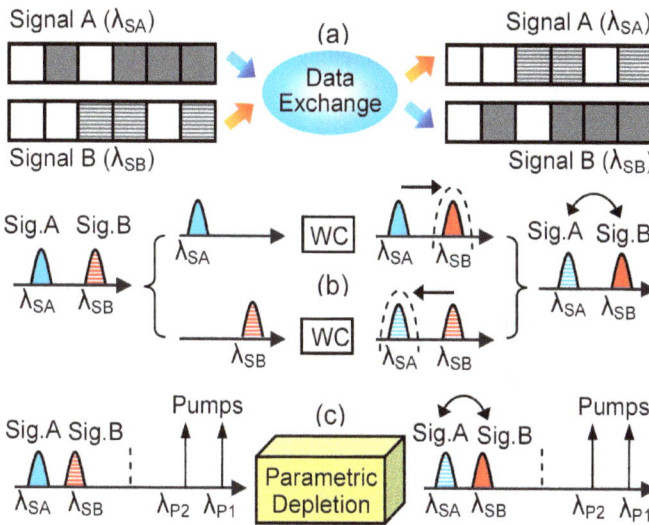

Figure 2. a) Concept of data exchange in the wavelength domain. (b) Data exchange by two separate wavelength converters (WCs). (c) An example of data exchange by signal depletion and wavelength conversion in a single nonlinear device.

3. Recent advances for robust data exchange

3.1. Data exchange using cSFG/DFG in a single PPLN waveguide [22-26]

As depicted in Fig. 2(c), data exchange based on signal depletion and wavelength conversion of cSFG/DFG involves two signals and two pumps, which can be described by the coupled-mode equations. To better understand the single-PPLN-based data exchange, under the slowly varying amplitude approximation, we can derive the following analytical solutions to the complex amplitudes of signal A ($A_{SA}(L)$) and signal B ($A_{SB}(L)$) after data exchange [22]

$$A_{SA}(L) = A_{SA}(0) + \omega_{SA}\omega_{SF}\kappa_1 \frac{1}{M^2} A_{P1}^*(0)[\kappa_1 A_{P1}(0)A_{SA}(0) + \kappa_2 A_{P2}(0)A_{SB}(0)][\cos(ML)-1] \text{ (a)}$$

$$A_{SB}(L) = A_{SB}(0) + \omega_{SB}\omega_{SF}\kappa_2 \frac{1}{M^2} A_{P2}^*(0)[\kappa_1 A_{P1}(0)A_{SA}(0) + \kappa_2 A_{P2}(0)A_{SB}(0)][\cos(ML)-1] \text{ (b)}$$

(1)

where $M = \sqrt{\omega_{SA}\omega_{SF}\kappa_1^2 P_{P1}(0) + \omega_{SB}\omega_{SF}\kappa_2^2 P_{P2}(0)}$. $A_{SA}(0)$, $A_{SB}(0)$, $A_{P1}(0)$ and $A_{P2}(0)$ are the input complex amplitudes of signal A, signal B, pump 1 and pump 2, respectively. $P_{P1}(0)$ and $P_{P2}(0)$ are the input power of pump 1 and pump 2. κ_1 (κ_2) refers to the coupling coefficient of the second-order nonlinear interaction involving signal A (signal B) and pump 1 (pump 2). ω_{SA}, ω_{SB} and ω_{SF} are the angular frequencies of signal A, signal B and sum-frequency (SF) wave, respectively. L is the waveguide length.

When ignoring the initial pump phase and setting the same power for two input pumps, we can further simplify Eqs. (1a)(1b) as follows

$$A_{SA}(L) = \frac{\cos(ML)+1}{2} A_{SA}(0) + \frac{\cos(ML)-1}{2} A_{SB}(0) \text{ (a)}$$

$$A_{SB}(L) = \frac{\cos(ML)-1}{2} A_{SA}(0) + \frac{\cos(ML)+1}{2} A_{SB}(0) \text{ (b)}$$

(2)

When satisfying the following relationship written by

$$ML = (2N+1)\pi, \quad N = 0, 1, 2, 3 \cdots \cdots$$

(3)

we can obtain

$$A_{SA}(L) = -A_{SB}(0) \text{ (a)}$$
$$A_{SB}(L) = -A_{SA}(0) \text{ (b)}$$

(4)

From Eq. (4) it can be clearly seen that data exchange between signal A and signal B is achieved under the exchange condition governed by Eq. (3). In particular, beyond the data exchange for OOK signal, the complex relationship in Eq. (4) also implies the modulation-format-transparency characteristic of PPLN-based data exchange.

Following the similar principle of PPLN-based data exchange using signal depletion and wavelength conversion of cSFG/DFG, we can further perform robust data exchange functions, including time- and channel-selective data exchange between WDM channels [23, 24] and low-speed tributary channel exchange of high-speed OTDM signals [25, 26].

The conceptual diagram of the proposed single-PPLN-based time- and channel-selective data exchange between WDM channels is illustrated in Fig. 3 [23, 24]. Multiple WDM channels (S1-S4) and two synchronized gated pumps (PA, PB) are coupled into a PPLN waveguide, in which cSFG/DFG processes take place. The wavelength selectivity of the quasi-phase matching (QPM) condition allows selection of channels for data exchange by proper choice of the two pump wavelengths. For proper QPM of both cSFG/DFG processes, the two pump wavelengths are nearly symmetric to the two exchanged data wavelengths with respect to the QPM wavelength. For instance, as illustrated in Fig. 3, within the gated pump pulse duration, PB mixes with S1 to produce an SF wave through the sum-frequency generation (SFG) process. Meanwhile, the SF wave interacts with PA to generate a new idler at the wavelength of S2 by the subsequent difference-frequency generation (DFG) process. During such nonlinear interactions, S1 can be depleted, and converted to S2 by means of proper control of the pump powers. Similarly, PA and S2 participate in the SFG process to create an SF wave, which simultaneously interacts with PB to yield an idler at the wavelength of S1 via the DFG process. Thus, S2 can also be consumed with its data copied onto S1. Consequently, it is expected to implement optical data exchange between S1 and S2 without the use of additional spectrum and touching other channels. Note that time- and channel-selective data exchange in specific time-slots (groups of bits) and between selective WDM channels can be accomplished by appropriately choosing the gated pump pulse duration and adjusting the pump wavelengths.

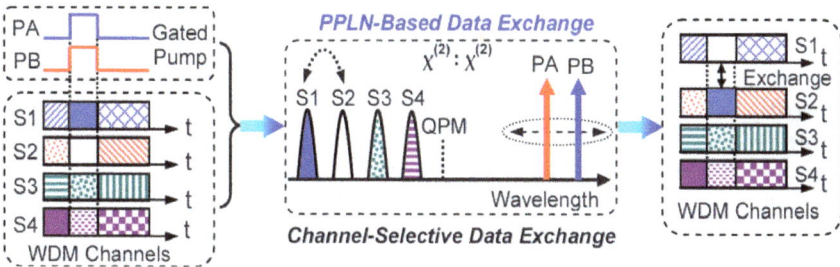

Figure 3. Concept and principle of single-PPLN-based time- and channel-selective data exchange between WDM channels.

We first demonstrate the data exchange between two 10-Gbit/s signals. Two gated pumps with a duty cycle of 1/127 and a pulse duration of ~3.2 ns are employed. The average power of each signal and peak power of each pump coupled into the PPLN waveguide are about 4 mW and 1 W, respectively. Figure 4 displays the observed temporal waveforms and eye diagrams of data exchange. The time-slots between the two straight lines correspond to the gated pump pulse duration, in which data exchange occurs. When S1 and the two pumps are on while S2 is off, the data of S1 within the gated pump pulse duration is depleted (a2) and converted to the wavelength of S2 (b3). Similarly, we can also observe the depletion of S2 (b2) and the conversion from S2 to S1 (a3) by switching S1 off and S2 on. In the case of simultaneously turning on the two signals and the two pumps, it is found that groups of bits data exchange between the two signals (S1 to S2: (b4), S2 to S1: (a4)) within the gated pump pulse duration is successfully achieved.

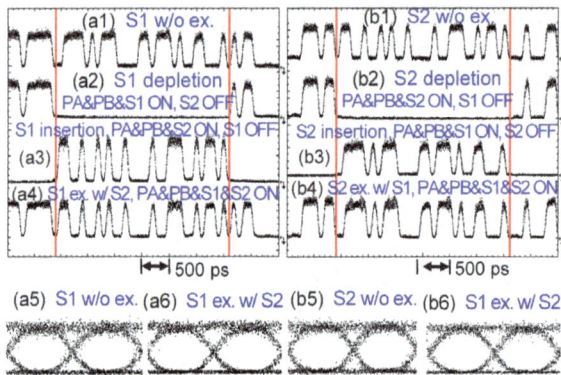

Figure 4. Measured (a1-a4)(b1-b4) temporal waveforms and (a5)(a6)(b5)(b6) eye diagrams of 10-Gbit/s groups of bits data exchange.

We further demonstrate the single-PPLN-based channel-selective data exchange for multiple WDM channels at 40 Gbit/s. Four WDM channels (S1: 1535.5 nm, S2: 1539.4 nm, S3: 1543.3 nm, S4: 1547.2 nm) are employed in the experiment. It is possible to perform a channel-selective data exchange by simply tuning the wavelength of the two pumps. Figure 5 displays the measured typical eye diagrams and bit-error rate (BER) performance for channel-selective data exchange between WDM channels. The power penalty of 40-Gbit/s channel-selective data exchange is estimated to be less than 4 dB at a BER of 10^{-9}.

Figure 6 illustrates the concept and principle of single-PPLN-based tributary channel exchange between two WDM high-speed OTDM signals [25, 26]. A PPLN waveguide is employed as the nonlinear device to perform the tributary channel exchange. Two WDM high-speed signals (S1, S2) each consisting of many low-speed time-division multiplexed tributary channels (e.g., 16 10-Gbit/s tributary channels for 160-Gbit/s signal), together with two synchronized subrate clock (e.g., 10 GHz) pumps, are launched into the PPLN waveguide for the tributary channel exchange. The wavelengths of two signals and two pumps are properly arranged to be symmetric (S1&P1, S2&P2) with respect to the QPM wavelength of PPLN. Inside the PPLN waveguide, two signals and two pumps participate in the cSFG/DFG nonlinear interactions, in which the photons of S1 (S2) and P1 (P2) are annihilated to produce the photons of SF wave, which are simultaneously consumed to generate the photons of S2 (S1) and P2 (P1). Due to the signal depletion and wavelength conversion effects, with the proper adjustment of pump powers, S1 can be depleted and converted to S2. Similarly, S2 can be extinguished to generate S1. As a result, data exchange between two signals (S1, S2) can be implemented. In particular, by exploiting two synchronized subrate (e.g., 10 GHz) clock pumps which are time aligned to one of the tributary channels of two WDM high-speed OTDM signals (e.g., 160 Gbit/s), it is possible to achieve the tributary channel exchange (e.g., 10 Gbit/s) between two WDM high-speed OTDM signals (e.g., 160 Gbit/s). As an example shown in Fig. 6, the tributary channel i (Ch. i) of two WDM high-speed OTDM signals is exchanged by using the signal depletion and wavelength conversion effects of cSFG/DFG in a PPLN waveguide.

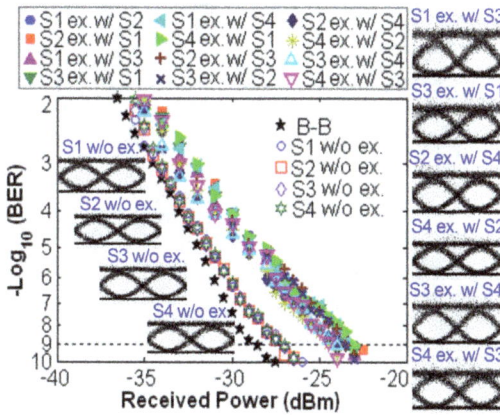

Figure 5. Measured eye diagrams and BER performance of 40-Gbit/s time- (groups of bits) and channel-selective data exchange between four WDM channels.

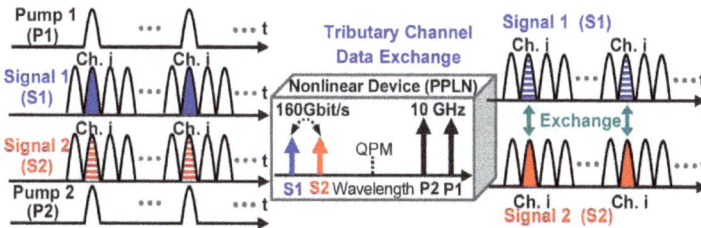

Figure 6. Concept and principle of single-PPLN-based tributary channel exchange between two WDM high-speed OTDM signals.

Figure 7 displays the eye diagrams for tributary channel exchange (Ch.1) measured by an optical sampling scope. Two 10-GHz clock pumps are time aligned to the tributary Ch.1 of two 160-Gbit/s signals. When the two pumps and S1 are present while S2 absent, Ch.1 of S1 is depleted and converted to the Ch.1 of S2 with the proper adjustment of pump powers and polarization states due to the signal depletion and wavelength conversion effects. Similarly, as the two pumps and S2 are turned on while S1 off, Ch.1 of S2 is extinguished with its data information copied onto the Ch.1 of S1. In the presence of two 10-GHz pumps and both two 160-Gbit/s signals, Ch.1 of S2 is exchanged to the Ch.1 of S1. Meanwhile, Ch.1 of S1 is swapped to the Ch.1 of S2, resulting in the implementation of 10-Gbit/s tributary channel exchange between two 160-Gbit/s signals. Moreover, it is convenient to further perform the 10-Gbit/s tributary exchange for all 16 tributary channels of two 160-Gbit/s signals simply by time shifting the 10-GHz clock pumps to be aligned with the corresponding tributary channel of interest.

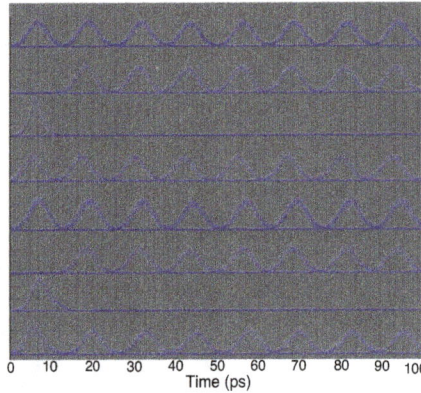

Figure 7. Measured eye diagrams for the tributary channel exchange (Ch. 1).

Figure 8 depicts power penalties at a BER of 10^{-9} of tributary exchange between two 160-Gbit/s signals for all 16 tributary channels. During the tributary channel exchange between two 160-Gbit/s signals, the average power penalty and the fluctuation of 16 tributary channels is around 3.7 and 1.1 dB for S1 (S2 to S1) and 3.9 and 1.1 dB for S2 (S1 to S2).

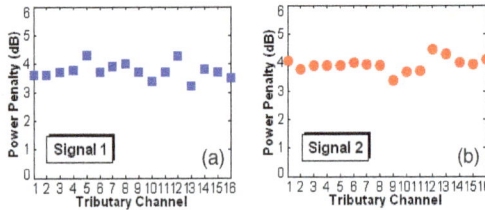

Figure 8. Power penalties of tributary exchange for 16 tributary channels. (a)(c) Signal 1. (b)(d) Signal 2.

3.2. Modulation-format-transparent data exchange using non-degenerate FWM in an HNLF [38-41]

In addition to cSFG/DFG ($\chi^{(2)} : \chi^{(2)}$) in a PPLN waveguide [22-26], signal depletion and wavelength conversion of non-degenerate FWM ($\chi^{(3)}$) in an HNLF can also enable the data exchange [29-41]. As shown in Fig. 9(a), when signal 1 (S1: λ_{S1}) and two continuous-wave (CW) pumps (P1: λ_{P1}, P2: λ_{P2}) are sent through the HNLF with S1 and P1 set symmetrically with respect to the zero-dispersion wavelength (ZDW) of the HNLF, S1 and P1 photons are consumed to produce photons of signal 2 (S2: λ_{S2}) and P2 during the non-degenerate FWM process. Thus the depletion of S1 is expected with its data information transparently copied onto a newly generated S2. Similarly, as shown in Fig. 9(b), the depletion of S2 accompanied by the generation of S1 can be achieved as S2 and two pumps are launched into the HNLF. As shown in Fig. 9(c), in the presence of two signals and two pumps at the input of HNLF with S1(S2) and P1(P2) symmetric relative to the ZDW of the HNLF, S1(S2) can be extinguished and converted to S2(S1), resulting in the implementation of data exchange between S1 and S2.

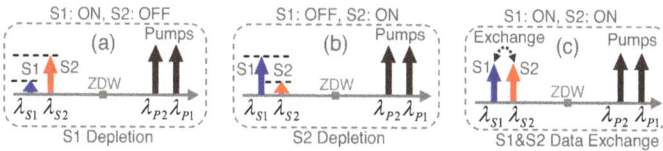

Figure 9. Concept and principle of non-degenerate FWM-based signal depletion and data exchange. (a) S1 depletion. (b) S2 depletion. (c) S1 & S2 data exchange.

For non-degenerate FWM-based data exchange, pump phase modulation is adopted in the experiment to suppress the stimulated Brillouin scattering (SBS) effect so that pump power can be efficiently utilized. Previous works of non-degenerate FWM-based data exchange have been reported for OOK signals [29-37], which are not affected by the phase modulation of two pumps. Shown in Fig. 10 is an example of data exchange (i.e., wavelength exchange) for 10-Gbit/s non-return-to-zero (NRZ) signals [32]. In order to perform phase-transparent data exchange for DPSK/DQPSK signals, it is desired that non-degenerate FWM-based data exchange has the characteristic of modulation-format-transparency.

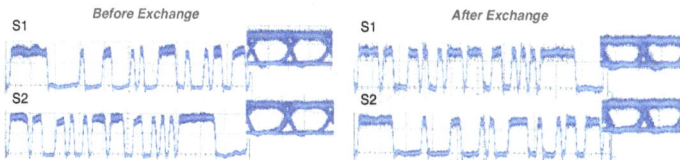

Figure 10. Results of data exchange (wavelength exchange) for 10-Gbit/s NRZ signals [32].

Under the non-depletion approximation, we derive the analytical solutions for the non-degenerate FWM involving two signals and two pumps written as [39]

$$g = \sqrt{4\gamma^2 P_{10} P_{20} + k^2 / 4} \tag{5}$$

where $g = \sqrt{4\gamma^2 P_{10} P_{20} + k^2 / 4}$ and $k = \Delta\beta + \gamma (P_{10} - P_{20})$ are constants related to the pump powers (P_{10}, P_{20}), nonlinear coefficient (γ), and phase mismatching ($\Delta\beta$). A_{SA0}, A_{SB0}, A_{P10} and A_{P20} are the complex amplitudes of input signals (SA, SB) and pumps (P1, P2) containing both amplitude and phase information. A'_{SA} and A'_{SB} are the complex amplitudes of output signals (SA, SB) after the data exchange. Under the exchange condition of phase matching (k =0) and $gz = (N + 1/2)\pi$ ($N = 0, 1, 2, ...$) enabled by the proper adjustment of pump powers, we can further simplify Eqs. (5a)(5b) as follows

$$A'_{SA} \propto A_{SB0} \tag{6}$$

Note that Eqs. (6a)(6b) indicate the linear relationship of complex amplitude between the output and input signals ($A'_{SA} \propto A_{SB0}$, $A'_{SB} \propto A_{SA0}$), implying the implementation of phase-transparent optical data exchange. We can further obtain the corresponding phase relationships of $\phi_{SA}' = \phi_{SB} + \phi_{P2} - \phi_{P1}$ and $\phi_{SB}' = \phi_{SA} + \phi_{P1} - \phi_{P2}$. Remarkably, the pump phase transfer ($\phi_{P1} - \phi_{P2} \neq 0$) to the exchanged signals does not impact on the OOK data exchange but could cause severe degradation on the DPSK/DQPSK data exchange. Fortunately, according to the deduced phase relationships, it is possible to cancel the pump phase transfer by applying the precisely identical phase modulation to the two pumps (i.e., $\phi_{P1} = \phi_{P2}$), which makes it applicable to implement the data exchange of DPSK/DQPSK signals [38-41].

A 1-km piece of HNLF is adopted in the experiment, which has a nonlinear coefficient of 9.1 $W^{-1} \cdot km^{-1}$, a ZDW of ~1552 nm, and a fiber loss of 0.45 dB/km. To suppress SBS, identical phase modulation is applied to the two pumps using a single phase modulator (PM) driven by a 10-Gbit/s pseudo-random binary sequence (PRBS). According to Eqs. (6a)(6b), the precisely identical phase modulation of the two pumps could be canceled in the output signals after data exchange.

We demonstrate the phase-transparent data exchange between two 100-Gbit/s 2^7-1 PRBS return-to-zero DQPSK (RZ-DQPSK) signals (S1: signal 1, S2: signal 2) [40, 41]. Figure 11(a) displays the measured temporal waveforms of the demodulated in-phase (Ch. I) and quadrature (Ch. Q) components for the 100-Gbit/s RZ-DQPSK data exchange. It can be clearly observed that the data information carried by two 100-Gbit/s RZ-DQPSK signals is successfully swapped after the non-degenerate FWM-based data exchange. Figure 11(b) and (c) plot BER curves for the 100-Gbit/s RZ-DQPSK data exchange. Less than 1.2-dB power penalty at a BER of 10^{-9} is obtained for the 100-Gbit/s RZ-DQPSK wavelength conversion (WC) with only one signal (S1 or S2) present. Less than 5-dB power penalty at a BER of 10^{-9} is observed for the 100-Gbit/s RZ-DQPSK data exchange. The extra power penalty of data exchange compared to wavelength conversion could be ascribed to the beating effect between the newly converted signal and the original residual signal.

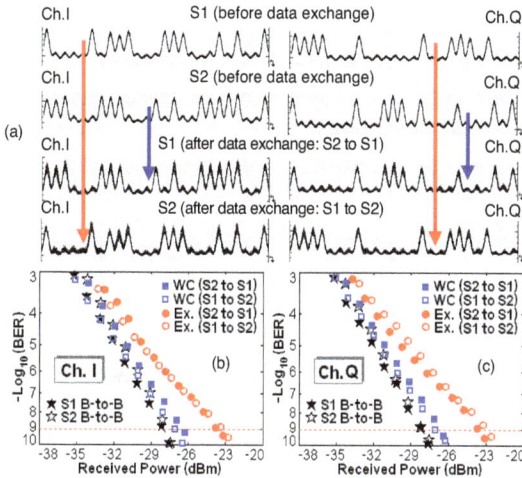

Figure 11. Measured (a) demodulated waveforms and (b)(c) BER curves for 100-Gbit/s RZ-DQPSK data exchange.

3.3. Multi-channel data exchange using bidirectional degenerate FWM in an HNLF [42-45]

The aforementioned signal depletion and wavelength conversion based schemes with two pumps enable the two-channel data exchange [22-26, 29-41]. However, the extended applications to simultaneous multi-channel data exchange might be limited. A laudable goal would be to explore the data exchange between multi-channel signals.

Figure 12 illustrates the concept and principle of multi-channel data exchange [42, 43]. Degenerate FWM with a single CW pump is utilized. Four-channel DQPSK signals (S1-S4) are symmetric with respect to the CW pump. Simultaneous data exchange between S1 and S4 as well as S2 and S3 is expected. In general, such exchange function is not applicable with the unidirectional degenerate FWM in a single HNLF since the newly converted signals cannot be separated from the original signals. A potential solution is to explore the bidirectional degenerate FWM in a single HNLF assisted by optical filtering. As shown in Fig. 12, for the input four-channel signals (S1-S4), the filtered S1, S2 and CW pump are sent to HNLF from the left side, yielding S4 and S3 via degenerate FWM. The newly generated S4 and S3 are selected at the right side of HNLF while the original S1, S2 and CW pump are blocked. Meanwhile, the filtered S3, S4 and CW pump are fed into HNLF from the right side, producing S2 and S1 by degenerate FWM. The newly converted S2 and S1 are selected at the left side of HNLF while the original S3, S4 and CW pump are removed. As a consequence, simultaneous four-channel data exchange (S1&S4, S2&S3) can be achieved using bidirectional FWM in a single HNLF assisted by optical filtering. The combined S1-S4 from both sides of HNLF are the output four-channel signals after data exchange. Note that the in-phase (Ch. I) and quadrature (Ch. Q) components of DQPSK signals are swapped after data exchange due to the phase-conjugation characteristic of degenerate FWM.

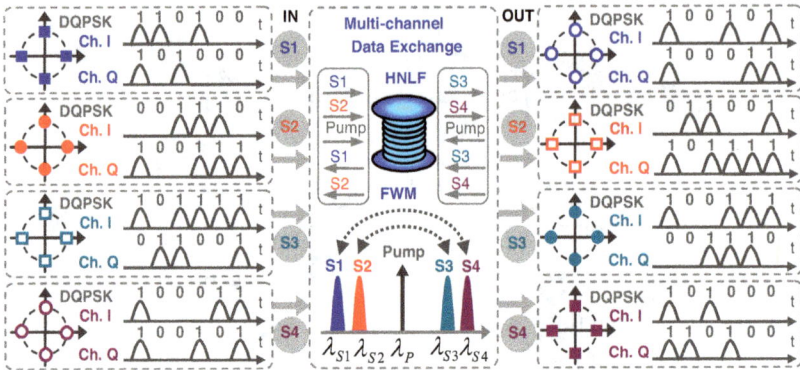

Figure 12. Concept and principle of simultaneous multi-channel DQPSK data exchange.

The proposed simultaneous multi-channel data exchange can be incorporated in a reconfigurable network switching element to enhance the efficiency and flexibility of optical networks. We construct a reconfigurable Tbit/s network switching element using double-pass liquid crystal on silicon (LCoS) technology accompanied by bidirectional degenerate FWM in a single HNLF. We demonstrate the LCoS+HNLF-based 2.3-Tbit/s multi-functional grooming switch which performs simultaneous selective add/drop, switchable data exchange, and power equalization, for 23-channel 100-Gbit/s RZ-DQPSK signals [44, 45].

ITU-grid-compatible 23-channel (from S1: 1531.12 nm to S23: 1566.31 nm) 100-Gbit/s RZ-DQPSK signals are employed in the experiment. Figure 13 shows the measured spectrum of the input unequalized 23-channel 100-Gbit/s RZ-DQPSK signals with a power fluctuation of ~9.1 dB. Shown in the insets are typical balanced eyes for the in-phase (Ch. I) and quadrature (Ch. Q) components.

Shown in Fig. 14 is the measured spectrum and balanced eyes after grooming switch with power equalization (<1 dB) for all 23 channels (input unequalization: ~9.1 dB), two-channel add/drop for S6 and S7, and simultaneous six-channel data exchange (S10, S11, S12, S21, S22, S23). The inset of Fig. 14 depicts the spectrum of dropped S6 and S7. The BER performance is plotted in Fig. 15 and power penalties less than 5 dB for six-channel data exchange are observed at a BER of 10^{-9}.

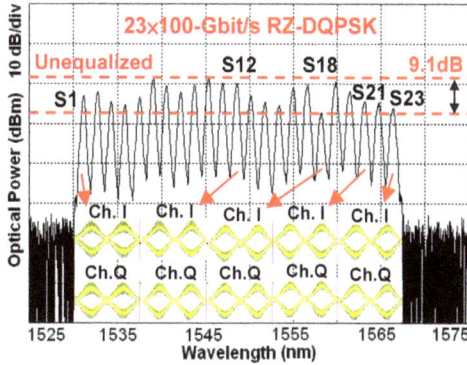

Figure 13. Measured spectrum and balanced eyes for input unequalized 23-channel 100-Gbit/s RZ-DQPSK signals.

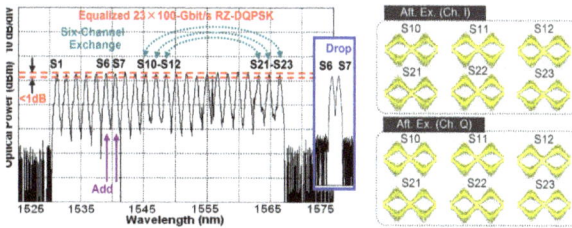

Figure 14. Measured spectrum and balanced eyes after multi-functional grooming switch (S6, S7: add/drop; S10, S11, S12, S21, S22, S23: data exchange; S1-S23: power equalization).

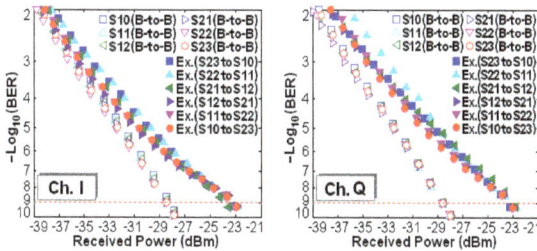

Figure 15. BER curves for simultaneous six-channel data exchange (S10, S11, S12, S21, S22, S23).

3.4. Data exchange between two orthogonal polarizations using kerr-induced nonlinear polarization rotation in an HNLF [48, 49]

In addition to the data exchange in the wavelength and time domains [22-45], it is also possible to perform data exchange between two orthogonal polarizations in the time and polarization domains [48-50]. We experimentally demonstrate the orthogonal tributary channel exchange between two pol-muxed DPSK OTDM data streams by using the Kerr effect-induced nonlinear birefringence in an HNLF [48, 49].

Figure 16 illustrates the concept and principle of the Kerr effect-based orthogonal tributary channel exchange of a pol-muxed DPSK OTDM signal. The strong subrate clock pump is 45° linearly polarized with respect to the two orthogonal polarizations of a pol-muxed DPSK OTDM signal. With the help of proper pump power control, the pump-induced nonlinear birefringence by Kerr effect could bring the selected tributary channel (aligned with the subrate clock pump) to a 90° polarization rotation for both of the two orthogonal polarizations of the pol-muxed signal, leading to the orthogonal tributary channel exchange when the pump is present. Other unselected orthogonal tributary channels with the pump absent will not experience the nonlinear polarization rotation and hence will be untouched. In addition, simply by shifting the subrate clock pump to be aligned with the tributary channel of interest, it is possible to implement orthogonal tributary channel exchange for all tributary channels of the pol-muxed DPSK OTDM signal.

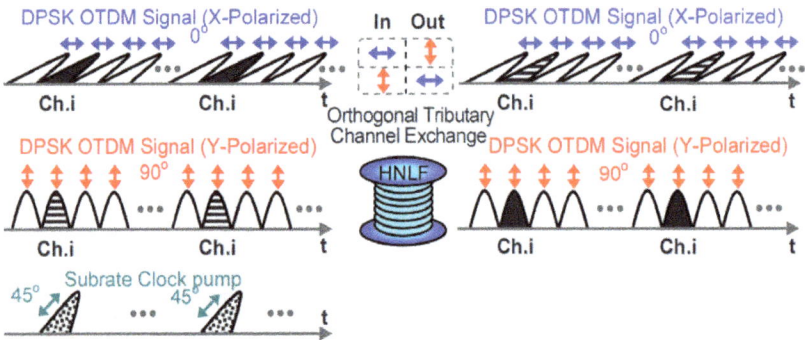

Figure 16. Concept and principle of Kerr effect-based orthogonal tributary channel exchange of a pol-muxed DPSK OTDM signal.

Figure 17 displays the eye diagrams measured by an optical sampling scope for the typical orthogonal tributary channel (Ch. 1) exchange of a 160-Gbit/s pol-muxed DPSK OTDM signal. As the 10-GHz clock pump is time aligned to tributary channel 1 (Ch. 1) of the X- and Y-polarized DPSK OTDM signal, in the absence of the Y-polarization, as shown in Fig. 17(b), Ch. 1 of the X-polarization is blocked by an X-polarizer after the HNLF due to the 90° rotation from the X- to the Y-polarization. When the Y-polarization is present but the X-polarization is absent, Ch. 1 of the Y-polarization is inserted to the X-polarization through the 90° rotation from the Y- to the X-polarization, as shown in Fig. 17(c). In the presence of both the X- and Y-polarizations, the tributary Ch. 1 of the Y-polarization is changed to the X-polarization, as shown in Fig. 17(d). Meanwhile, the original tributary Ch. 1 of the X-polarization is also changed to the Y-polarization, as shown in Fig. 17(h), resulting in the orthogonal tributary channel exchange of a pol-muxed DPSK OTDM signal.

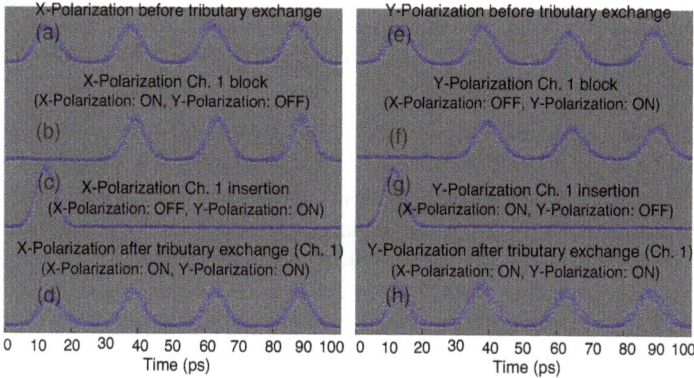

Figure 17. Eye diagrams of orthogonal tributary channel (Ch. 1) exchange of a 160-Gbit/s pol-muxed DPSK OTDM signal.

Figure 18 plots the power penalties of the orthogonal tributary exchange for 8 tributary channels. Less than 4-dB power penalty at a BER of 10^{-9} is obtained for all 8 tributary channels with a fluctuation of <1.5 dB.

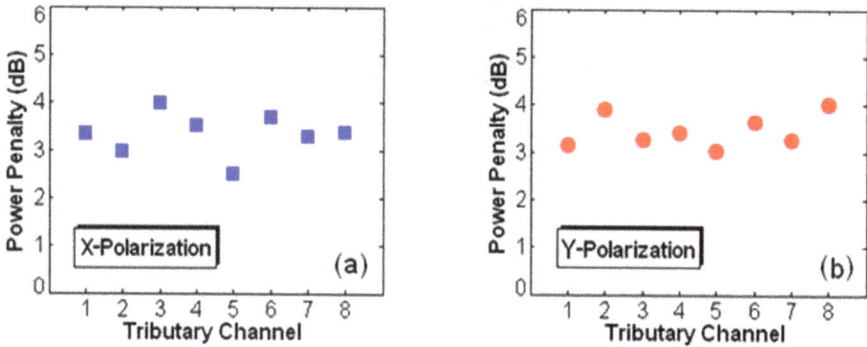

Figure 18. Power penalties of orthogonal tributary exchange for 8 tributary channels.

3.5. Time-slot-exchange using conversion-dispersion-based tunable delays [46, 47]

The demonstrated data exchange of groups of bits or tributary channels manipulates data in multiple degrees of freedom, such as time- (groups of bit) and channel-selective data exchange between WDM channels [23, 24], tributary channel exchange between two WDM high-speed OTDM signals [25, 26], and orthogonal tributary channel exchange of a pol-muxed OTDM signal [48-50]. Another important traffic grooming function, known as time-slot exchange or time-slot interchange, is to manipulate data only in the time domain to enable contention resolution and increase throughput efficiency in time-based networks. Time-slot exchange, occurring on the bit or packet level, can afford the network enhanced flexibility. For packet-switched networks, exchanging full data packets in the time domain requires optical delays that are tunable.

Figure 19 shows concept and principle of conversion-dispersion-based time-slot exchange of two separate packets in the time domain [46]. Three clocked pumps (λ_{P1}, λ_{P2}, λ_{P3}) are fed into an HNLF, along with a packetized input signal (λ_S) located near the ZDW of HNLF. Degenerate FWM between the clocked pumps and signal generates replicas of the input signal at new converted wavelengths (λ_1, λ_2, λ_3) which contain only the information of the input signal at times when the clocked pumps are on. The three pumps are clocked to convert: (i) only packet A to λ_1, (ii) all information but packets A and B to λ_2, and (iii) only packet B to λ_3. The three converted signals (λ_1, λ_2, λ_3) then pass through a dispersion module, such as dispersion compensation fiber (DCF), and experience a wavelength-dependent delay via inter-channel chromatic dispersion. Due to the conversion-dispersion-based tunable delays, packet A is advanced, while packet B is retarded, relative to the reference (all information but packets A and B), resulting in the swapping of packets A and B in the time domain. After the delays, all three converted signals are converted back to the original signal wavelength using a PPLN waveguide followed by the compensation for intra-channel chromatic dispersion.

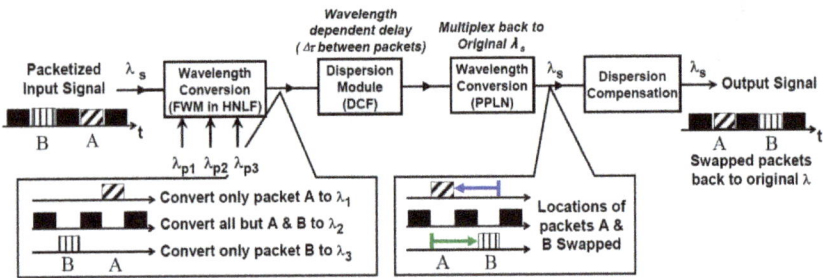

Figure 19. Concept and principle of time-slot exchange of two separate packets using conversion-dispersion-based tunable delays.

Shown in Figs. 20 and 21 are experimental results of time-slot exchange of 40-Gbit/s optical data packets using conversion-dispersion-based tunable delays. Separate 182-bit packets are converted to separate wavelengths, delayed relative to one another using conversion-dispersion-based tunable delays, and then recombined together to achieve new packets with time slots exchanged.

Similarly, time-slot exchange between odd and even data packets is also achievable using conversion-dispersion-based tunable delays [47]. As illustrated in Fig. 22, odd and even data packets are extracted from an input signal via wavelength multicasting with clocked pumps, delayed relative to one another in a dispersion module, and then multiplexed back together using wavelength conversion in a PPLN waveguide followed by dispersion compensation. Shown in Fig. 23 are experimental results of time-slot exchange of 40-Gbit/s odd and even packets. The conversion-dispersion-based tunable delays enable time-slot exchange of variable length optical packets (182 and 288 bits/packet).

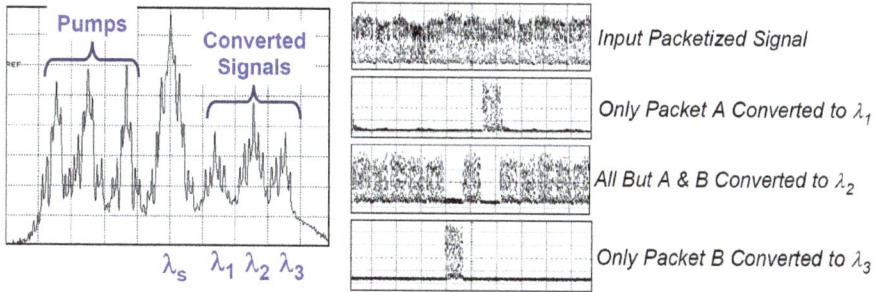

Figure 20. Optical spectra (left) and temporal waveforms (right) of converted signals after multicasting using clocked pumps in HNLF.

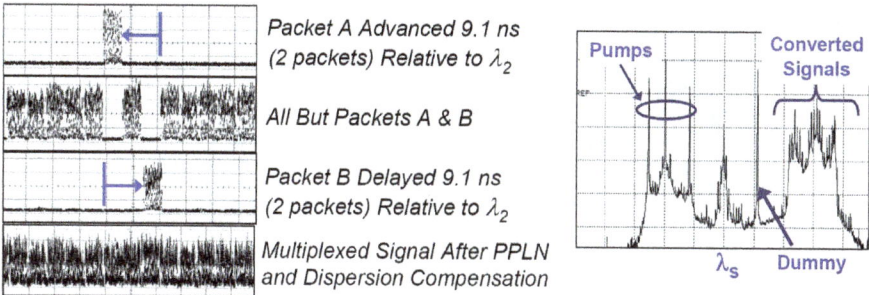

Figure 21. Temporal waveforms of optical signals following delay via dispersion, conversion back to the original signal wavelength and dispersion compensation (left) and optical spectra after PPLN (right).

3.6. Data exchange between "twisted" light beams carrying Orbital Angular Momentum (OAM) [51, 52]

In optical communications, beyond well-known existing degrees of freedom such as wavelength, time and polarization, other degrees of freedom are encouraged to be explored to break "capacity crunch". For example, OAM which is related to the helical phase front of "twisted" light beams [59-61], can be considered as an additional degree of freedom [62, 63], where the multiplexing of data-carrying OAM beams provides yet another dimension in the ever-continuing effort to increase the capacity and spectral efficiency of communication links [63]. When employing OAM beams to carry data information, a desirable function for flexible data processing would be the data exchange between "twisted" OAM beams.

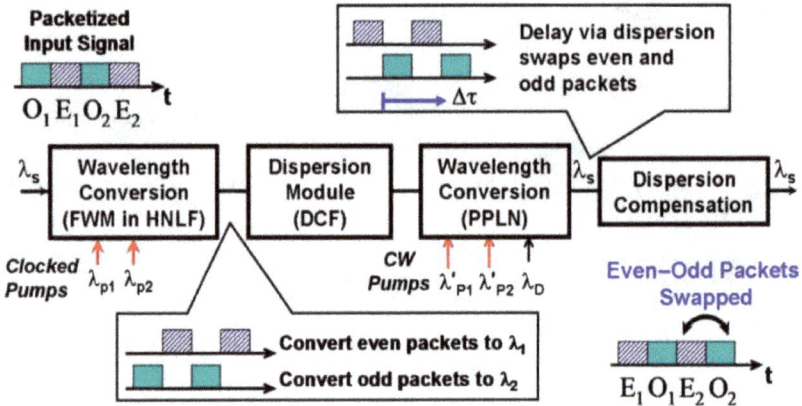

Figure 22. Concept and principle of time-slot exchange of odd and even packets.

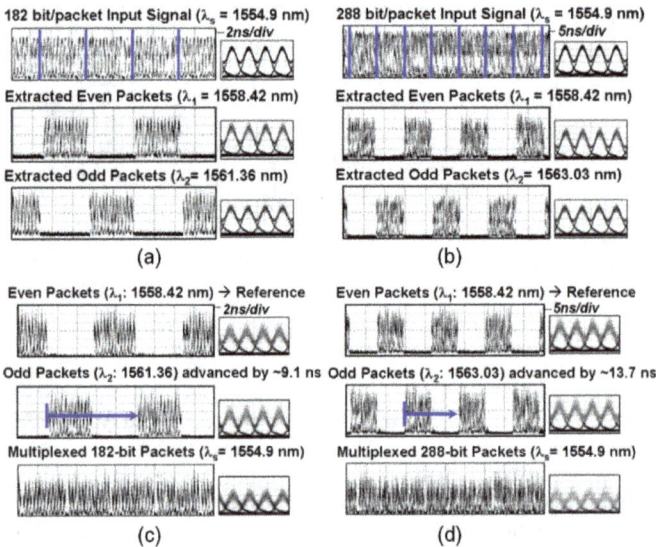

Figure 23. Temporal waveforms of time-slot exchange of 40-Gbit/s odd and even packets with two variable packet lengths. (a)(c) 182 bits/packet. (b)(d) 288 bits/packet.

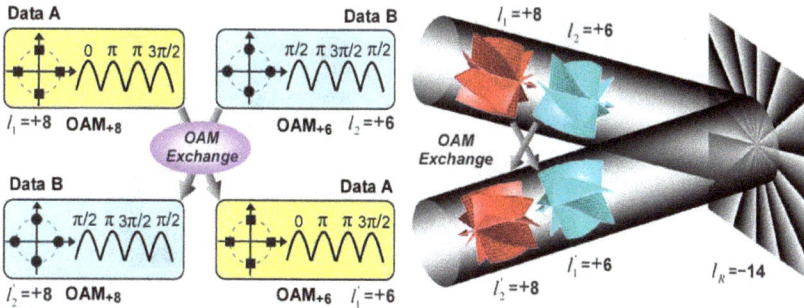

Figure 24. Concept and principle of data exchange between "twisted" OAM beams.

Figure 24 shows concept and principle of data exchange between OAM beams [51, 52]. Superposed two OAM beams (OAM_{ℓ_1}, OAM_{ℓ_2}), each carrying different data information (Signal A, Signal B), shine at a reflective-type spatial light modulator (SLM) loaded with a spiral phase mask with a charge of $\ell_R = -(\ell_1 + \ell_2)$. After reflecting off the SLM, this phase mask adds an azimuthal phase term $\exp(i\ell_R\theta)$ to the two OAM beams and converts them into $OAM_{-\ell_2}$ and $OAM_{-\ell_1}$, which are further transformed into OAM_{ℓ_2} and OAM_{ℓ_1} due to reflection of the SLM which flips the charge sign. As a result, data exchange between two OAM beams is implemented. For the input of two OAM beams with varied charges, reconfigurable data exchange is available by updating the phase mask loaded into the reflective-type SLM. Shown in Fig. 24 is an example of data exchange between DQPSK-carrying "twisted" OAM beams (OAM_{+8}, OAM_{+6}).

The measured interferograms (i.e., interference between OAM beams and a reference Gaussian beam), as shown in Fig. 25(a) and (b), indicate that the charges of two OAM beams before exchange are +8 and +6. After exchange, the measured interferograms, as shown in Fig. 25(c) and (d), verify that the charges of two OAM beams after exchange become +6 and +8 (see Ref. 52 for details). Figure 25(e)-(h) show measured interferograms of reconfigurable data exchange between another two OAM beams (OAM_{+10} and OAM_{+6}) by updating the spiral phase mask loaded into the SLM.

We measure temporal waveforms and balanced eyes of demodulated in-phase (Ch. I) and quadrature (Ch. Q) components of 100-Gbit/s RZ-DQPSK signals. As shown in Fig. 26, the observed temporal waveforms confirm the successful implementation of data exchange between two OAM beams (OAM_{+10} and OAM_{+6}). Shown in Fig. 27 are measured BER curves for 100-Gbit/s RZ-DQPSK data exchange between OAM_{+10} and OAM_{+6} beams with power penalty <1.9-dB at a BER of 1e^{-9}.

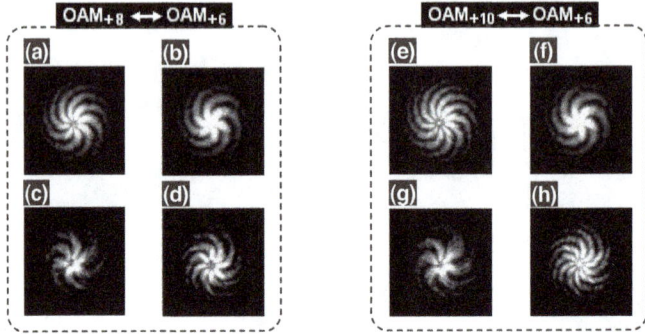

Figure 25. Measured interferograms. (a) OAM$_{+8}$ and (b) OAM$_{+6}$ beams become (c) OAM$_{+6}$ and (d) OAM$_{+8}$ ones after exchange. (e) OAM$_{+10}$ and (f) OAM$_{+6}$ beams become (g) OAM$_{+6}$ and (h) OAM$_{+10}$ ones after exchange.

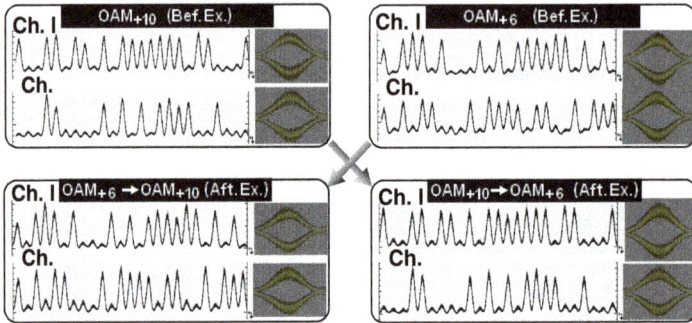

Figure 26. Measured waveforms and balanced eyes of demodulated in-phase (Ch. I) and quadrature (Ch. Q) components of 100-Gbit/s RZ-DQPSK signals for data exchange between OAM$_{+10}$ and OAM$_{+6}$ beams. Bef. Ex., before exchange; Aft. Ex., after exchange.

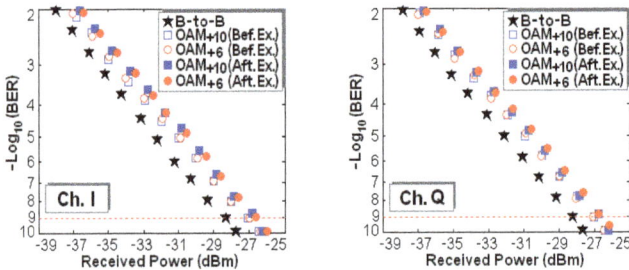

Figure 27. Measured BER curves for 100-Gbit/s RZ-DQPSK data exchange between OAM$_{+10}$ and OAM$_{+6}$ beams.

4. Discussions

The demonstrated miscellaneous data exchange functionalities provide great potential for facilitating flexible networks. With future improvement, several aspects could be considered as follows.

1. In practical applications, some supplementary functionalities might be required to construct complete and independent data exchange modules. Taking tributary channel exchange as an example (Figs. 6-8), subrate clock pumps should be synchronized with high-speed OTDM signals. Note that the incoming signals and locally generated pumps are usually independent with each other. Hence, a supplementary functionality of clock recovery is required in real situations to get synchronized subrate clock pumps from incoming signals. Fortunately, various optical clock recovery methods have been developed [64]. In particular, recent promising demonstrations have shown the successful synchronization and sub-clock recovery for ultra-high speed OTDM signals up to 640 Gbit/s [65, 66]. As a consequence, it is possible develop a complete and independent data exchange module by incorporating a synchronization and clock recovery unit.

2. Beyond reported functionalities, data exchange can be further extended in terms of degrees of freedom, modulation formats, and granularities. For example, some additional degrees of freedom have recently attracted increasing interest in high-speed optical fiber communications to break the "capacity crunch", such as space [67, 68] and mode [69, 70]. A valuable goal would be achieve data exchange in these degrees of freedom. Also, some high-level modulation formats have been used in fiber transmission, such as 16-ary quadrature amplitude modulation (16-QAM), 32-QAM, etc [67, 68, 71]. According to the demonstrated characteristic of modulation-format-transparency, most of the presented data exchange should be, in principle, available for these advanced modulation formats. However, high-level modulation formats show reduced tolerable performance degradation, and therefore accurate manipulation of amplitude and phase would be expected. Additionally, data exchange with fine granularity in the time domain requires accurate control of time delay, which could be achievable assisted by the fine tuning of conversion-dispersion-based optical delays [7].

3. In addition to PPLN and HNLFs, there would be some other alternative candidates applicable for data exchange, including the use of third-order nonlinearities in semiconductor optical amplifiers (SOAs) [3], As_2S_3 waveguides [19], and silicon waveguides [18].

5. Conclusion

In this chapter, we have reviewed recent research efforts towards robust data exchange. Various kinds of optical nonlinearities, i.e., cSFG/DFG in a PPLN waveguide, non-degenerate FWM in an HNLF, bidirectional degenerate FWM in an HNLF, Kerr-induced nonlinear polarization rotation in an HNLF, and conversion-dispersion-based tunable delays, together with simple linear optics, are exploited to enable robust data exchange in different degrees of freedom (wavelength, time, polarization, phase front), for different modulation formats (OOK, DPSK, DQPSK, pol-muxed), and at different granularities (entire data, groups of bits, tributary channels).

First, analytical solutions to the single-PPLN-based data exchange are derived showing the exchange condition. 40-Gbit/s time- (groups of bits) and channel-selective data exchange between four WDM channels is implemented. 10-Gbit/s tributary channel exchange between two WDM 160-Gbit/s OTDM signals is demonstrated. Second, analytical solutions to the non-degenerate FWM-based data exchange are derived indicating the exchange condition and implying the characteristic of modulation-format-transparency. Phase-transparent data exchange (entire data) of 100-Gbit/s RZ-DQPSK signals is demonstrated. Third, a simple approach is proposed to perform simultaneous multi-channel data exchange using bidirectional degenerate FWM in an HNLF. A reconfigurable Tbit/s network switching element is constructed using double-pass LCoS technology, together with bidirectional degenerate FWM in a single HNLF. LCoS+HNLF-based 2.3-Tbit/s (23X100-Gbit/s RZ-DQPSK) multi-functional grooming switch (e.g., simultaneous add/drop, six-channel data exchange, and power equalization) is implemented. Fourth, 10-Gbit/s tributary channel exchange between two orthogonal polarizations of a 160-Gbit/s pol-muxed DPSK OTDM signal is demonstrated based on the Kerr-induced nonlinear polarization rotation. Fifth, time-slot exchange of 40-Gbit/s optical data packets is demonstrated using conversion-dispersion-based tunable delays. Finally, reconfigurable 100-Gbit/s RZ-DQPSK data exchange between "twisted" OAM beams is demonstrated using simple linear optics.

The obtained theoretical and experimental results of data exchange in the wavelength, time, polarization and phase front domains, show that robust data exchange for different modulation formats and at different granularities could potentially enhance the efficiency and flexibility of optical networks.

Acknowledgements

We acknowledge Jeng-yuan Yang, Xiaoxia Wu, Scott R. Nuccio, Omer F. Yilmaz, Zahra Bakhtiari, Hao Huang, Xue Wang, Nisar Ahmed, Irfan Fazal, Yan Yan, Yang Yue, Lin Zhang, Yinying Xiao-Li, Bishara Shamee, Yongxiong Ren, Amanda Bozovich, Robert W. Hellwarth, Moshe Tur, Kevin Birnbaum, John Choi, Baris Erkmen and Samuel Dolinar for the helpful discussions, and the generous support of the National Natural Science Foundation of China (NSFC) under grants 61077051, 11274131, 61222502, the Program for New Century Excellent Talents in University (NCET-11-0182), the Defense Advanced Research Projects Agency (DARPA) under contract FA8650-08-1-7820, and the DARPA under InPho (Information in a Photon) program.

Author details

Jian Wang[1*] and Alan E. Willner[2]

*Address all correspondence to: jwang@hust.edu.cn

1 Wuhan National Laboratory for Optoelectronics, College of Optoelectronic Science and Engineering, Huazhong University of Science and Technology, Wuhan, Hubei, China

2 Department of Electrical Engineering, University of Southern California, Los Angeles, California, USA

References

[1] Saruwatari M. All-optical signal processing for terabit/second optical transmission. IEEE J. Sel. Top. Quantum Electron. 2000; 6(6): 1363-1374.

[2] Yoo S.J.B. Wavelength conversion technologies for WDM network applications. J. Lightwave Technol. 1996; 14(6): 955-966.

[3] Chan K, Chan CK, Chen LK, Tong F. Demonstration of 20-Gb/s all-optical XOR gate by four-wave mixing in semiconductor optical amplifier with RZ-DPSK modulated inputs. IEEE Photon. Technol. Lett. 2004; 16(3): 897-899.

[4] Wang J, Sun JQ, Zhang XL, Huang DX, Fejer MM. All-optical format conversions using periodically poled lithium niobate waveguides. IEEE J. Quantum Electron. 2009; 45(2): 195-205.

[5] Okawachi Y, Sharping JE, Xu C, Gaeta AL. Large tunable optical delays via self-phase modulation and dispersion. Opt. Express 2006; 14(25): 12022-12027.

[6] Wang Y, Yu CY, Yan LS, Willner AE, Roussev R, Langrock C, Fejer MM, Sharping JE, Gaeta AL. 44-ns continuously tunable dispersionless optical delay element using a PPLN waveguide with two-pump configuration, DCF, and a dispersion compensator. IEEE Photon. Technol. Lett. 2007; 19(11): 861-863.

[7] Nuccio SR, Yilmaz OF, Wu X, Willner AE. Fine tuning of conversion/dispersion based optical delays with a 1 pm tunable laser using cascaded acousto-optic mixing. Opt. Lett. 2010; 35(4): 523-525.

[8] Dai Y, Okawachi Y, Turner-Foster AC, Lipson M, Gaeta AL, Xu C. Ultralong continuously tunable parametric delays via a cascading discrete stage. Opt. Express 2010; 18(1): 333-339.

[9] Salem R, Foster MA, Turner AC, Geraghty DF, Lipson M, Gaeta AL. Signal regeneration using low-power four-wave mixing on silicon chip. Nature Photonics 2008; 2(1): 35-38.

[10] Kataoka N, Sone K, Wada N, Aoki Y, Kinoshita S, Miyata H, Miyazaki T, Onaka H, Kitayama K. Field trial of 640-Gbit/s-throughput, granularity-flexible optical network using packet-selective ROADM prototype. J. Lightwave Technol. 2009; 27(7): 825-832.

[11] Wang, J., Fu, H. Y., Geng, D. Y., & Willner, A. E. All-optical wavelength-/time-selective switching/dropping/swapping for 100-GHz-spaced WDM signals using a periodically poled lithium niobate waveguide. ECOC2012, paper Th.1.A.5, 2012.

[12] Wu, X. X., Bogoni, A., Yilmaz, O. F., Nuccio, S., Wang, J., & Willner, A. E. Eightfold 40-320 Gbit/s phase-coherent multiplexing and 320-40 Gbit/s demultiplexing using highly nonlinear fibers. Opt. Lett. (2010). , 35(11), 1896-1898.

[13] Brès, C. S., Boggio, J. M. C., Alic, N., & Radic, S. 1-to-40 10-Gb/s channel multicasting and amplification in wideband parametric amplifier. IEEE Photon. Technol. Lett. (2008). , 20(16), 1417-1419.

[14] Biberman A, Lee BG, Turner-Foster AC, Foster MA, Lipson M, Gaeta AL, Bergman K. Wavelength multicasting in silicon photonic nanowires. Opt. Express 2010; 18(17): 18047-18055.

[15] Hamza HS, Deogun JS. Wavelength-exchanging cross connects (WEX)—a new class of photonic cross-connect architectures. J. Lightwave Technol. 2006; 24(3): 1101-1111.

[16] Winzer PJ, Essiambre RJ. Advanced optical modulation formats. Proc. IEEE 2006; 94(5): 952-985.

[17] Winzer PJ, Essiambre RJ. Advanced modulation formats for high-capacity optical transport networks. J. Lightwave Technol. 2006; 24(23): 4711-4728.

[18] Oxenløwe LK, Ji H, Galili M, Pu MH, Hu H, Mulvad, HCH, Yvind K, Hvam JM, Clausen AT, Jeppesen P. Silicon photonics for signal processing of Tbit/s serial data signals. IEEE J. Sel. Top. Quantum Electron. 2012; 18(2): 996-1005.

[19] Pelusi MD, Ta'eed VG, Fu LB, Mägi E, Lamont MRE, Madden S, Choi DY, Bulla DAP, Luther-Davies B, Eggleton BJ. Applications of highly-nonlinear chalcogenide glass devices tailored for high-speed all-optical signal processing. IEEE J. Sel. Top. Quantum Electron. 2008; 14(3): 529-539.

[20] Chowdhury A, Hagness SC, McCaughan L. Simultaneous optical wavelength interchange with a two-dimensional second-order nonlinear photonic crystal. Opt. Lett. 2000; 25(11): 832-834.

[21] Chowdhury A, Staus C, Boland BF, Kuech TF, McCaughan L. Experimental demonstration of 1535–1555 nm simultaneous optical wavelength interchange with a nonlinear photonic crystal. Opt. Lett. 2001; 26(17): 1353-1355.

[22] Wang, J., Sun, Q. Z. Theoretical analysis of power swapping in quadratic nonlinear medium. Appl. Phys. Lett. (2010). , 96(8), 081108.

[23] Wang, J., Nuccio, S. R., Wu, X., Yilmaz, O. F., Zhang, L., Fazal, I., Yang, J. Y., Yue, Y., & Willner, A. E. 40-Gbit/s optical data exchange between WDM channels using second-order nonlinearities in PPLN waveguides. NLO 2009, paper PDPA1, 2009.

[24] Wang, J., Nuccio, S. R., Wu, X., Yilmaz, O. F., Zhang, L., Fazal, I., Yang, J. Y., Yue, Y., & Willner, A. E. 40 Gbit/s optical data exchange between wavelength-division-multiplexed channels using a periodically poled lithium niobate waveguide. Opt. Lett. (2010). , 35(7), 1067-1069.

[25] Wang, J., Bakhtiari, Z., Xiao-Li, Y., Yilmaz, O. F., Nuccio, S. R., Wu, X., Huang, H., Yang, J. Y., Yue, Y., Fazal, I., Hellwarth, R., & Willner, A. E. Experimental demonstration of data traffic grooming of a single 10-Gbit/s TDM tributary channel between two 160-Gbit/s WDM channels. OFC 2010, paper OWF1, 2010.

[26] Wang, J., Bakhtiari, Z., Yilmaz, O. F., Nuccio, S. R., Wu, X., & Willner, A. E. 10 Gbit/s tributary channel exchange of 160 Gbit/s signals using periodically poled lithium niobate. Opt. Lett. (2011). , 36(5), 630-632.

[27] Mori K, Takara H, Saruwatari M. Wavelength interchange with an optical parametric loop mirror. Electron. Lett. 1997; 33(6): 520-522.

[28] Gao Y, Dai YH, Shu C, He SL. Wavelength interchange of phase-shift-keying signal. IEEE Photon. Technol. Lett. 2010; 22(11): 838-840.

[29] Wong, K. K. Y., Marhic, M. E., Uesaka, K., & Kazovsky, L. G. Demonstration of wavelength exchange in a highly nonlinear fiber. ECOC 2001, pp. 272-273, 2001.

[30] Uesaka, K., Wong, K. K. Y., Marhic, M. E., & Kazovsky, L. G. Polarization-insensitive wavelength exchange in highly-nonlinear dispersion-shifted fiber. OFC2002, paper ThY3, 2002.

[31] Uesaka K, Wong KKY, Marhic ME, Kazovsky LG. Wavelength exchange in a highly nonlinear dispersion-shifted fiber: theory and experiments. IEEE J. Sel. Topics Quantum Electron. 2002; 8(3): 560-568.

[32] Fung, R. W. L., Cheung, H. K. Y., & Wong, K. K. Y. Widely tunable wavelength exchange in anomalous-dispersion regime. IEEE Photon. Technol. Lett. (2007). , 19(22), 1846-1848.

[33] Cheung, H. K. Y., Fung, R. W. L., Kwok, C. H., & Wong, K. K. Y. All-optical packet switching by pulsed-pump wavelength exchange in a highly nonlinear dispersion-shifted fiber. OFC2007, paper OTuB4, 2007.

[34] Shen M, Xu X, Yuk TI, Wong KKY. Byte-level parametric wavelength exchange for narrow pulsewidth return-to-zero signals. IEEE Photon. Technol. Lett. 2009; 21(21): 1591-1593.

[35] Kwok, C. H., Kuo, B. P. P., & Wong, K. K. Y. Pulsed pump wavelength exchange for high speed signal de-multiplexing. Opt. Express (2008). , 16(15), 10894-10899.

[36] Shen, M., Xu, X., Yuk, T. I., & Wong, K. K. Y. A 160-Gb/s OTDM demultiplexer based on parametric wavelength exchange. IEEE J. Quantum Electron. (2009). , 45(11), 1309-1316.

[37] Shen, M., Cheung, H. K. Y., Fung, R. W. L., & Wong, K. K. Y. A comprehensive study on the dynamic range of wavelength exchange and its impact on exchanged signal performance. J. Lightwave Technol. (2009). , 27(14), 2707-2716.

[38] Wang, J., Bakhtiari, Z., Xiao-Li, Y., Nuccio, S. R., Yilmaz, O. F., Wu, X., Yang, J. Y., Yue, Y., Fazal, I., Hellwarth, R., & Willner, A. E. Phase-transparent optical data exchange of 40-Gbit/s DPSK signals using four-wave-mixing in a highly nonlinear fiber. OFC 2010, paper OMT6, 2010.

[39] Wang J, Bakhtiari Z, Nuccio SR, Yilmaz OF, Wu X, Willner AE. Phase-transparent optical data exchange of 40 Gbit/s differential phase-shift keying signals. Opt. Lett. 2010; 35(17): 2979-2981.

[40] Wang, J., Nuccio, S. R., Huang, H., Wang, X., Yilmaz, O. F., Wu, X., Yang, J. Y., Yue, Y., & Willner, A. E. Demonstration of 100-Gbit/s DQPSK data exchange between two different wavelength channels using parametric depletion in a highly nonlinear fiber. ECOC 2010, paper Mo.1.A.4, 2010.

[41] Wang J, Nuccio SR, Huang H, Wang X, Yang JY, Willner AE. Optical data exchange of 100-Gbit/s DQPSK signals. Opt. Express 2010; 18(23): 23740-23745.

[42] Wang, J., Huang, H., Wang, X., Yang, J. Y., & Willner, A. E. Optical phase-transparent data grooming exchange of multi-channel 100-Gbit/s RZ-DQPSK signals. IEEE 23rd Photonics Society Annual Meeting 2010, paper WN2, 2010.

[43] Wang, J., Huang, H., Wang, X., Yang, J. Y., & Willner, A. E. Multi-channel 100-Gbit/s DQPSK data exchange using bidirectional degenerate four-wave mixing. Opt. Express (2011). , 19(4), 3332-3338.

[44] Wang, J., Huang, H., Wang, X., Yang, J. Y., Yilmaz, O. F., Wu, X., Nuccio, S. R., & Willner, A. E. 2.3-Tbit/s (23X100-Gbit/s) RZ-DQPSK grooming switch (simultaneous add/drop, data exchange and equalization) using double-pass LCoS and bidirectional HNLF. OFC 2011, paper OTuE2, 2011.

[45] Wang, J., Huang, H., Wang, X., Yang, J. Y., & Willner, A. E. Reconfigurable 2.3-Tbit/s DQPSK simultaneous add/drop, data exchange and equalization using double-pass LCoS and bidirectional HNLF. Opt. Express (2011). , 19(19), 18246-18252.

[46] Christen, L., Yilmaz, O. F., Nuccio, S., Wu, X. X., Fazal, I., & Willner, A. E. Tunable time-slot-interchange of 40-Gb/s optical packets using conversion/dispersion-based tunable 100-ns delays. OFC2008, paper OThA4, 2008.

[47] Yilmaz, O. F., Christen, L., Wu, X. X., Nuccio, S. R., Fazal, I., & Willner, A. E. Time-slot interchange of 40 Gbit/s variable length optical packets using conversion-dispersion-based tunable delays. Opt. Lett. (2008). , 33(17), 1954-1956.

[48] Wang, J., Yilmaz, O. F., Nuccio, S. R., Wu, X. X., Bakhtiari, Z., Xiao-Li, Y., Yang, J. Y., Huang, H., Yue, Y., Fazal, I., Hellwarth, R., & Willner, A. E. Data traffic grooming/ exchange of a single 10-Gbit/s TDM tributary channel between two pol-muxed 80-Gbit/s DPSK channels. CLEO 2010, paper CFJ5, 2010.

[49] Wang J, Yilmaz OF, Nuccio SR, Wu XX, Willner AE. Orthogonal tributary channel exchange of 160-Gbit/s pol-muxed DPSK signal. Opt. Express 2010; 18(16): 16995-17008.

[50] Suzuki J, Taira K, Fukuchi Y, Ozeki Y, Tanemura T, Kikuchi K. All-optical time-division add-drop multiplexer using optical fibre Kerr shutter. Electron. Lett. 2004; 40(7): 445-446.

[51] Wang, J., Willner, A. E. Review of robust data exchange using optical nonlinearities. International Journal of Optics (2012). , 2012, Article ID 575429. doi: 10.1155/2012/575429.

[52] Wang, J., Yang, J. Y., Fazal, I. M., Ahmed, N., Yan, Y., Willner, A. E., Dolinar, S., & Tur, M. Experimental demonstration of 100-Gbit/s DQPSK data exchange between orbital-angular-momentum modes. OFC2012, paper OW1I.5, 2012.

[53] Wang, J., Yang, J. Y., Fazal, I. M., Ahmed, N., Yan, Y., Huang, H., Ren, Y. X., Yue, Y., Dolinar, S., Tur, M., & Willner, A. E. Terabit free-space data transmission employing orbital angular momentum multiplexing. Nature Photonics (2012). , 6(7), 488-496.

[54] Willner AE, Yilmaz OF, Wang J, Wu XX, Bogoni A, Zhang L, Nuccio SR. Optically efficient nonlinear signal processing. IEEE J. Sel. Topics Quantum Electron. 2011; 17(2): 320-332.

[55] Tian Y, Xiao XS, Gao SM, Yang CX. All-optical switch based on two-pump four-wave mixing in fibers without a frequency shift. Appl. Opt. 2007; 46(23): 5588-5592.

[56] Parameswaran KR, Fujimura M, Chou MH, Fejer MM. Low-power all-optical gate based on sum frequency mixing in APE waveguides in PPLN. IEEE Photon. Technol. Lett. 2000; 12(6): 654-656.

[57] Wang J, Sun JQ, Sun QZ. Experimental observation of a 1.5 μm band wavelength conversion and logic NOT gate at 40 Gbit/s based on sum-frequency generation. Opt. Lett. 2006; 31(11): 1711-1713.

[58] Wang, J., Sun, J. Q., & Sun, Q. Z. Single-PPLN-based simultaneous half-adder, half-subtracter, and OR logic gate: proposal and simulation. Opt. Express (2007). , 15(4), 1690-1699.

[59] Allen L, Beijersbergen MW, Spreeuw RJC, Woerdman JP. Orbital angular momentum of light and the transformation of Laguerre–Gaussian laser modes. Phys. Rev. A 1992; 45(11): 8185-8189.

[60] Franke-Arnold S, Allen L, Padgett M. Advances in optical angular momentum. Laser Photon. Rev. 2008; 2(4): 299-313.

[61] Yao, A. M., Padgett, M. J. Orbital angular momentum: origins, behavior and applications. Adv. Opt. Photon. (2011). , 3(2), 161-204.

[62] Gibson G, Courtial J, Padgett M, Vasnetsov M, Pas'ko V, Barnett S, Franke-Arnold S. Free-space information transfer using light beams carrying orbital angular momentum. Opt. Express 2004; 12(22): 5448-5456.

[63] Djordjevic IB, Arabaci M, Xu L, Wang T. Spatial-domain-based multidimensional modulation for multi-Tb/s serial optical transmission. Opt. Express 2011; 19(7): 6845-6857.

[64] Lerber TV, Honkanen S, Tervonen A, Ludvigsen H, Küppers F. Optical clock recovery methods: Review (Invited). Opt. Fiber Technol. 2009; 15(4): 363-372.

[65] Mulvad, H. C. H., Tangdiongga, E., Waardt, H., & Dorren, H. J. S. 40 GHz clock recovery from 640 Gbit/s OTDM signal using SOA based phase comparator. Electron. Lett. (2008). , 44(2), 146-147.

[66] Oxenløwe, L. K., Gómez-Agis, F., Ware, C., Kurimura, S., Mulvad, H. C. H., Galili, M., Nakajima, H., Ichikawa, J., Erasme, D., Clausen, A. T., & Jeppesen, P. 640-Gbit/s data transmission and clock recovery using an ultrafast periodically poled lithium niobate device. J. Lightwave Technol. (2009). , 27(3), 205-213.

[67] Takara, H., Ono, H., Abe, Y., Masuda, H., Takenaga, K., Matsuo, S., Kubota, H., Shibahara, K., Kobayashi, T., & Miaymoto, Y. 1000-km 7-core fiber transmission of 10X96-Gb/s PDM-16QAM using Raman amplification with 6.5 W per fiber. Opt. Express (2012). , 20(9), 10100-10105.

[68] Liu, X., Chandrasekhar, S., Chen, X., Winzer, P. J., Pan, Y., Taunay, T. F., Zhu, B., Fishteyn, M., Yan, M. F., Fini, J. M., Monberg, E. M., & Dimarcello, F. V. 1.12-Tb/s 32-QAM-OFDM superchannel with 8.6-b/s/Hz intrachannel spectral efficiency and space-division multiplexed transmission with 60-b/s/Hz aggregate spectral efficiency. Opt. Express (2011). , 19(26), B958-B964.

[69] Al, Amin. A., Li, A., Chen, S., Gao, G., & Shieh, W. Dual-LP11 mode 4X4 MIMO-OFDM transmission over a two-mode fiber. Opt. Express (2011). , 19(17), 16672-16679.

[70] Ryf R, Randel S, Gnauck AH, Bolle C, Sierra A, Mumtaz S, Esmaeelpour M, Burrows EC, Essiambre RJ, Winzer PJ, Peckham DW, McCurdy AH, Lingle R. Mode-division multiplexing over 96 km of few-mode fiber using coherent 6 × 6 MIMO processing. J. Lightwave Technol. 2012; 30(4): 521-531.

[71] Koizumi, Y., Toyoda, K., Yoshida, M., & Nakazawa, M. 1024 QAM (60 Gbit/s) single-carrier coherent optical transmission over 150 km. Opt. Express (2012). , 20(11), 12508-12514.

Chapter 4

All-Optical Quaternary Logic Based Information Processing: Challenges and Opportunities

Jitendra Nath Roy and Tanay Chattopadhyay

Additional information is available at the end of the chapter

1. Introduction:

Science and Technology is providing people all over the world with much better ways of communicating than ever before, and the winds of change have whipped up the desire to exchange more of everything from messages to movies. The field of computation and signal processing is growing day by day [1-7]. In last three to four decades, the philosophy, science and technical prospects enriched the scientific communities a lot. Massive parallelism, speed of operation, increased spatial density attracts in many ways the scientists, researchers and technologists. Very Large Scale Integration (VLSI) technology has revolutionized the electronics industry and established the 20th century as the computer age. But, VLSI technology is approaching its fundamental limits in the sub-micron miniaturization process. Therefore an alternative technological solution to the problem of high speed information processing is needed, and unless we gear our thoughts toward a totally different pathway, we will not be able to further improve our information processing performance for the future. Conservative and reversible logic gates are widely known to be compatible with revolutionary computing paradigms. At the same time the Multi-valued logic (MVL) is also positioned as a coming generation technology that can execute arithmetic functions faster and with less interconnect than binary logic [8-48].

In order to overcome the electronic bottlenecks and fully exploit the advantages of optics, it is necessary to move towards networks, where the transmitted data will remain exclusively in all optical domains without optical electrical optical (OEO) conversions. Ultra high-speed optical network is developing rapidly as growing capacity demand in telecommunication system is increasing. In these networks, it is desired to carry out switching, routing and processing in optical domain to avoid bottlenecks of optoelectronic conversions. The dream of photonics is to have a completely all-optical technology. All-optical switching is an essential technology for transparent fiber optic networks and for all forms of optical signal processing as the optical interconnections and optical integrated circuits is immune to electromagnetic interference, and free from electrical short circuits. In a pursuit to probe into cutting-edge research areas, the development of different ultra-fast all-optical switches has received considerable interest in recent years all over the world for future optical information processing [49-59]. As photon is the ultimate unit of information with unmatched speed and with data package in a signal of zero mass, the techniques of computing with light may provide a way out of the limitations of computational speed and complexity inherent in electronics computing.

The fundamentals of digital signal processing are straightforward. To send something as simple as a phone message or as complicated as a picture, we digitize it by breaking it up into a series of binary bits, transmit the bits, and decode them at the other end to re-create the message or picture. The ones or zeroes in the bits are encoded by turning some signal on or off. In the past, the signal has been electrical, but increasingly it is composed of light pulses. We use a laser to produce the light, and then add information to it with a modulator, transmit it through optical fibers, amplify it if needed, receive it with a photo detector and re-create the message with a demodulator. An optical signal is better than an electrical one, with less attenuation, faster switching, and more signals traveling together. In everyday we have to handle enormous and ever increasing, amounts of information. Binary number (0 and 1) is insufficient in respect to the demand of the coming generation. The application of multi-valued (non-binary) signals can provide a considerable relief in transmission, storage and processing of large amount of information in digital signal processing. Quaternary logic (4-valued) is one type of MVL [60-82].

In this chapter, all-optical scheme for designing some polarization encoded quaternary logic gates (quaternary min and quaternary delta literal) with the help of nonlinear material based interferometric switches have been discussed. Design of all-optical quaternary multivalued multiplexer and demultiplexer circuits have also been described with the help of these basic gates. For the quaternary data processing in optics, the quaternary number (0, 1, 2, 3) have been represented by four discrete polarized state of light. In optical implementation we can consider the set of Quaternary logic states {0, 1, 2, 3} as : 0= No light, 1 = vertically polarized light (↕), 2 = horizontally polarized light (•), 3 = partially polarized light (↔).This chapter is organized as follows. Section-1.1 to 1.3 gives a brief overview of multivalued logic (MVL) i.e. What is MVL? Why do we need it? How it can be implemented and where MVL can be applied? Section-2.1 describes the basic principle of all-optical interferometric switches which is the cornerstone of all logic based signal processing. Section-2.2 and section 2.3 describes the design and operational principle of some basic all-optical quaternary logic circuits (QMIN, Delta Literal). All-optical circuit for quaternary multiplexer and demultiplexer are described in section-2.4 and section 2.5. Also quaternary T-gate is discussed in section 3. Challenges in design issues that is to be considered for experimentally achieve result from the proposed scheme is mentioned in section section-4. Chapter ends with Conclusions and Future Scopes given in section-5.

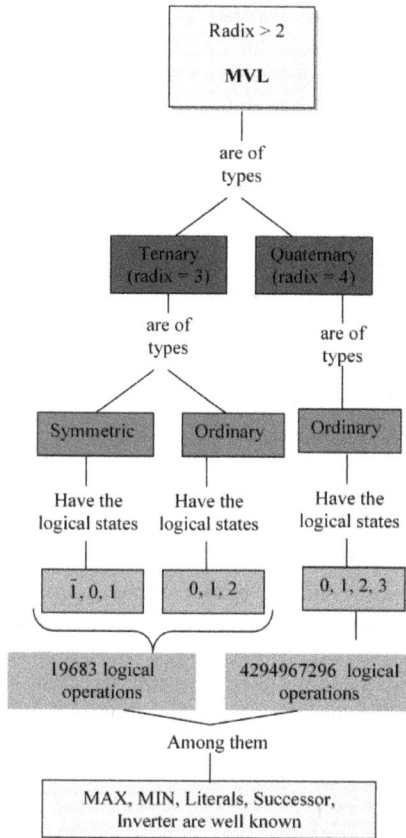

Figure 1. Different field in multi-valued logic.

1.1. What is Quaternary Logic?

Multi-valued logic (MVL) is a non binary logic with radix >2. Binary logic is limited to only two states 'True' (1) and 'False' (0), MVL replaces these with finite and infinite number of values. MVL system is defined as system operating on higher radix than two. In the base-R number system, a numerical value of N-bit data$(a_{N-1}a_{N-2}\cdots a_2a_1a_0)_R$; [where $0 \le a \le (R-1)$] can be written as [56] :

$$(a_{N-1}.R^{N-1} + a_{N-2}.R^{N-2} + \cdots + a_1.R^1 + a_0.R^0) = \sum_{i=0}^{N-1} a_i.R^i \qquad (1)$$

For example Ternary logic ($R=3$) has three logical states {0, 1, 2} or {$\bar{1}$, 0, 1} [18]. These are known as ordinary ternary and symmetric ternary logic respectively. Quaternary logic ($R = 4$) has four logical states {0, 1, 2, 3}. Like binary world there are also numbers of basic gates in multi-valued logic world. Depending on the radix and number of variables used, different logic functions can be generated. The numbers of possible functions are [37].

$$f(R,n) = R^{\left(R^n\right)} \tag{2}$$

Where R = radix, n = number of variables. In ternary logic of two variable ($R = 3$, $n = 2$) there are $f(3, 2) = 3^{3^2} = 19683$ possible functions. For quaternary ($R = 4$, n = 2) there are $f(4, 2) = 4^{4^2} = 4294967296$ logical operations. Hence, huge numbers of logical operation can be possible for higher radix (Fig. 1). Among them, some basic gates are the MAX, MIN, Complement, Cycle or successor, Literals etc [6, 7, 38-41], which is indicated in Fig. 2.

1.2. Why do we need All-optical Quaternary Logic based signal processing?

The most pressing problems in present-day binary systems are interconnection problems, both on-chip and between chip. On chip the difficulties of placement and routing of the digital logic elements which go to make up the complete chip are escalating with increase in capability per chip, and the silicon area used for interconnections may be greater than that used for the active logic elements. Similarly, the difficulties of bringing an increasing number of connections off-chip is promoting -a new consideration of packaging concepts in an attempt to overcome problems which are becoming mechanically, thermally, and electrically extreme. All these factors point to the attraction of raising the information content per interconnection from the present lowest-possible (binary) level. Multiple-valued logic, in which the number of discreet logic levels is not confined to two, has been the subject of much research over many years. The practical objective of this work is to increase the information content of the digital signals in a system to a higher value than that provided by binary operation. To increase the transmission capacity of future communication the present binary system is becoming very critical. A more formal approach would be an n-valued logic which has n different states, each state having a unique identifier. Multi-valued logic (MVL) is defined as a non-binary logic and involves the switching between more than two states. Multi-valued logic can be viewed as an alternative approach to solve many problems in transmission, storage and processing of large amount of information in digital signal processing [22]. The main advantages of multi-valued logic systems and circuits are greater speed of arithmetic operations realization, greater density of memorized information, better usage of transmission paths, decreasing of interconnections complexity and interconnections area, decreasing of pin number of integrated circuits and printed boards, possibilities for easier testing.

In the field of data communication, the quaternary codes are preferred because four-valued (i.e. quaternary) logic signals easily interface with the binary world. They may be decoded directly into their two binary-digit equivalent. Quaternary logic world can easily be interfaced with binary logic in all-optical domain with the help of our suggested DEC and ENC schemes [57, 59]. The block diagram of this interfacing circuit is shown in the Fig. 3. Here input & outputs are 4-valued and the internal circuitry is binary (radix=2). Decoder circuit converts quaternary input into its binary equivalent. After performing the logical operation in binary system, it is then encoded to its quaternary equivalent by encoder circuit. Hence, it can be said that this scheme requires no major modifications of the existing transmitter, receiver, or transmission link. Quaternary digits are two major types: ordinary quaternary digit (OQD) and quaternary signed digit (QSD) [75-76]. QSD is useful for carry free arithmetic operations [19, 31-35]. Fig. 4 indicates how quaternary and binary are interfaced.

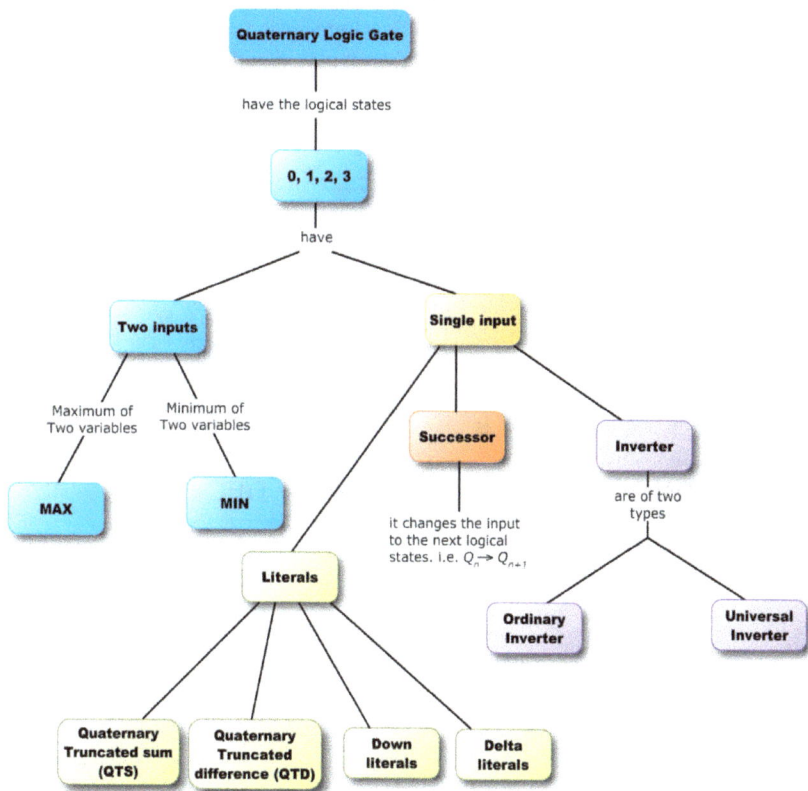

Figure 2. Different field in quaternary logic.

1.3. How All-optical Quaternary Logic can be implemented?

Consideration of different logical states is a challenge. It can be done in different ways and given in Fig. 5. In electronics, efforts have already been made to design four-valued logic

[37-48] with charge couple device (CCD). In I²L circuits 0mA, 10mA, 20mA and 30mAg are four different logical states, in νMOS (neuron-MOS) have the logic levels 0.0V, 1.1V, 2.2V and 3.3V [39], also in CMOS MVL have the logic levels 0 V, 1V, 2V, 3V respectively [37]. Quantum computation and information is the study of the information processing tasks that can be accomplished using quantum mechanical systems. Just as the classical computation is built upon bits, quantum computation also has an analogous concept called qubits. Analogous to classical computation, the operations on qubits are carried out using quantum logic gates. Of late, renewed interest in optical computing has been witnessed due to the emergence of novel photonics structures that includes nano-photonics, silicon photonics, bio-photonics and plasmonics etc. Optical quaternary logical operation is an interesting and challenging field of research for future optical signal processing where we can expect much innovation [58-82]. Polarization properties of light can play significant role here.

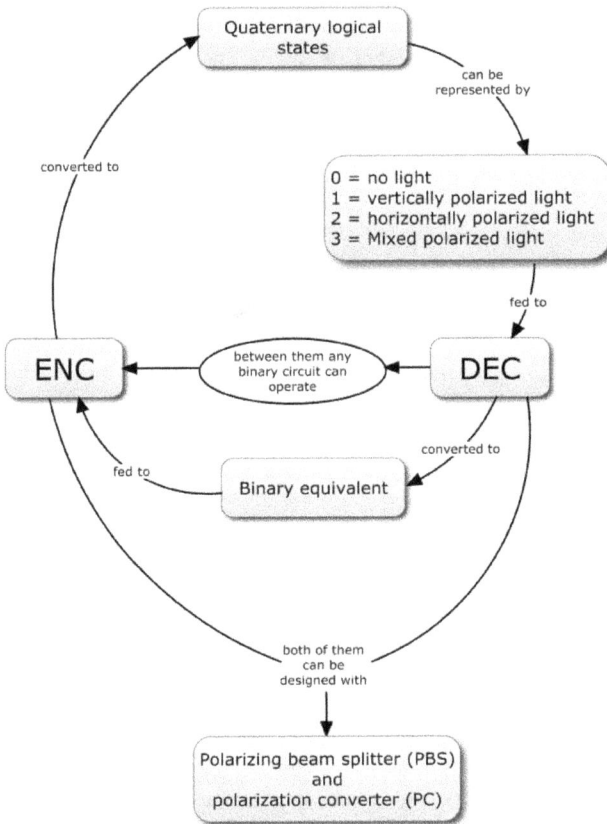

Figure 3. Binary-to-quaternary Encoder and Quaternary-to-binary decoder.

Polarization may be a good choice for representing different logical states in all-optical quaternary (4-valued) logic operations because [7, 15],

- Nature of polarization does not change due to absorption of light like intensity. Therefore the strength or weakness of the beam plays no role in the operation of the devices.
- The sate of polarization can be changed easily by Polarization converter.
- No photon energy is wasted.

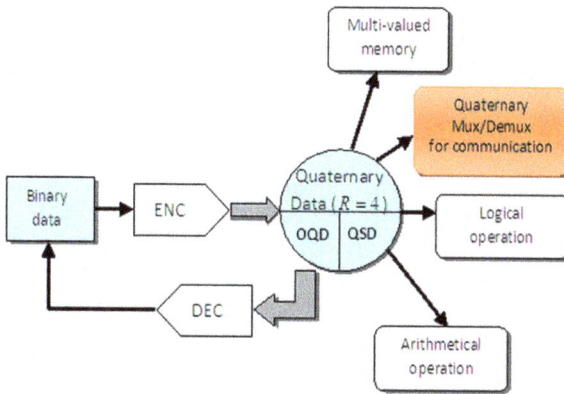

Figure 4. Interfacing Binary and Quaternary world by the help of ENC and DEC.

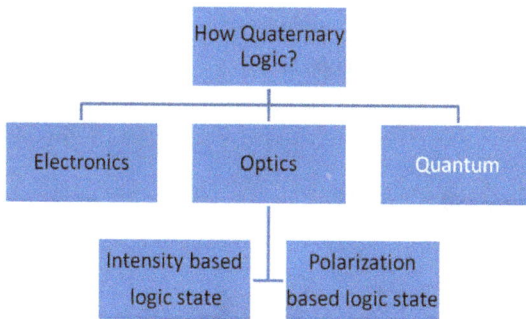

Figure 5. Quaternary (4-valued) logic implementation.

For the quaternary data processing in optics, the quaternary logic states {0, 1, 2, 3} can be represented by four discrete polarized state of light as mentioned below:

0 = No light.

1 = vertically polarized light ($\updownarrow$)

2 = horizontally polarized light ($\bullet$)

3 = mixed polarized light or un-polarized light ($\leftrightarrow$).

2. Designing of Polarization encoded all-optical Quaternary multiplexer / De-multiplexer

Multiplexing and de-multiplexing are two essential features in almost all the signal communication systems, where a lot of information is being handled without any mutual disturbances. The principles and possibilities of designing of all-optical quaternary multi-valued multiplexer and de-multiplexer circuits are described with the help of quaternary MIN and quaternary delta literal gates (Fig. 6). Nonlinear material based interferometric switches can take an important role here. Working principle of Terahertz Optical Asymmetric Demultiplexer (TOAD) based all-optical switch is discussed in Section-2.1. Section-2.2 and Sec-2.3 describes the design and operational principle of some basic all-optical quaternary logic circuits (QMIN, Delta Literal). All-optical circuit for quaternary multiplexer and demultiplexer are proposed and described in Sec-2.4 and sec-2.5 respectively.

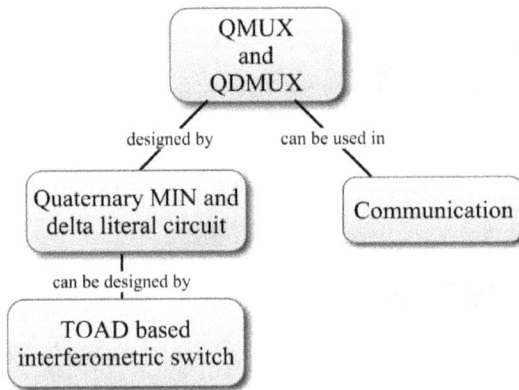

Figure 6. Overview of Quaternary Mux/Demux.

2.1. Interferometer based optical switch:

Interferometric devices for optical processing have been of great interest in the recent years [50-55]. Optical switch using a nonlinear interferometer makes it possible for one optical signal to control and switch another optical signal through the nonlinear interaction in a material. The incoming signal to be switched is split between the arms of the interferometer. The interferometer is balanced so that, in the absence of a control signal, the incoming signal emerges from one output port. The presence of a strong control pulse changes the refractive index of the medium given by

$$\Delta n = n_2 I \tag{3}$$

Where Δn is the change in the refractive index of the medium, n_2 is the nonlinear refractive coefficient and I is the intensity of the light incident on the medium. A change in the index adds a phase shift between the two arms of the interferometer, so that the incoming signal is switched over to another output port. This method of switching is based on cross phase modulation (XPM). In much resent years, Semiconductor optical amplifier (SOA) makes a revolution in designing high speed (>100 Gb/s) interferometric switches in all-optical information processing system. The technology of SOA has been evolving rapidly during the re-sent years and has become mature enough so that it is now key factor in implementation of modern optical communication networks. SOA are commercially available device and have different important properties. Such as, fast and strong nonlinearities, short latency, thermal stability, low power consumption, large dynamic range, short response time, broadband and versatile operation and capability of large scale integration with chip level design.

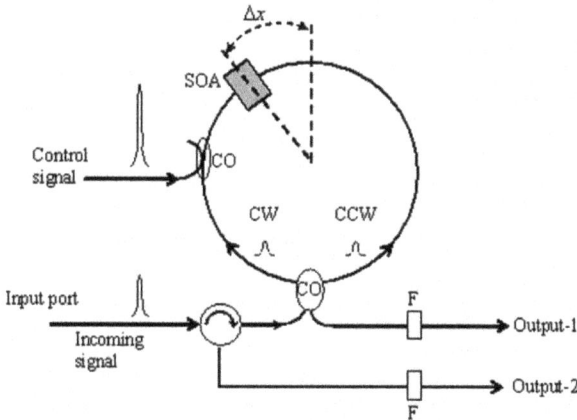

Figure 7. A TOAD based optical switch, where SOA: Semiconductor optical amplifier, CW: Clockwise pulse, CCW: Counter clockwise pulse, CO: coupler, F: Filter which blocks control pulse.

The Fig. 7 is a Sagnac interferometer that uses an SOA offset from the midpoint of the loop and is known as a terahertz optical asymmetric demultiplexer (TOAD). It can operate at frequencies in terahertz range [50-51]. There are two couplers; 1) the control coupler provides an input path for the control pulses to enter the fiber loop in order to saturate the SOA, and 2) the input coupler (50:50) where the incoming pulse signal train entering the loop splits equally into clockwise (CW) and counter clockwise (CCW) pulses. CW and CCW pulses arrive at the SOA at slightly different times as determined by the offset Δx of the SOA from the midpoint of the loop. Another strong light pulse is also injected to the loop. It is called control signal (CS). When CS=ON, then SOA changes its index of refraction. As a result, the two counter-propagation data signal will experience a differential gain saturation profiles. Therefore cross phase modulation (XPM) take place when they recombine at the input coupler. Then relative phase difference between CW and CCW pulse become $\sim \pi$ and the data will exit from the transmitted port / T-port (output-1 according to the Fig. 7). In the absence of a control signal (CS=OFF), the incoming signal enters the fiber loop, pass through the SOA at different times as they counter-propagate around the loop, and experience the nearly same unsaturated amplifier gain of SOA, recombine at the input coupler. Then, relative phase difference between CW and CCW is zero (0). Then no data is found at the T-port. Then data is reflected back toward the source and isolated by optical circulator (CR). The port through which it comes is called reflected port /R-port (output-2 according to the Fig. 7). A filter (F) may be used at the output to reject the control and pass the incoming pulse. 'F' can be polarization filter of band pass filter.

The output power of transmitted port (T-port) and reflected port (R-port) of a TOAD based switch can be expressed as [52-53],

$$P_T(t) = \frac{P_{in}(t)}{4} \cdot \left\{ G_{cw}(t) + G_{ccw}(t) - 2\sqrt{G_{cw}(t) \cdot G_{ccw}(t)} \cdot \cos\left(\Delta\varphi\right) \right\} \tag{4}$$

$$P_R(t) = \frac{P_{in}(t)}{4} \cdot \left\{ G_{cw}(t) + G_{ccw}(t) + 2\sqrt{G_{cw}(t) \cdot G_{ccw}(t)} \cdot \cos\left(\Delta\varphi\right) \right\} \tag{5}$$

where, $G_{cw}(t)$, $G_{ccw}(t)$ are the power gain of CW and CCW pulse, $\Delta\varphi = (\varphi_{cw} - \varphi_{ccw})$ is the phase difference between CW and CCW pulse, can be expressed as $\Delta\varphi = -\alpha/2 \cdot \ln(G_{cw}/G_{ccw})$. The temporal duration of the switching window (τ_{win}) that depends on the offset position of the SOA in the loop (Δx) is given by $T_{off} = 2\Delta x / c_{fiber}$, where c_{fiber} is the velocity of light inside the optical fiber. More specifically eccentricity of the loop must be less than half the bit period, otherwise the two counter-propagating halves of incoming signal (IS) being processed will not experience the gain dynamics caused by their synchronized control pulses but instead by others resulting in incomplete switching. T_{FWHM} of the control pulse must be as short as possible and ideally less than the switching window so that when CCW pulse is inserted in the SOA the CW pulse already passed through and the SOA gain has started to recover after saturation by the control pulse or else the two components of IS will overlap inside the SOA perceiving its nonlinear properties only partially altered, thus obstructing the creation of the required differential phase shift [52-53].

$$\sigma < T < 0.5\xi < \tau_e < 1.5\xi \tag{6}$$

ξ is bit period. For low switching window eccentricity of the loop (T) should be small. One data when transmit through the switching window, next data cannot pass until the gain recovery of the SOA takes place.

In summary we can say, in the absence of control signal, the incoming signal exits through input port of TOAD and reaches to the output port-2 as shown in Fig. 8(a).In this case no light is present in the output port-1. But in the presence of control signal, the incoming signal exits through output port of TOAD and reaches to the output port-1 as shown in Fig. 8(a).In this case no light is present in the output port-2. In the absence of incoming signal, port-1 and port-2 receives no light as the filter blocks the control signal. Only incoming signal is passed through filter. Truth table is given in Fig. 8(b).The above principle of the switch is used to design basic quaternary logic circuits.

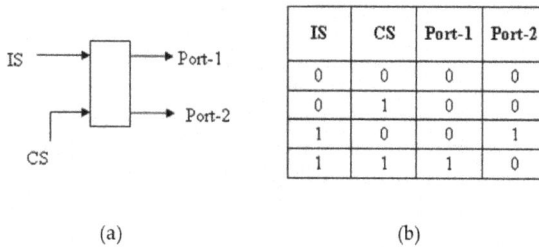

IS	CS	Port-1	Port-2
0	0	0	0
0	1	0	0
1	0	0	1
1	1	1	0

(a)   (b)

Figure 8. (a) Schematic block diagram TOAD based switch (b) the corresponding truth table.

2.2. All-optical two input Quaternary MIN Gate

Quaternary MIN gate is equivalent AND gate in binary world [6,39]. It is an important multi-valued logic function. The QMIN operation is shown in the equation no (7), the operator $\wedge$ is QMIN operation. Truth table is shown in Table 1.

$$x_1 \wedge x_2 \wedge \cdots \wedge x_n = QMIN\left(x_1, x_2, \cdots, x_n\right) \tag{7}$$

X	Y			
	0	1	2	3
0	0	0	0	0
1	0	1	1	1
2	0	1	2	2
3	0	1	2	3

Table 1. Truth Table of Quaternary MIN(x,y).

Polarization encoded all- optical QMIN Gate is shown in the Fig. 9. Here light from inputs X and Y fall into two PBS (PBS$_1$& PBS$_2$), where it split into two polarized light, one is vertically polarized ($\uparrow$) and other is horizontally polarized ($\bullet$).Hence, X_1& Y_1 are vertically polarized ($\uparrow$) and X_2& Y_2 are horizontally polarized ($\bullet$). Light from X_2 and Y_2 are fed to two Interferometric switches (here TOAD) S_1 and S_2 as incoming signal and also their control signals are taken from Y_2 and X_2 respectively. The lower outputs of S_1 and S_2 are passed through a PC (polarization converter, which is preferably half wave plate; converts vertically polarized light to horizontal one and vice versa). It is indicated as S_{1L} and S_{2L} respectively in the Fig. 9. X_1 and S_{1L} is combined by a BC-1, the combined ray (C_1) is connected to another switch S_3 as incoming signal. Also Y_1 and S_{2L} are combined by BC-2 and the combined ray (C_2) is connected to S_3 as control signal. The upper output channel of S_3 (S_{3U}) is feed to BC-3. Again X_2 and Y_2 are feed to another switch S_4 as incoming and the control signal respectively. All the control signals are amplified by EDFA (Erbium Doped Fiber Amplifier). When incoming light signal is incident on wavelength converter (WC) then the wavelength converter (WC) converts the wavelength of incoming signal to wavelength of control signal. The upper output channel of this switch S_4 (S_{4U}) is connected to BC-3. The combined ray is the final output. Let us describe the operational principles in detail [66].

1. When X=0, X_1=X_2=0. X_2 which act as a incoming signal of S_1 and S_4 is zero. So S_{1L} and S_{4U} receive no light. So the BC-1 receives no light and hence the output of BC-1 is zero therefore S_{3U} receives no light. Hence the final Output after BC-3 is 0. This result cannot be changed by any value of Y.

2. Similarly when Y=0, Y_1=Y_2=0 so all the incoming signals of S_2 and control signal of S_4 and S_3 is zero hence S_{4U}=S_{2L}=C_2=S_{3U}=0 i.e. receive no light. So the final Output after BC-3 is 0. This result cannot change by any value of X.

3. When X=1($\uparrow$), X_1=1 and X_2=0. So S_{1L}& S_{4U} receives no light (as incoming signal of S_1 and S_4 is absent). And C_1=1 ($\uparrow$) i.e. vertically polarized light.

    •   Now as Y=1($\uparrow$) then Y_1=1, Y_2=0. So S_{2L}=0 (as incoming signal of S_2 is absent), C_2=1. So S_{3U}=1 (as C_2 is the incoming signal of S_3). So the final Output after BC-3 is 1 ($\uparrow$).

    •   When Y=2 ($\bullet$) i.e. horizontally polarized light, then Y_1=0 (no light) and Y_2=2. So S_{2L} and hence C_2 receives vertically polarized light (1 i.e. $\uparrow$). Hence S_{3U}=1. So the final Output after BC-3 is 1 ($\uparrow$).

    •   Now when Y=3 ($\leftrightarrow$), then Y_1=1,Y_2=2. So S_{2L}=1 (as incoming signal is present but control signal is absent at S_2), C_2=1. Hence S_{3U}=1. So the final Output after BC-3 is 1 ($\uparrow$).

4. When X is 2 ($\bullet$) and Y is 1 ($\uparrow$), then X_1& Y_2 receives no light. That means here, X_1 =0 & Y_2 =0 and Y_1=1, X_2=2. Hence S_{4U}=S_{2L}=0 (as the control signal of S_4 and incoming signal of S_2 is absent) and S_{1L}=C_1=C_2=1 i.e. vertically polarized light. So S_{3U}=1 as both the incoming and control signal of S_3 are present. So the final Output after BC-3 is 1 ($\uparrow$).

5. When X=Y=2 ($\bullet$) i.e. both of them are horizontally polarized light, then X_1& Y_1 receives no light (0) and X_2& Y_2 receives horizontally polarized light (2). Hence S_{4U}=2 and S_{2L}=0

(as both the incoming and control is present of S_4 and S_2) As $Y_1=S_{2L}=0$. So $S_{3U}=0$. So the final Output after BC-3 is 2 ($\bullet$).

6. When X takes horizontally polarized light i.e. 2 ($\bullet$) and Y is partially polarized light i.e.3 ($\leftrightarrow$) then X_1 receives no light (0) and Y1=1, X2=Y2=2. Hence $S_{4U}=2$ and $S_{2L}=0$, $C_2=1$. Again as $X_2=Y_2=2$, then $S_{1L}=C_1=0$ (as both the incoming and control signal are present at S_1). As C_1 is the incoming signal of S_3, hence $S_{3U}=0$. And the final Output 2 ($\bullet$).

7. When X=3 ($\leftrightarrow$), then $X_1=1$ and $X_2=2$. When Y=1, then $Y_1=1$ and $Y_2=0$. So $S_{4U}=S_{2L}=0$ (as the control signal is absent in S_4 and the incoming signal is absent in S_2) and $S_{1L}=1$ (as incoming is present but control signal is absent in S_1). So $C_1=C_2=1$, hence $S_{3U}=1$. So final Output is 1 ($\updownarrow$).

8. When X=3 ($\leftrightarrow$), then $X_1=1$ and $X_2=2$. When Y=2 ($\bullet$), $Y_1=0$ and $Y_2=2$. So $S_{4U}=2$ and $S_{2L}=S_{1L}=0$. So $C_1=1$ and $C_2=0$, hence $S_{3U}=0$. So the final output is 2 ($\bullet$).

9. When X & Y both are partially polarized light i.e. 3 ($\leftrightarrow$), Then $X_1=Y_1=1$ and $X_2=Y_2=2$. So S_{4U} receives horizontally polarized light (2 i.e. $\bullet$) and $S_{2L}=S_{1L}=0$. Hence $C_1=C_2=1$, hence $S_{3U}=1$. So final Output is 3 ($\leftrightarrow$).

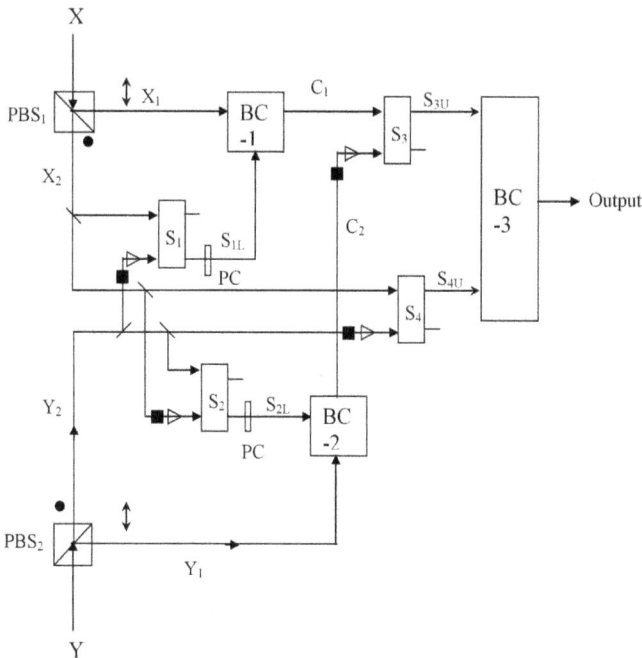

Figure 9. All-optical Quaternary QMIN(X,Y) Circuit. S (Switch): PBS : Polarizing Beam Splitter BC : Beam Combiner PC : Polarization Converter, ▶ EDFA : Erbium Doped Fiber Amplifier, ■ : WC Wavelength Converter.

2.3. All-optical Quaternary delta LITERALS

Literals are very important function in multi-valued logic based information processing [67]. The truth table of Delta literal circuit [66] is in the Table 2 and the circuit diagram is shown in the Fig. 10. Here, X is the quaternary input, which can take any one of the four quaternary logic states and the output is x^0, x^1, x^2 and x^3 respectively.

Input X	Output			
	X⁰	X¹	X²	X³
0	3	0	0	0
1	0	3	0	0
2	0	0	3	0
3	0	0	0	3

Table 2. Truth table of quaternary delta Literals.

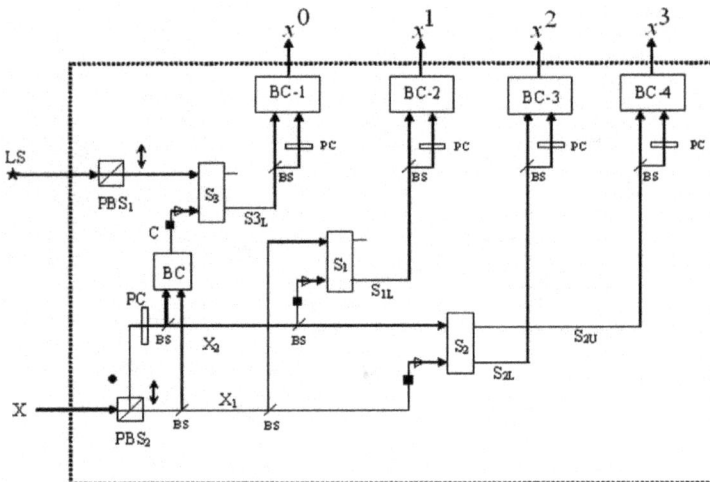

Figure 10. All-optical Quaternary Delta Literal Circuit.

1. When X=0 (absent of light), X_1 & X_2 receives no light and the other outputs of the switch S_1, S_2 are 0 as they receives no light. Here only vertically polarized light (↕), which comes from LS through PBS_1, falls on S_3. This act as incoming signal. Here as control signal is absent (because of C=0) the light comes out through lower channel of S_3 i.e. through S_{3L}. A part of this directly enters in the beam combiner (BC) and another part is passed through PC, which converts vertically polarized (↕) to horizontally polarized

light ($\bullet$). And hence the output of BC-1 is x^0 receives partially polarized light i.e. 3 ($\leftrightarrow$). Hence the final outputs are $x^0=3(\leftrightarrow)$ i.e. partially polarized light and others receives no light i.e. $x^1=0$, $x^2=0$ and $x^3=0$.

2.   When X=1 ($\updownarrow$), then $X_1=1$, $X_2=0$ and C=1. So $S_{3L}=0$ as control signal of S_3 is present and $S_{1L}=1$ ($\updownarrow$) as control signal of S_1 is absent. Other outputs (S_{2U} & S_{2L}) of the switch S_2 are absent, as the incoming signal of S_2 is absent. So the final outputs are $x^0=0$, $x^1=3(\leftrightarrow)$, $x^2=0$ and $x^3=0$.

3.   When X=2 ($\bullet$), then $X_1=0$, $X_2=1$ and C=1. So $S_{3L}=0$, $S_{1L}=0$ as the incoming signal of S_1 is absent. $S_{2U}=0$ and $S_{2L}=1$ ($\updownarrow$), because control signal of S_2 is absent. Hence the final outputs are $x^0=0$, $x^1=0$, $x^2=3(\leftrightarrow)$ and $x^3=0$.

4.   When X is partially polarized light i.e.3 ($\leftrightarrow$), then all the X_1, X_2 & C receives vertically polarized light i.e.1. So $S_{3L}=S_{1L}=S_{2L}=0$ and $S_{2U}=1$ ($\updownarrow$) as both the incoming and control signal of S_3, S_1 and S_2 are present. So the final outputs are $x^0=0$, $x^1=0$, $x^2=0$ and $x^3=3$ ($\leftrightarrow$).

2.4. Design of All-optical Quaternary Multiplexer (4:1) :

From the truth Table 3 (truth table of QMIN gate) we can say that,

$$\left.\begin{array}{l} 3 \wedge A = A \\ 0 \wedge A = 0 \end{array}\right\} \quad where \ A \in \{0,1,2,3\} \tag{8}$$

Now we design quaternary multiplexer (QMUX) and demultiplexer (QDEMUX) using the basic gates QMIN and Delta Literal, made by switching character of the non-liner material based switch [66].

Control input signal (X)	QMUX Output (Y)	QDEMUX Outputs			
		Y_0	Y_1	Y_2	Y_3
0	A	A	0	0	0
1	B	0	B	0	0
2	C	0	0	C	0
3	D	0	0	0	D

Table 3. Truth Table of Quaternary Multiplexer (QMUX) and Demultiplexer (QDEMUX).

Multiplexing means many into one. A multiplexer is system dealing with many inputs and only with single output. A quaternary multiplexer with n-control inputs can be used to route one of 4^n data inputs (it may be any one of the four logical states) to the output. Fig. 11

is the design of 4:1 all-optical quaternary multiplexer. Here four inputs A, B, C and D [which can be any one of the 4-logical state i.e. 0(no light), 1($\updownarrow$), 2 ($\bullet$), 3 ($\leftrightarrow$)] are connected to four 2-input QMIN gates (QMIN$_0$, QMIN$_1$, QMIN$_2$, and QMIN$_3$). Other input of the QMIN is fed from one of the Delta Literal outputs (i.e. x^0, x^1, x^2 and x^3) respectively as shown in Fig. 11. These inputs of QMIN are act as a select line.

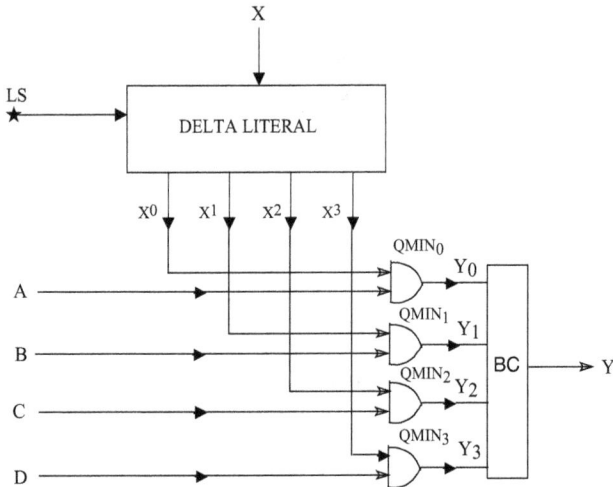

Figure 11. All optical Quaternary 4:1 Multiplexer (QMUX).

1.   When X (delta literal input) is zero i.e. with no signal (0), then x^0 receives the logical state 3 ($\leftrightarrow$) i.e. partially polarized light and others outputs of the delta literal (x^1, x^2 and x^3) receive no light (0) as discussed in earlier section. As x^0 is connected with QMIN$_0$, then according to the equation no 7, only QMIN$_0$ is active and others QMIN gates (QMIN$_1$, QMIN$_2$ and QMIN$_3$) are inactive. Hence the corresponding input A is at the output i.e. Y$_0$ = A. and Y$_1$ = Y$_2$ = Y$_3$ =0. Hence after combining in BC we receives Y = A at the outputs.

2.   When X is vertically polarized light i.e. 1 ($\updownarrow$), then only x^1 receives the logical state 3 ($\leftrightarrow$). & $x^0 = x^2 = x^3 = 0$ (no light). As x^1 is connected with QMIN$_1$, then according to the equation no (2) only QMIN$_1$ is active and QMIN$_0$, QMIN$_2$ and QMIN$_3$ are inactive. Hence Y$_1$ = B & Y$_0$ = Y$_2$ = Y$_3$ =0 and at the final output we receives Y = B.

3.   When X is horizontally polarized i.e. 2 ($\bullet$), then only x^2 receives the logical state 3 ($\leftrightarrow$). And $x^0 = x^1 = x^3 = 0$ (No light). Hence only QMIN$_2$ is active and QMIN$_0$, QMIN$_1$ and QMIN$_3$ are inactive. Hence Y$_2$ = C & Y$_0$ = Y$_1$ = Y$_3$ =0 and at the final output, we receives Y = C.

4. When X is partially polarized light i.e. 3 ($\leftrightarrow$), then only x^3 receives the logical state 3 ($\leftrightarrow$). And $x^0 = x^1 = x^2 = 0$ (no light). Hence only $QMIN_3$ is active and others $QMIN_0$, $QMIN_1$ and $QMIN_2$ are inactive. Hence $Y_3 = D$ & $Y_0 = Y_1 = Y_2 = 0$ and combining, at the final output we receives $Y = D$. The truth table of this circuit is shown in the Table 3 (second column).

2.5. Design of All-optical Quaternary Demultiplexer (1:4):

A quaternary demultiplexer has the opposite function of QMUX. Here one input data is passed to one of the outputs according to the selection of the control. Fig. 12 is the design of 1:4 all-optical quaternary demultiplexers. Here one Input A is fed to every four 2-input QMIN gate as one input light (light from A is split by three beam splitters (BS) and fed to four 2-input QMIN gates) and other input of the QMIN is fed from one of the Delta Literal outputs (i.e. x^0, x^1, x^2 and x^3) respectively. These inputs of QMIN gates act as select line. This circuit act like same way as multiplexer circuit i.e. one QMIN is active (depends on the selection of the control line) and others QMIN gates are inactive. The active QMIN gate passes the input data from A which may be one of the four logical sate. The final outputs are taken from the combination of four QMIN output line (Y_0, Y_1, Y_2 and Y_3) as shown in Fig. 12. The truth table is shown in the Table 3 (third column).

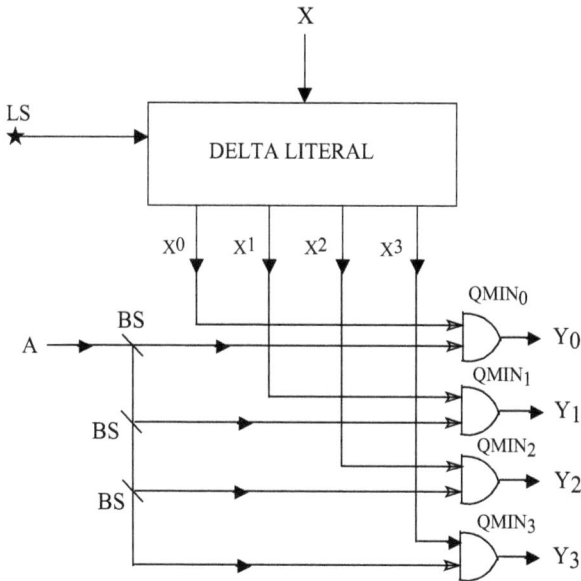

Figure 12. All optical Quaternary 1:4 Demultiplexer (QDEMUX).

3. Quaternary T-gate:

In section 2.4 we have reported all-optical 4:1 all-optical quaternary multiplexer. It is also known as 'T-Gate' [82]. The schematic diagram for quaternary T-gate is shown in Fig. 13. some logic operations are given in table 4.

Name of the functions	Symbol & mathematical expression	Inputs (logical states)				Outputs
		A	B	C	D	
Compliment / Inverter	$\bar{x}=(R-1)-x$	3	2	1	0	$\langle3210\rangle$
Successor	$Suc(x)=(x+1)\ \mathrm{mod}\ 4$	1	2	3	0	$\langle1230\rangle$
Clockwise Cycle	$\hat{X}_b=(x+b)\mathrm{mod}\ 4$	1	2	3	0	$\hat{X}_1=\langle1230\rangle$
		2	3	0	1	$\hat{X}_2=\langle2301\rangle$
		3	0	1	2	$\hat{X}_3=\langle3012\rangle$
Counter Cycle	$\hat{X}_{bc}=(x-b)\mathrm{mod}\ 4$	3	0	1	2	$\hat{X}_{1c}=\langle3012\rangle$
		2	3	0	1	$\hat{X}_{2c}=\langle2301\rangle$
		1	2	3	0	$\hat{X}_{3c}=\langle1230\rangle$
Literal	$^a x^b=\begin{cases}(R-1) & \text{if } a\le x\le b\\ 0 & \text{otherwise}\end{cases}$	0	3	3	0	$^1x^2=\langle0330\rangle$
		0	3	3	3	$^1x^3=\langle0333\rangle$
		0	0	3	3	$^2x^3=\langle0033\rangle$
Truncated Sum	$X\boxplus a=\begin{cases}(X+a)\ \text{if, } X<(R-1)\\(R-1)\qquad \text{otherwise}\end{cases}$	1	2	3	3	$X\boxplus1=\langle1233\rangle$
Truncated difference	$X\boxminus a=\begin{cases}(X-a)\ \text{if, } X\ge a\\0\qquad \text{otherwise}\end{cases}$	0	0	1	2	$X\boxminus1=\langle0012\rangle$
Threshold literals (up)	$U_a(x)=\begin{cases}1 & \text{if } x\ge a\\0 & \text{else}\end{cases}$	0	1	1	1	$U_1(x)=\langle0111\rangle$
		0	0	1	1	$U_2(x)=\langle0011\rangle$
Step literals (down)	$D_a(x)=\begin{cases}1 & \text{if } x\le a\\0 & \text{else}\end{cases}$	1	1	0	0	$D_1(x)=\langle1100\rangle$
		1	1	1	0	$D_2(x)=\langle1110\rangle$

Table 4. Some well-known one input quaternary logical functions (radix R=4) and design process with quaternary T-gate.

The mathematical expression for all-optical quaternary T-gate using MIN & delta literals can be written as:

$$O = \left(A \wedge x^{\langle 3000 \rangle} + B \wedge x^{\langle 0300 \rangle} + C \wedge x^{\langle 0030 \rangle} + D \wedge x^{\langle 0003 \rangle} \right) \qquad (9)$$

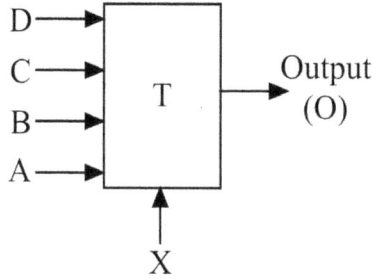

Figure 13. All optical Quaternary T-gate.

Where, '$\wedge$' is MIN operator ($x \wedge y$= minimum of (x, y)) and δ- literals function is $x^a = (R-1)$ if $x=a$, else 0] The four incoming data transmission lines are 'A', 'B', 'C' and 'D' [which can be any one of the 4-logical state i.e. 0 (no light), 1($\updownarrow$), 2 ($\bullet$), 3 ($\leftrightarrow$)] and 'X' is the selection input. By using proper section we can get any data (A, B, C or D) at the output. If X=0, the output is A, when X=1 then the output is B, for X=2 the output is C and when X=3 then the output is D respectively i.e. [82]:

$$T(A,B,C,D;x) = \begin{cases} A & \text{if } x = 0 \\ B & \text{if } x = 1 \\ C & \text{if } x = 2 \\ D & \text{if } x = 3 \end{cases} \qquad (10)$$

This T-gate can successfully used for designing any quaternary circuits. So it is called 'universal' element of quaternary logic. Some quaternary logical operations with T-gate is shown in Table 4. Here inputs of T-gate A, B, C, and D are shown in colomn-3 of that table 4. X is the select input =$\langle 0123 \rangle$. Here the quaternary multiplexer or T-gate is all-optical in nature. Hence all the quaternary circuits are all-optical.

4. Challenges in designing the polarization encoded all-optical system:

Here, in this proposed scheme, we have proposed and described an all-optical circuit for designing quaternary (four-valued) multiplexer & de-multiplexer with the help of some polarization encoded basic quaternary logic gates (quaternary min and quaternary delta literal). It is important to note that the above discussions are based on simple model. In order to experimentally achieve result from the proposed scheme, some design issues have to be considered. For example, polarization properties of fiber, predetermined values of the intensities, wavelength of laser light for control and incoming signals, introduction of filter, intensity losses due to beam splitters/fiber couplers etc. The output logical states of every ternary circuit can be determined by, stokes vector [S] measurement. Stokes vector can be calculated from the measurement of six intensities ($I_{i,j}$) in the photo detector (PD) by use of a linear analyzer (LA) followed by a quarter wave plate ($\lambda/4$plate), which is shown in the Fig. 14. The formula for calculating stokes vector is [83]:

$$[S] = \begin{bmatrix} S_0 \\ S_1 \\ S_2 \\ S_3 \end{bmatrix} = \sqrt{\frac{\mu_0}{\varepsilon_0}} \begin{bmatrix} I_{(0,0^\circ)} + I_{(0,90^\circ)} \\ I_{(0,0^\circ)} - I_{(0,90^\circ)} \\ I_{(0,45^\circ)} - I_{(0,135^\circ)} \\ I_{\left(\frac{\lambda}{4},45^\circ\right)} - I_{\left(\frac{\lambda}{4},135^\circ\right)} \end{bmatrix} \qquad (11)$$

Where first subscript ('i') index lack or presence of $\lambda/4$ plate and the second ('j') gives the azimuth of the analyzer. μ_0 and ε_0 is free space permeability and permittivity. Degree of polarization (DOP) is also calculated by the equation:

$$DOP = \frac{\sqrt{S_1^2 + S_2^2 + S_3^2}}{S_0} \qquad (12)$$

The value of DOP can be plotted in Poincaŕe sphere in point 'P' and we found that, for vertically ($\updownarrow$) and horizontally ($\bullet$) polarized light OP=DOP=1 and lies on the equator of the Poincaŕe sphere (at point y and x respectively).

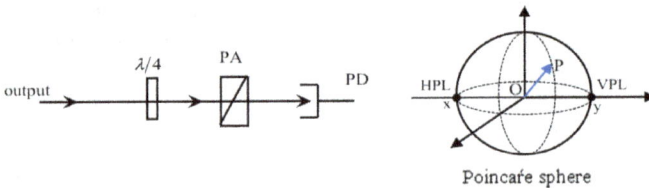

Poincaŕe sphere

Figure 14. Measurement technique of output logical states.

In high speed data communication (50 GB/s or TB/s) random change of polarization in a short time can produce power fluctuation at the output. So polarization dependent loss

(PDL) degrades the optical signal to noise ratio (OSNR) and also degrades the extinction ratio. PDL of 3 dB could cause 1 dB power penalty [84]. Optical depolarizers can be used to reduce the polarization-induced noise in optical sensing and measurement systems [85]. Again random birefringence in optical fibers induces an unpredictable rotation of the sate of polarization (SOP); this can be adjusted by using polarization controller and PM fiber. Intrinsic cross talk between two polarization states, imperfection of polarized tracking after transmission link etc may cause polarization mode dispersion (PMD). This may cause the delay among the different states of polarization. The effects of PMD are expected to be similar to those of other approaches that have been studied in the literature [86]. Optical amplifiers degrade the signal-to-noise ratio (SNR) of the amplified signal because of spontaneous emission added to the signal during its amplification (ASE). The OSNR-errors arise in this process. For polarized signal PHB will cause ASE polarization orthogonal to the signal polarization. Bruyere et al [87] have shown that the DOP of ASE could exceed 70% in transoceanic links with low PMD. Of course, the most of ASE light does not traverse the entire light path, and then OSNR-errors become less as < 0.6 dB. Polarization related problem discussed above would occur inside the considered circuit. The said problem will not occur in optical communications system once the signal comes out the output. State of polarization may be changed if it is passed through bi-refringent crystals or optically active substances. The significant advantage of this proposed scheme is that the schemes are all-optical in nature and can be easily and successfully be extended for higher order multiplexer and demultiplexer. As an example, for 16:1 multiplexer, 16-select lines can be constructed by two Delta literals outputs and 16-QMIN gates. Now select lines are to be fed to again 16-QMIN gate as first input and the second input is to be taken from the input signals. By selecting proper select line we can transfer any one of 16-input signals to the output. This scheme is easily practicable as the components of our design are technically highly developed and widely used in optical communication. The proposed scheme will work with other 2x2 Interferometric switches (like Mach–Zehnder interferometer) also.

5. Conclusions and Future Scopes:

In present day digital signal processing is based on conventional binary number system (radix = 2). It has two logical states 'LOW' and 'HIGH'. Binary logic (or logic for that matter) is NOT a law of nature. The reason why binary logic seems more natural is because we have been more exposed to it. The perspective MVL is more like an extension of binary logic and very conventional, though broader in possibilities. In a wide sense, a signal may be anything that can be observed to have states that change in time and space. In narrow sense, a signal is a physical quantity that can be measured, usually by an electronic device. Signals, as conveyers of information about the state of a system, should be processed to extract and understand the information content encoded. Nowadays, digital systems, and sometimes mixed-signal systems, are prevalent in information transmission, storage, and processing. However, enormous, and ever increasing, amounts of information that can be handled, even in everyday life, focus attention to multiple-valued (MV) logic, which permit more compact encoding of information within the same amount of digits. Although, having certain considerable demerits, multiple-valued logic is viewed as promising alternatives in many practical solutions. Many contemporary logic design technologies are oriented towards supporting an efficient implementing of various signal processing algorithms. In order to entirely exploit all the available resources, sophisticated methods are required. Humans count by tens, machines count by twos, these sums up the way we do arithmetic today. However, there are countless other ways to count. Challenges and opportunities are wide.

Author details

Jitendra Nath Roy[1*] and Tanay Chattopadhyay[2]

*Address all correspondence to: jnroys@yahoo.co.in

1 Department of Physics, National Institute of Technology, Agartala, Jirania, Tripura, India

2 Mechanical operation (stage-II), Kolaghat Thermal Power station, WBPDCL, West Bengal, India

References

[1] Woods, D., & Naughton, T. J. (2012). Photonic neural networks. *Nature Physics*, 8, doi: 10.1038/nphys2283, 257-259.

[2] Caulfield, H. J., & Dolev, S. (2010). Why future supercomputing requires optics. *Nature Photonics*, 4, doi:10.1038/nphoton.2010.94, 262-263.

[3] Caulfield, H. J., Dolev, S., & Green, W. M. J. (2009). Appl. Opt. A, 48(Optical High-Performance Computing feature issue), http://dx.doi.org/10.1364/AO.48.0OHPC1.

[4] Roy, S., Prasad, M., Topolancik, J., & Vollmer, F. (2010). All optical switching with bacterirhodopsin protein coated microcavities and its application to low power computing circuits. *Journal of Applied Physics*, 107, 053115, http://dx.doi.org/10.1063/1.3310385.

[5] Reed, T., Mashanovich, G., Gardes, F. Y., & Thomson, D. J. (2010). Silicon optical modulators. *Nature Photonics*, 4, doi:10.1038/nphoton.2010.179, 518-526.

[6] Hurst, S. L. (1984). Multiple-Valued Logic-Its Status and its Future. *IEEE Transactions on computers*, C-33(12), 1160-1179, Doi:10.1109/TC.1984.1676392.

[7] Lohmann, A. W. (1988). Polarization and optical logic. *Applied Optics*, 25, 1594-1597.

[8]   Fagotto, E. A. M., & Abbade, M. L. F. (2009). All-optical demultiplexing of 4-ASK op-
       tical signals with four-wave mixing optical gates. *Optics Communications*, 283,
       1102-1109, Doi: 10.1016/j.optcom.2009.10.094.

[9]   Khan, M. H. A., & Perkowski, M. A. (2007). GF(4) based synthesis of quaternary re-
       versible / quantum logic gate. *Proc. of the 37th international symposium on multiple-val-
       ued logic, ISMVL'07 IEEE*, Doi:10.1109/ISMVL.2007.26.

[10] Obiniyi, A. A., Absalom, E. E., & Adako, K. (2011). Arithmetic Logic Design with
       Color-Coded Ternary for Ternary Computing. *International Journal of Computer Appli-
       cations*, 26(11), 0975-8887, Doi: 10.5120/3162-2929, 31-37.

[11] Garai, S. K., Pal, A., & Mukhopadhyay, S. (2010). All-optical frequency-encoded in-
       version operation with tristate logic using reflecting semiconductor optical am-
       plifiers. *Optik*, 121, DOI:10.1016/j.ijleo.2009.02.011, 1462-1465.

[12] Stakhov, A. (2002). Brousentsov's ternary principle Bergman's number system and
       ternary mirror-symmetrical arithmetic. *The Computer Journal*, 45(2), 221-236, doi:
       comjnl/45.2.221.

[13] Walklin, S., & Conradi, J. (1997). A 10 Gb/s 4-ary ASK lightwave system. *ECOC 97,
       IEE, Conference publication*, 448(448), 255-258, DOI: cp:19971538.

[14] Liu, S., Li, C., Wu, J., & Lin, Y. (1989). Optoelectronic multiple-valued logic imple-
       mentation. *Optics letters*, 14(14), 713-715, http://dx.doi.org/10.1364/OL.14.000713.

[15] Awwal, A. A. S., Karim, M. A., & Cherri, A. K. (1987). Polarization-encoded optical
       shadow-casting scheme: design of multi-output trinary combinational logic units.
       Applied optics, 26(22), 4814-4818, http://dx.doi.org/10.1364/AO.26.004814.

[16] Imai, Y., & Ohtsuka, Y. (1987). Optical multiple-output and multiple-valued logic op-
       eration based on fringe shifting techniques using a spatial light modulator. *Applied
       optics*, 26(2), 274-277, doi: AO.26.000274.

[17] Li, C., & Yan, J. (2011). Design Method and Implementation of Ternary Logic Optical
       Calculator. *Advances in Information and Communication Technology*, 347, Doi:
       10.1007/978-3-642-18369-0_17, 147-166.

[18] Chattopadhyay, T. (2010). All-optical symmetric ternary logic gate. *Optics and Laser
       Technology*, 42, DOI: 10.1016/j.optlastec.2010.01.023, 1014-1021.

[19] Song, Kai, & Yan, Liping. (2012). Design and implementation of the one step MSD
       adder of optical computer. *Applied Optics*, 51(7), 917, doi: AO.51.000917.

[20] Chattopadhyay, T., Maity, G. K., & Roy, Jitendra Nath. (2008). Designing of all-opti-
       cal tri-state logic system with the help of optical nonlinear material. *Journal of Nonlin-
       ear Optical Physics & Materials*, 17(3), 315-328, Doi: S0218863508004159.

[21] Chattopadhyay, T., & Roy, J. N. (2011). Semiconductor optical amplifier (SOA)-assist-
       ed Sagnac switch for designing of all-optical tri-state logic gates. *Optik International*

Journal for Light and Electron Optics, 122(12), DOI: 10.1016/j.ijleo.2010.06.045, 1073-1078.

[22] Chattopadhyay, T., & Roy, J. N. (2011, March 26-27). All-optical multi-valued computing: the future challenges and opportunities. Kolkata. *International conference on convergence of Optics and Electronics (COE 11)*, 94-101, 978-8-19064-011-4.

[23] Vasundara Patel, K. S., & Gurumurthy, K. S. (2010). Quaternary Sequential Circuits. *IJCSNS International Journal of Computer Science and Network Security*, 10(7), 110, DOI: 10.1109/ICCIT.2009.5407139.

[24] Yi, J., Huacan, H., & Yangtian, L. (2005). Ternary computer architecture. *Physica Scripta*, T 118(98), 101, doi:10.1238/Physica.Topical.118a00098.

[25] Roy, J. N. (2009). Mach-Zehnder interferometer-based tree architecture for all-optical logic and arithmetic operations. *Optik-International Journal for Light and Electron Optics*, 120(7), 218-324, http://dx.doi.org/10.1016/j.ijleo.2007.09.004.

[26] Khan, M. H. A., Siddika, N. K., & Perkowski, M. A. (2008). Minimization of quaternary galois field sum of products expression for multi-output quaternary logic function using quaternary Galois field decision diagram. *Proceedings of the 38th international symposium on Multiple-valued logic (DOI 10.1109/ISMVL.2008.31) IEEE*, 125-130, Doi: 10.1109/ISMVL.2008.31.

[27] Li, G., Qian, F., Ruan, H., & Liu, L. (1999). Compact Parallel Optical Modified-Signed-Digit Arithmetic-Logic Array Processor with Electron-Trapping Device. *Applied Optics*, 38(23), 5039-5045, http://dx. doi.org/10.1364/AO.38.005039.

[28] Chattopadhyay, T., & Sarkar, T. (2012). All-optical switching by Kerr nonlinear prism and its application to of binary-to-gray-to-binary code conversion. *Optics & Laser Technology*, 44(6), 1722-1728, http://dx.doi.org/10.1016/j.optlastec.2012.02.007.

[29] Jung, Y. J., Lee, S., & Park, N. (2008). All-optical 4-bit gray code to binary coded decimal converter. *Proc. of SPIE*, 6890, 68900S, doi:10.1117/12.762429.

[30] Nakamura, T., Kani, J. I., Teshima, M., & Iwatsuki, K. (2004). A quaternary amplitude shift keying modulator for suppressing initial amplitude distortion. *IEEE Journal of Lightwave Technology*, 22(3), 733-738, DOI: 10.1109/JLT.2004.824465.

[31] Awwal, A. A. S., & Ahmed, J. U. (1993). Fast Carry free Adder Design Using QSD Number System. *IEEE*, Doi:10.1109/NAECON.1993.290791, 1085-1088, CH3306-8/93/0000-1085.

[32] Awwal, A. A. S., & Ahmed, J. U. (1994). Two-bit restricted signed-digit quaternary full adder. *Proc. Of IEEE*, DOI:10.1109/NAECON.1994.332917, 1119-1125.

[33] Cherri, A. K., & Al-Zayed, A. S. (2009). Circuit designs of ultra-fast all-optical modified signed-digit adders using semiconductor optical amplifier and Mach-Zehnder interferometer. *Optik*, doi:10.1016/j.ijleo.2009.02.029, in press.

[34] Ghosh, A. K., Bhattacharya, A., Raul, M., & Basuray, A. (2012). Trinary arithmetic and logic unit (TALU) using savart plate and spatial light modulator (SLM) suitable for optical computation in multivalued logic. *Optics & Laser Technology*, 44(5), DOI: 10.1016/j.optlastec.2011.11.044, 1583-1592.

[35] Jahangir, I., Hasan, D. M. N., & Reza, M. S. (2009, 23-26 Jan.) Design of Some Quaternary Combinational Logic Blocks Using a New Logic System. Singapore. *TENCON '09, IEEE Region 10 Conference*, Doi: 10.1109/TENCON.2009.5396095, 1-6.

[36] Smith, K. C. (1988). Multiple-valued logic: a tutorial and appreciation. *IEEE computers*, 21(4), 1160-1179, doi:10.1109/2.48.

[37] Cunha, R., Boudinov, H., & Carro, L. (2007). Quaternary look-up tables using voltage mode CMOS logic design. *Proceedings of the 37th International Symposium of multiple valued logic (ISMVL'07) IEEE*, 0-7695-2831-7, Doi:10.1109/ISMVL.2007.47.

[38] Shanbhag, N. R., Nagchoudhuri, D., Ferd, R. E. S., & Visweswaran, G. S. (1990). Quaternary logic circuits with 2-jum CMOS technology. *IEEE Journal of solid-state circuits*, 25(3), 790-798, doi: 10.1109/4.102677.

[39] Park, S. J., Yoon, B. H., Sub Yoon, K., & Kim, H. S. (2004). Design of Quaternary Logic Gate Using Double Pass-transistor Logic with neuron MOS Down Literal Circuit. *Proceedings of the 34 th International Symposium on Multiple-valued logic (ISMVL2004)*, 198-203, doi: 10.1109/ISMVL.2004.1319941.

[40] Kerkho, H. G., & Tervoert, M. L. (1981). Multiple-valued logic charge-coupled devices. *IEEE Transactions computers*, C-30(9), 644-652, http://doi.ieeecomputersociety.org/10.1109/TC.1981.1675862.

[41] Yasuda, Y., Tokuda, Y., Zaima, S., Pak, K., Nakamura, T., & Yoshida, A. (1986). Realization of Quaternary Logic Circuits by n-Channel MOS Devices. *IEEE Journal of solid-state circuits*, sc-21(1), 162-168, DOI: 10.1109/JSSC.1986.1052493.

[42] Kerkhoff, H. G., & Tervoert, M. L. (1981). Multiple-Valued logic Charge-Coupled Devices. *IEEE Transactions computers*, C-30(9), 644-652, http://doi.ieeecomputersociety.org/10.1109/TC.1981.1675862.

[43] Brilman, M., Etiemble, D., Oursel, J. L., & Tatareau, P. (1982). A 4-valued ECL encoder and decoder circuit. *IEEE Journal of solid-state circuits*, sc-17(3), 547-552, Doi: 10.1109/TC.1986.1676733.

[44] Mangin, J. L., & Current, K. W. (1986). Characteristics of prototype CMOS quaternary logic Encoder-Decoder circuits. *IEEE Transactions on Computers*, C-35(2), 157-161, Doi: 10.1109/TC.1986.1676733.

[45] Etiemble, D., & Israël, M. (2002). A current-mode folding / interpolating CMOS analog to quaternary encoding block. *Proc. Of 32nd IEEE International Symposium on Multiple-valued logic (ISMVL2002)*, 0-7695-2831-7, http://doi.ieeecomputersociety.org/10.1109/ISMVL.2002.1011099.

[46] Chan, H. L., Mohan, S., & George, I. Haddad. (1996). Compact Multiple-Valued Multiplexerers Using negative Differential Resistance Devices. *IEEE Journal of Solid-State circuits*, 31(8), 1151-1155, doi=10.1.1.136.8146.

[47] Chattopadhyay, T., Bhowmik, P., & Roy, J. N. (2012). Polarization encoded optical N-valued inverter. *JOSA B*, Accepted.

[48] Keshavarzian, P., & Mirzaee, M. M. (2012). A novel, efficient CNTFET Galois design as a basic ternary-valued logic field. *Nanotechnology, Science and Applications*, 5, 1-11, http://dx.doi.org/10.2147/NSA.S27550.

[49] Khorasaninejad, M., & Saini, S. S. (2011). All-optical logic gate in silicon nanowire optical waveguides. *IET Circuits, Devices and system*, 5(2), 115-122, 10.1049/iet-cds.2010.0142.

[50] Glesk, I., Runser, R. J., & Prucnal, P. R. (2001). New generation of devices for all-optical communication. *Acta physica slovaca*, 51(2), 151-162, http://dx.doi.org/10.1117/12.498224.

[51] Wang, B. C., Baby, V., Tong, W., Xu, L., Friedman, M., Glesk, I., Runser, R. J., & Prucnal, P. R. (2002). A novel fast optical switch based on two cascaded terahertz optical asymmetric demultiplexers (TOAD). *Optics Express*, 10(1), 15-23, http://www.opticsinfobase.org/oe/abstract.cfm?URI=oe-10-1-15.

[52] Houbavlis, T., & Zoiros, K. E. (2004). Numerical simulation of semiconductor optical amplifier assisted Sagnac gate and investigation of its switching characteristics. *Optical Engineering*, 43(7), 1622-1627, http://dx.doi.org/10.1117/1.1751132.

[53] Zoiros, K. E., Chasioti, R., Koukourlis, C. S., & Houbavlis, T. (2007). On the output characteristics of a semiconductor optical amplifier driven by an ultrafast optical time division multiplexing pulse train. *Optik*, 118(3), 134-146, DOI: 10.1016/j.ijleo.2006.01.012.

[54] Shen, Z. Y., & Wu, L. L. (2008). Reconfigerable optical logic unit with a teraheartz optical asymmetric demultiplexer and electro-optic switches. *Applied Optics*, 47(21), 3737-3742, DOI: 10.1364/AO.47.003737.

[55] Roy, J. N., Maity, A. K., & Mukhopadhyay, S. (2006). Designing of an all-optical time division multiplexing scheme with the help of non linear material based tree-net architecture. *Chinese Optics letters*, 4(8), 483-486.

[56] Karim, M. A., & Awwal, A. A. S. (2003). *Optical Computing: an introduction*, Wiley, New York, Chap-6.

[57] Chattopadhyay, T., & Roy, J. N. (2009). An all-optical technique for a binary-to-quaternary encoder and a quaternary-to-binary decoder. *J.Opt.A: Pure Appl. Opt.*, 11, doi: 10.1088/1464-4258/11/7/075501.

[58] Shen, Z. Y., Wu, L., & Yan, J. (2012). The reconfigurable module of ternary optical computer. *Optik International Journal for Light and Electron Optics,* http://dx.doi.org/10.1016/j.ijleo.2012.03.081.

[59] Chattopadhyay, T., & Roy, J. N. (2009). All-optical conversion scheme: binary to quaternary and quaternary to binary number. *Optics & Laser Technology,* 41(3), 289-294, http://dx.doi.org/10.1016/j.optlastec.2008.06.003.

[60] Taraphdar, C., Chattopadhyay, T., & Roy, J. N. (2009). Polarization Encoded All-optical Ternary Max Gate. Kolkata. *International conference on computers and devices for communication CODEC-09,* Paper ID OLT-3512. Print, 978-1-42445-073-2, INSPEC Accession Number 11136770.

[61] Taraphdar, C., Chattopadhyay, T., & Roy, J. N. (2011). All-optical integrated ternary MIN and MAX gate. Kolkata, India. *Proccedings of international conference on Trends in optics and photonics (IconTOP 2011),* 476-481, 978-81-908188-1-0.

[62] Bhowmik, P., Roy, J. N., & Chattopadhyay, T. (2011). Designing of all-optical two input ternary MAX logical operation. *National conference of photonics and nano sciences, Dept of physics, Garhbeta college,* 51-58.

[63] Roy, J. N., Chattopadhyay, T., Manna, S., & Maity, G. K. (2008). Polarization encoded all-optical quaternary max gate. IIT Delhi, India. *PHOTONICS '08, International Conference on Fiber Optics and Photonics,* 1-4.

[64] Chattopadhyay, T., & Roy, J. N. *Quaternary MAX gate and its applications in all-optical domain,* unpublished.

[65] Bhowmik, P., Chattopadhyay, T., Taraphdar, C., & Roy, J. N. (2011, March 26-27). Designing of All Optical Circuit for Two Input Ternary MIN Logical Operation. Kolkata. *International conference on convergence of Optics and Electronics, (COE 11),* 94-101, 978-8-19064-011-4.

[66] Chattopadhyay, T., & Roy, J. N. (2009). Polarization encoded all-optical quaternary multiplexer and demultiplexer-a proposal. *Optik International Journal for Light and Electron Optics,* 120, 941-946, http://dx.doi.org/10.1016/j.ijleo.2008.03.030.

[67] Chattopadhyay, T., & Roy, J. N. (2010). Polarization encoded TOAD based all-optical quaternary Literals. *Optik International Journal for Light and Electron Optics,* 121, 617-622, http://dx.doi.org/10.1016/j.ijleo.2008.09.014.

[68] Taraphdar, C., Chattopadhyay, T., & Roy, J. N. (2011). Designing of an all-optical scheme for single input Ternary logical operations. *Optik International Journal for Light and Electron Optics,* 122(1), 33-36, http://dx.doi.org/10.1016/j.ijleo.2009.09.016.

[69] Chattopadhyay, T., & Roy, J. N. Polarization encoded four valued ordinary inverter. *XXXVI OSI Symposium on Frontiers in Optics and Photonics, (FOP 11) IIT,* P 99, 978-81-309-1964-5 .

[70] Chattopadhyay, T., & Roy, J.N. (2012). All-optical ordinary quaternary inverter (QNOT) using binary NOT gate. *Optik International Journal for Light and Electron Optics*, in press, doi: 10.1016/j.ijleo.2012.01.035.

[71] Chattopadhyay, T., & Roy, J.N. (2010). Polarization Encoded All-optical Quaternary Universal Inverter and Designing of Multi-valued Flip-flop. *Optical Engineering*, 49(3), 035201, DOI: 10.1117/1.3362897.

[72] Chattopadhyay, T., & Roy, J.N. (2011). Polarization encoded all-optical quaternary successor with the help of SOA assisted Sagnac switch. *Optics communication*, 284(12), 2755-2762, DOI:10.1016/j.optcom.2011.02.005.

[73] Chattopadhyay, T., & Roy, J.N. (2011). All-optical quaternary Galois field sum of product (GFSOP) circuits. *Optik International Journal for Light and Electron Optics*, 122(9), 758-763, http://dx.doi.org/10.1016/j.ijleo.2010.06.002.

[74] Chattopadhyay, T., & Roy, J. N. (2010, 27-28 March). All-optical quaternary half-adder circuit with the help of Terahertz optical asymmetric demultiplexer (TOAD). Burdwan. *National conference on materials, devices and circuits in communication Tech. (MDCCT'2010)*, TS. 4.12, 50.

[75] Chattopadhyay, T., Das, M.K., Roy, J.N., Chakraborty, A.K., & Gayen, D.K. (2011). Interferometric switch based all optical scheme for Conversion of Binary number to its Quaternary Signed Digit form. *IET Circuits, Devices and system, (special issue on 'Optical Computing Circuits, Devices and Systems')*, 5(2), 132-142, doi: 10.1049/iet-cds.2010.0056.

[76] Chattopadhyay, T., & Roy, J.N. (2011). Easy conversion technique of Binary to Quaternary Signed Digit and vice versa. *Physics Express*, 1(3), 165-174.

[77] Chattopadhyay, T., & Roy, J.N. (2009, March 1-4). All-optical conversion of binary number to quaternary signed digit (QSD) number. Kolkata, India. *Proceedings of international conference on Trends in optics and photonics (IconTOP 2009)*, 130-137.

[78] Chattopadhyay, T., & Roy, J.N. (2011, 28th February- 1st March). All-optical carry free addition using quaternary signed digit (QSD). *18th West Bengal state science & Technology congress*, 1(2), 3-4.

[79] Chattopadhyay, T., Taraphdar, C., & Roy, J.N. (2009). Quaternary Galois field adder based all-optical multivalued logic circuits. *Applied Optics, (feature issue on 'optical high-performance computing')*, 48(22), E35-E44, http://dx.doi.org/10.1364/AO.48.000E35.

[80] Chattopadhyay, T., Roy, J.N., & Chakraborty, A.K. (2009). Polarization encoded all-optical quaternary R-S flip-flop using binary latch. *Optics Communications*, 282, 1287-1293, DOI:10.1016/j.optcom.2008.12.022.

[81] Taraphdar, C., Chattopadhyay, T., & Roy, J.N. (2011). Designing of Polarization encoded all-optical ternary multiplexer and Demultiplexer. *Recent Patents on Signal Processing*, 1(2), 143-155, doi:10.2174/1877612411101020143.

[82] Chattopadhyay, T. (2010). All-optical quaternary circuits using quaternary T-gate. *Optik International Journal for Light and Electron Optics*, 121, 1784-1788, DOI:10.1016/j.ijleo.2009.04.014.

[83] Domanski, A.W. (2005). Polarization degree fading during propagation of partially coherent light through retarders. *Opto-Electronics Review, 7th International Workshop on Nonlinear Optics applications*, 13(2), 171-176.

[84] Mecozzi, A., & Shtaif, M. (2002). The statistics of polarization dependent loss in optical communication systems. *IEEE photonics tech. letter*, 14(3), 313-315, DOI: 10.1109/68.986797.

[85] Nelson, L.E, Nielson, T.N., & Kogelnik, H. (2001). Observation of PMD-induced coherent crosstalk in polarization-multiplexed transmission. *IEEE photonics tech. letter*, 13(7), 738-740, DOI:10.1109/68.930432.

[86] Tang, J.M., & Shore, K.A. (1998). Strong picosecond optical pulse propagating in semiconductor optical amplifiers at transparency. *IEEE Journal of Quantum Electronics*, 34(7), 1263-1269, doi: 10.1109/3.687871.

[87] Bruyere, F., & Andouin, O. (1994). Penalties in long-haul optical amplifiers systems due to polarization dependent loss and gain. *IEEE Photonics Tech. Letters*, 6(5), 654-656, doi: 10.1109/68.285570.

Image and Video Processing

Chapter 5

Video Encoder Implementation on Tilera's TILEPro64™ Multicore Processor

José Parera-Bermúdez, Javier Casajús-Quirós and
Igor Arambasic

Additional information is available at the end of the chapter

1. Introduction

The Moore's law states that the transistor number on integrated circuits approximately doubles every two years. This trend has been met since its description in 1965. But this exponential growth in transistor count does not always translate into similar growth of CPU performance; some issues such as power density, total power and intra-chip distances are preventing clock speeds above 4.5 GHz. During the past decades advances in semiconductor technology and architecture have overcome the obstacles, but at present there is no alternative technology and all the possibilities of micro-parallelism seem to have been explored. Another major issue is that the speed of dynamic memory has not grown with the same strength as the CPU's speed, while static memory is prohibitively expensive for widespread use.

The solution being put into practice is the use of the so called multicore CPU, i.e. the integration of multiple cores on a single chip. Today nearly all computers, including desktops and laptops, are equipped with CPUs with at least 2 cores and it is not uncommon to see servers with 8 or 16 cores.

The evolution and the steady decrease in the price of technology have enabled the digital video to be a media component included on any device from small pocket players to professional projection equipment on movie theaters. Today the *de facto* standard for video coding is ITU-T/ISO H.264 /MPEG-4 Part 10 or AVC (Advanced Video Coding) [1]. Since its first publication, back in 2003, it has become one of the most commonly used formats due to its flexibility to be applied to a wide variety of applications on a wide variety of networks and systems, including low and high bit rates, low and high resolution video, broadcast, DVD storage, RTP/IP packet networks, multimedia telephony systems, etc. In 2004 the standard was extended to enable higher quality video coding by adding several new features (increased sample bit depth precision, higher-resolution color information, adaptive switching between 4x4 and 8x integer transforms...) required by professional applications.

H.264 performs significantly better than any prior standard under a wide variety of circumstances in a wide variety of application environments, and outperforms MPEG-2 video, the DVD standard for movies, typically obtaining the same quality at half the bit rate or less, especially on high bit rate and high resolution situations.

Like other ITU-T standards, H.264 only specifies the syntax of the bitstream and the decoding procedures for reconstructing the video images; the encoding process is not specified at all allowing the use of different approaches, algorithms and optimizations as long as the bitstream is syntactically correct. Unlike previous standards, it is designed bearing in mind its implementation avoiding complex calculations and favoring the use of just adders and shifters; nevertheless encoding is far more involved than decoding.

It is easy to find lots of papers and books dealing with almost every aspect of H.264; there are also countless proprietary and open source software libraries and custom hardware implementations particularly for the consumer market. Therefore, what is special about the implementation described in the following paragraphs? In brief, the remarkable aspects are:

- It is targeted to very high quality with very low latency,

- It is a software-only solution, and

- The hardware is based on a commercial off-the-shelf multicore processor: the TILE*Pro*64™ from Tilera Corporation.

The performance achieved allows encoding 4K (*DLP Cinema Technology*, 4096x1716 pixels, 24 frames per second) video, the current standard for digital cinema, in real time with just one frame latency.

The study has been undertaken as part of a project that develops optimized hardware for those applications in which real time analysis-synthesis of high definition image streams is needed. The project was led by Datatech (www.datatech-sda.com) and it focused on a particular case of search & track applications for the aerospace segment, namely automatic refueling of flying military aircrafts.

2. System architecture

Figure 1 shows the hardware building blocks of the system. As can be seen the TILE processor is the very heart of the system and only some adapting logic is required to deal with the camera output; the I/O capabilities of the processor, including the Gigabyte Ethernet interface, make the rest.

From a logical point of view, the system behaves as a standalone RTSP (*Real-Time Streaming Protocol*) server [2] that packetizes the video encoded data for RTP delivering [3] over an IP network.

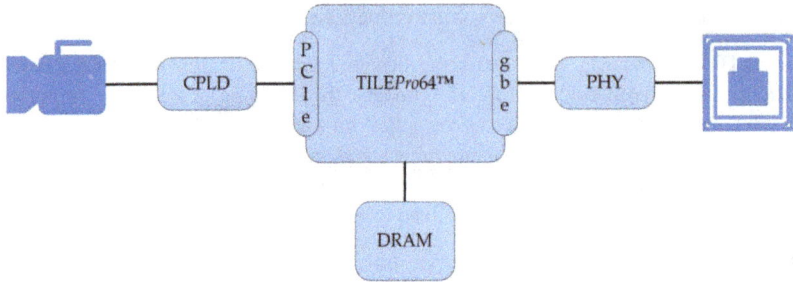

Figure 1. Hardware System Architecture

2.1. The TILE*Pro*64™ processor

The TILE*Pro*64™ [4], the second generation of Tilera's processors, is a fully programmable 64-core processor organized as a two-dimensional array (8x8) of processing elements (each referred to as a tile), connected through the iMesh™, a bunch of two-dimensional mesh networks. The processor also integrates external memory and I/O interfaces connected to the tiles via the iMesh™ interconnect fabric.

Each tile contains a Processor Engine, a Cache Engine, and a Switch Engine, which combine components to make a powerful, full-featured compute engine.

- The Processor Engine is a conventional 32 bit VLIW (Very Long Instruction Word) processor with three instructions per bundle and full memory management, protection, and OS support configuring a powerful, full-featured computing system that can independently run a Linux operating system. The Tile Processor includes special instructions to support commonly-used embedded operations in DSP, video and network packet processing, including: hashing and checksums, instructions to accelerate encryption, SIMD (Single Instruction Multiple Data) instructions for sub-word parallelism, saturating arithmetic, multiply-accumulate (MAC) instructions, sum of absolute differences (SAD), and unaligned access acceleration. All arithmetic instructions are of integer type because there is not a floating point unit.

- The Cache Engine contains the tile's Translation Lookaside Buffers (TLBs), caches, and cache sequencers. Each tile has 16KB L1 instruction cache; 8KB L1 data cache, and a 64KB unified L2 cache. This delivers a total of 5.5 MB of on chip memory. The cache can be configured as coherent or incoherent; in the first case, the hardware automatically maintains the consistency of data between processors, converting all on chip memory in a sort of L3 unified cache. Each tile also contains a DMA engine that works together with the cache engine for orchestrating memory data streaming between tiles and external memory, and among the tiles.

- The Switch Engine implements six independent networks. The Switch Engine switches scalar data between tiles through the Static Network (STN) with very low latency. Five dynamic networks (UDN, TDN, MDN, CDN and IDN) aid the Switch Engine by routing packet-based data among tiles, tile caches, external memory, and I/O controllers. Of the five dynamic networks, only the User Dynamic Network (UDN) is user-visible; the others are used to satisfy cache misses from external memory and other tiles, and for various system-related functions. The Static Network in addition to the five Dynamic Networks comprise the interconnect fabric of the Tilera iMesh™. The user does not explicitly need to manage these networks; rather they are used by the system software to efficiently implement the application-level API abstractions, such as user-generated inter-process socket-like streams.

It is noteworthy that all the cores are identical forming a homogeneous architecture that contrasts with other notable multicore processors such as the Cell Broadband Engine (from Sony, Toshiba and IBM) or the DaVinci from Texas Instruments. As a result, programming is easier, more portable and more easily scalable. Furthermore, the combination of cores and interconnecting network enable different kinds of parallelism: fine or coarse grain, shared-memory multithreading, message passing multitasking, etc. making the architecture suitable for a broad range of parallel problems.

The TILE*Pro*64™ supports the following primary external interfaces:

- Memory: four memory interface channels, each supporting 64-bit DDR2 DRAM up to 800 Mbps, for a peak total bandwidth of 25.6 GB/s. The memory controllers are on-chip.

- 10Gb Ethernet: Two full-duplex XAUI-based 10Gb ports with integrated MACs.

- PCIe: Two 4-lane PCI Express ports configurable as 4-lane, 2-lane or 1-lane (4x, 2x, 1x) with integrated MACs, supporting both root complex and endpoint modes.

- 10/100/1000 Ethernet: Two on-board RGMII 10/100/1000 Ethernet MACs.

- HPI: 16-bit host port interface.

- Flexible I/O: 64 bits of dedicated Flexible I/O for programmable I/O and interrupt support, with frequency up to 150 MHz and streaming capability.

- UART, I2C and SPI ROM.

Figure 2 shows a block diagram of the processor:

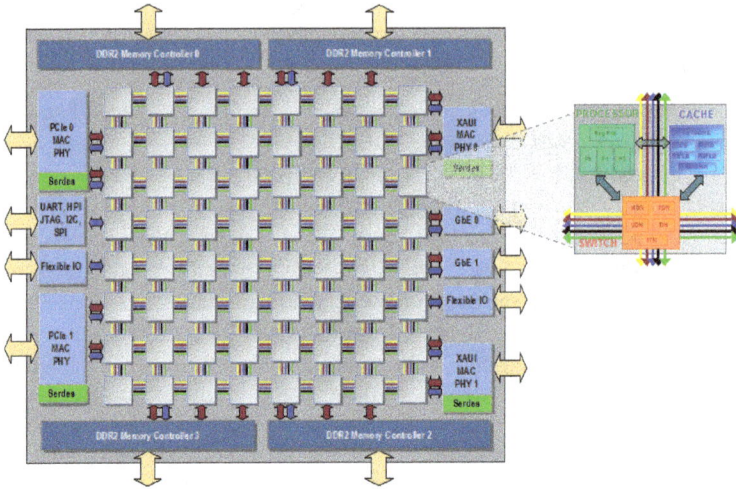

Figure 2. TILE*Pro*64™ Block Diagram

There are two versions of the processor that differ only in the operating frequency: 700 or 866 MHz. A few performance metrics at 866 MHz are listed in the next table:

Operations Per Second	8-bit	443 BOPs
	16-bit	222 BOPs
	32-bit	166 BOPs
Data I/O	40+ Gbps	
Memory I/O	205 Gbps	
Bisection Bandwidth	2660 Gbps	
On-chip Cache Memory Bandwidth	1774 Gbps	

Table 1. TILE*Pro*64™ Performance Metrics

2.2. The multicore development environment

The Tilera MDE [5] provides a complete software environment, including the system software stack, a variety of helpful software libraries, and standard Linux command-line utilities. The execution environment includes three layers: the hypervisor, the client operating system (Linux), and user space:

- The hypervisor is the lowest layer of the software stack. It abstracts hardware details of the processor, manages communication between tiles, from tiles to I/O controllers, and provides a low-level virtual-memory. This layer also provides I/O drivers that run on dedicated tiles and, therefore, do not run Linux or user space applications. The running drivers, the tiles on which they run and their parameters can be configured at boot time.

- The supervisor layer, composed of SMP Linux, provides system calls and I/O devices for user-space applications and libraries. This layer enables multi-process applications and multi-threaded processes to exploit multiple tiles for increased performance. The OS software manages hardware resources and provides higher-level services, such as processes and virtual memory allocation.

- The application layer runs user space programs that can invoke Linux system calls and link against standard libraries just as on any other Linux platform. Tilera provides the standard C/C++ run-time and other processor specific libraries.

The MDE also provides a complete suite of tools for all phases of program development, starting with authoring or porting an application, through debugging, and into performance evaluation. These tools include:

- C/C++ compiler, assembler and linker, and other standard Unix tools. This tool chain is compatible with that of GNU; specifically, the compiler supports ANSI C99 standard as well as GNU extensions. The tools enable the use of portable source code, easy to program and with the same concurrency support as that available for Intel x86 processors in a Linux box.

- A standard, open-source gdb debugger with support for the Tile Processor architecture.

- A software simulator that provides cycle-accurate profiling and tracing.

- A custom version of the open-source Eclipse IDE providing a GUI interface for all stages of program development: authoring, building, running, debugging and profiling.

3. Encoder parallelization

Programming a parallel application is not an easy job. The design space is enormous: different kinds of parallelism, data granularity, tools... The algorithm being implemented, the performance objectives and the computing platform impose some constraints but do not determine the design choices. Fortunately, the Tilera's platform supports a broad range of possibilities. See, for example [6], which explores several alternatives for the H.264 encoder.

The design of the application has completed all four typical stages established by best practices: task decomposition, assignment, orchestration and distribution [7]. The following paragraphs detail the course of action.

3.1. Task decomposition

3.1.1. H.264 encoder procedures

An H.264 encoder [8] [9] [10] consists of a few basic procedures; besides these, the standard defines a wide range of ancillary techniques, many of them optional, designed to provide enough flexibility to be applied to multiple scenarios. The encoder structure (see figure 3) does not differ substantially from that of other encoders but its many details and subtleties allow for a much more efficient compression.

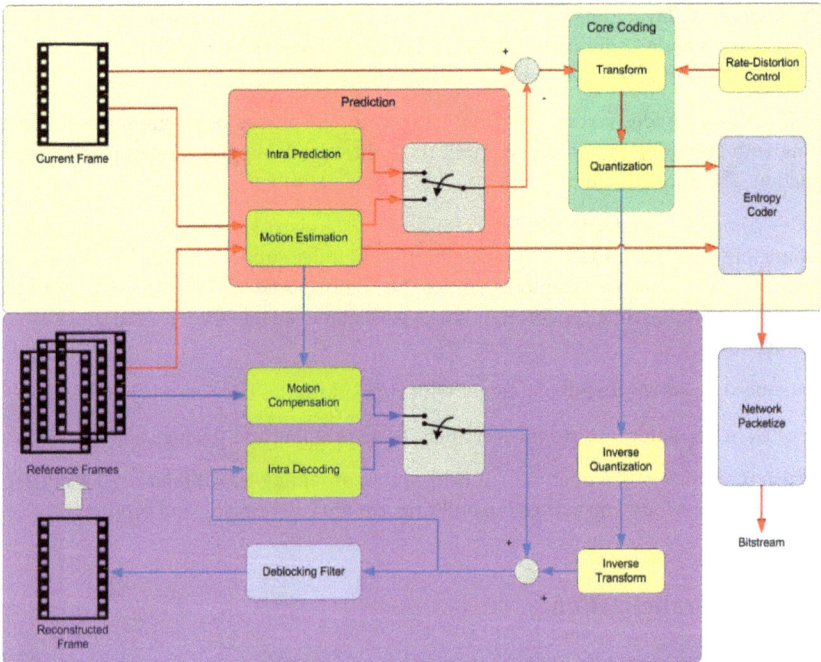

Figure 3. Block Diagram of Basic Encoder Procedures in H.264

The upper part of the figure (yellow) shows the encoding process. The frame being encoded, in YCbCr color space format, is divided into macroblocks, i.e. chunks of 16x16 luminance (Y) pixels and their corresponding chrominance (Cb, Cr) pixels, whose size varies according to the subsampling method (8x8 if 4:2:0, 8x16 if 4:2:2 or 16x16 if 4:4:4). The luminance and chrominance channels are processed separately using the same techniques.

In the prediction phase the encoder builds a macroblock using previously encoded data, either from the current frame (spatial or intra prediction) or from other frames (temporal or inter prediction). Intra prediction can be carried out for the whole macroblock in 4 modes or dividing it into 8x8 or 4x4 blocks in 9 modes. Each mode, related to a spatial direction, is just an extrapolation computed as averages of the neighboring pixels. Inter prediction is more involved since it tries to find a description of the macroblock by estimating its motion with respect to similar regions of previous frames. The search is performed on a variable number of reference frames, in several rectangular section sizes and with increased pixel accuracy to allow for sub pixel motion. Finally, the predicted macroblock can be expressed in terms of motions vectors from regions of the reference frames.

Once a macroblock is predicted the encoder subtracts the predicted pixel values from the input macroblock to form a residual. The prediction methods supported by H.264 make it possible to accurately predict the input macroblock, thus resulting in an outstanding video compression because, being a differential encoder, the residual values are small and very often nulls. Unfortunately, the computational complexity is too high, which poses a severe trouble for a real-time implementation.

The residual data is transformed using an approximate 4x4 or 8x8 2D DCT (Discrete Cosine Transform). This transform has the particular feature of compacting the energy of the input in the low frequency coefficients and thus the transformed residual usually contains a few non-zero values close to the matrix's upper left corner. H.264 does not use standard DCTs but modified so that their kernels consist solely of integer numbers eliminating the need for floating point calculations.

The modified DCT output should be quantized by dividing the coefficients by an integer, but the derivation of the transform left pending a scaling factor. Both numbers are combined to avoid division and, incidentally, to accommodate the quality parameter QP. The overall computation reduces the precision of the coefficients according to the desired quality governed by the value of QP: the larger the value, the poorer the quality but the higher the compression and vice versa. The rate-distortion control procedure can update the value of QP for each frame or macroblock in order to balance the opposing goals of high quality and low bit rate. Usually, the rate-distortion control procedure aims a maximum or constant bit rate but it is also possible to encode aiming constant quality in which case this procedure does nothing.

The quantized DCT coefficients are scanned in zigzag to sort them according to increasing spatial frequency; then they are converted into binary codes along with other signaling information (macroblock partitioning, prediction modes or motion vectors...) and inserted into the output bitstream. The binary codes are computed by the entropy coder procedure using variable length or arithmetic coding, both adapted to the context to further reduce the number of bits. Hence, their names are context-adaptive variable length coding (CAVLC) and context-adaptive binary arithmetic coding (CABAC).

The quantization output also feeds the decoding process (the purple box in figure 3). This process is needed at the encoder because it reconstructs the macroblocks and frames to be used for prediction. The first procedure is the inverse quantization; actually it is a rescaling as quantization cannot be inverted. Afterwards the coefficients are inverse transformed to form a residual which is added to the prediction to get the reconstructed macroblock. If using inter prediction, the result could be optionally filtered for reducing blocking distortion by smoothing the macroblock edges. Note that typically the reconstructed macroblock will differ from the original due to the loss of precision caused by quantization. Ultimately, H. 264 is a lossy compressor.

3.1.2. Implementation tradeoffs

In order to fulfill the requirements of real-time, low-latency and high quality the encoder implements only a subset of the standard features and techniques available in the standard. This selection does not prevent the encoder to comply with the standard because H.264 allows a high degree of flexibility in the techniques used. Specifically, the implemented encoder has undergone the two following main tradeoffs:

- Intra only prediction, i.e. all predicted pixels are computed using only the current frame; otherwise, the very low latency requirement couldn't be achieved.

- CAVLC entropy coding; the alternative method, CABAC, is much more efficient from the standpoint of the bit rate but it cannot be parallelized due to its recursive nature.

These tradeoffs, and some others discussed below, adversely affect the compression ratio of the encoder resulting in an increased bit rate. Fortunately, neither the best compression ratio nor constant bit rate are requirements of the implementation. These goals are distinct from those required for consumer applications for which there are lot of solutions, but they are essential in many professional applications: remote monitoring, remote assistance, content generation, broadcast, video surveillance. The H.264 standard dedicates some specific profiles for this kind of applications (High, High10 and High10 Intra), that in some cases the industry has adopted, such as Panasonic's AVC-Intra.

Undoubtedly the temporal prediction improves the bit rate but not for free; in order to obtain a high quality a large number of reference frames are needed, scene changes can lead to devastating effects, especially if a constant bit rate is desired, and latency, measured from end to end, i.e. camera to monitor, increases linearly with the number of reference frames, and can reach 1 second, only for decoding.

In our implementation, the encoding of a frame begins as soon as the first 16 lines of pixels are available and once the row is encoded it is sent so that the decoding can start even before the whole frame is encoded. Such an extremely low latency is only possible using spatial prediction. Additional advantages of the intra-only prediction are: 1) ease of frame by frame video editing, 2) the resilience against transmission errors, since an error affects only one frame, and 3) a significant saving of memory, which is especially important for very large resolutions.

3.1.3. Amdahl's law

The very first step in parallelizing an application is to determine if it is worth. Regardless of costs, running platform, software architecture and any other constraint a parallel application can run faster only a limited amount compared to its sequential version. Amdahl's law states that if P is the fraction of code that can be made parallel and $S = (1 - P)$ is the fraction not parallelizable then the maximum speed up that can be achieved by using N processors is

$$speedup(N) = \frac{1}{S + \frac{P}{N}} \qquad (1)$$

The results of applying Amdahl's law to a given problem are just a rough approximation to reality but serve to get an estimate of the maximum parallel performance and to focus attention on potential bottlenecks and hot spots that can be found in the algorithm under development. So, the essential starting point in parallelization is to get an optimized sequential code, in order to determine the value of S.

The available literature dealing with H.264 encoding focuses on algorithmic description or performance improvement but usually forget to emphasize the inherently non-parallelizable part: the composition of the bitstream. Once the input raw video is encoded the resulting data must be packaged into NAL (Network Abstraction Layer) units which are byte aligned structures with header and trailing data. The data, known as RBSP (Raw Byte Sequence Payload), is written into the NAL units using a strict syntax in which the macroblock raster order must be preserved. The number of bits generated depends on the image, making it impossible to compose the NAL units without complying with the order. Furthermore, the Annex B of the standard states that RBSP data must be checked against patterns of bytes that can confuse the framing alignment while decoding. Those patterns must be disambiguated by byte stuffing the RBSP, i.e. inserting a fixed 0x03 byte each time they occur. Again, this procedure is neither predictable nor parallelizable.

The execution of the optimized sequential code on a Linux box equipped with an Intel Core2 Duo (T7700) CPU @ 2.40 GHz reveals that the fraction of time spent handling the composition of NAL units is 0.45%, yielding the value of $S = 0.0045$. The encoded frames per second (fps) for a 4096x1716 video are 0.75. These figures have been obtained without taking into account the input or output in order to accurately measure the time spent in the algorithm. Solving equation (1) for $N = 60$ processors results in a speed up of 47.4x that applied to the throughput gives 35.56 fps, enough for the digital cinema format (24 fps). A similar run on the Tilera platform @ 866 MHz yields $S = 0.015$ (1.5%) and a throughput of 0.67 fps. Solving again for $N = 60$ processors results in a speed up of 31.8x and a final throughput of 21.33 fps, less than the requirement for the above mentioned format. The different values of S are mainly due to the unequal facilities fitted in the CPUs for handling bytes and bit fields. The result of this analysis indicates that the NAL unit management is clearly a hot spot in the code that could ruin the overall performance. Obviously, an optimization is needed in the Tilera side to fulfill the goal.

3.1.4. Data dependencies

No parallel program can be built without knowing the data dependencies that the algorithm imposes. As previously stated, the basic procedures of the encoder are the pixels prediction and the entropy encoding of the residuals; in our case, intra prediction and CAVLC. It is clear that the second procedure must follow the first one, since it's not feasible to encode any data without having calculated it. Aside from this obvious fact, an analysis of the H.264 encoder algorithm from the data flow standpoint shows:

- The input image is partitioned for processing into so called macroblocks, square chunks of 16x16 pixels.

- The macroblocks are processed in raster scan order, i.e. from left to right and from top to bottom.

- Each macroblock is predicted using some data from previously encoded macroblocks, specifically the boundary pixels of the upper, upper right and left macroblocks. The only exception to this rule is when the neighbouring macroblocks are not available; for example, the first macroblock of an image does not use any additional information because their neighbours do not exist.

- In order to compute the entropy encoding each macroblock needs a quantization parameter, QP. There is no provision to determine how a macroblock selects this parameter, but usually this job is entrusted to the block labelled as "Rate Distortion Control" in figure 3, because it affects the number of bits generated in the entropy encoder and ultimately the bit rate of the whole encoder. The quantization parameter can be seen as a quality parameter: the lower the value the better the quality but also the higher the bit rate. Our implementation allows to select between constant quality (fixed QP) and constant bit rate (adaptable QP), but for the ease of parallelizing the latter option is applied in a frame by frame basis.

In summary, the data dependencies at this algorithm level are the boundary pixels from neighboring macroblocks and a frame constant quantization parameter.

3.1.5. Tasks

After analyzing the extent of parallelization and data dependencies it is the time to analyze the tasks that make up the algorithm. Here we mean by task not the usual computing term but any procedure of the algorithm that could be accomplished in parallel.

The core encoding algorithm assuming the above mentioned tradeoffs could be described as a kind of streaming with frames as elements and macroblocks as the units of computation. At the system level there must be a single task that implements the RTSP service, waiting for a client connection and then delivering the RTP packets. At the frame level the following tasks can be identified:

- Compute the Rate-Distortion procedure (usually known as RDO with the O meaning Optimization). The result is the quantization parameter to be applied to the frame.

- Open and initialize a Network Abstraction Layer (NAL) unit

- Read the raw input pixels of the frame.

- Encode the frame in macroblock chunks in raster scan order.

- Write the encoded data to the NAL unit.

- Close the NAL unit.

- Update RDO with the frame information.

- Deliver the NAL unit.

At the macroblock level, the task list is as follows:

- Read the macroblock raw input pixels.

- Get the boundary pixels from neighbouring macroblocks.

- Select the best prediction.

- Compute the residual error.

- Transform and quantize the residual error.

- Inverse transform and quantize the transformed data.

- Encode the residual transformed data using CAVLC.

Note that some tasks at macroblock level can be interspersed with those at the frame level, e.g. once a macroblock is CAVLC encoded the resulting data could be written to the NAL unit, and therefore there is no need of collecting the data from all macroblocks and writing it afterwards. The rearrangement of task order and the intermixing at the described or even finer levels broaden the parallelization options as long as the data dependencies are met.

Two potential hot spots can be found at the input and output of data. A digital cinema camera with a chroma sampling of 4:2:0 produces 10,543,104 bytes of raw video data per frame totaling more than 240 Mbytes/s. If we assume a compression ratio of 10, the total output bit rate will exceed 24 Mbytes/s. These figures are not unmanageable, but indicate that the input and output procedures should be treated with special care and, as far as possible, run them overlapped with the rest of tasks in the algorithm.

Another hot spot is concerned with prediction. The luminance part of each macroblock can be predicted in three pixel sizes: 16x16, the full macroblock, four 8x8 blocks or sixteen 4x4 blocks; the chrominance is always predicted in full size blocks (8x8 if using 4:2:0 chroma format) for each component. Each block is explored in several modes related to different spatial directions: four modes for 16x16 luminance and chrominance and nine modes for the rest. Summing up all modes by iterating all the luminance prediction modes for each possible chrominance prediction mode yields a total search space of 736 combinations, each with its associated metric. The standard says nothing about how to compute these metrics and, therefore, how to select the best prediction mode for each block. There are two main approaches to assess this measure: 1) in the spatial domain, calculating the cumulative sum of absolute differences between actual and predicted pixels and (SAD); and 2) calculating the same sum but using the data in the DCT transformed domain (SATD). The latter provides, in general, better results but the quality or bit rate difference is not significant when the video resolution is high. By means of a test suite we have determined that using SAD instead of SATD on high-definition (HD) and above formats, the bit rate increases by only 1% whereas the computational load is 30% lower. Needless to say, the approach chosen is the use of SAD. Luckily, it also allows taking advantage of some of the more specific and powerful instructions of the TILEPro64™ processor: the "sum of absolute difference" SIMD group.

In whatever case, these computations are very time consuming; note that the prediction of 4x4 and 8x8 blocks requires the reconstructed neighboring blocks since this will be the information available at the decoder. Therefore, once a mode is selected as best for a given block, it must be reconstructed emulating the decoder procedure in order for the neighbors to use its boundaries. This circumstance has promoted a lot of research over the last years [11] aimed to diminish the search space making available fast methods to "predict" the best predictor from among a substantially reduced set modes without compromising too much the bit rate. We have chosen for our implementation a simple yet effective fast mode decision algorithm called "Selective Intra Prediction" [12]. The key idea of this algorithm stems from the fact that the dominating direction of a bigger block is similar to that of a smaller block and therefore it is feasible to avoid the computation of the unlikely modes after the determination of the best 16x16 mode. The algorithm has been combined with the usual early-termination technique, but in spite of this the fraction of time dedicated to the selection of the predictor exceeds 60% after manually optimizing the code.

3.2. Task assignment

3.2.1. Parallel pattern

Previous sections have explored the opportunities for parallelism highlighting the hot spots of the encoder. Now it is time to choose the most appropriate type of parallelism and to logically organize the tasks.

The best parallelization pattern for achieving high throughput is the pipeline; if in addition its number of stages is not large latency can be low enough. However, a video encoder is not a good candidate for pipelining because, among other considerations, the computational burden of tasks is very dissimilar, the flow of control is not regular as it depends on data, data must be shared or copied and, specifically for the TILE processor, the number of stages should be no less than 60.

If we reject the pipeline approach, the remaining choices to consider are multiprocessing, multithreading or a mix of both. The main difference concerns virtual memory space; a process has its own non shared virtual space while a thread shares it with all other threads. Multithreading demands a more elaborated synchronization among threads but facilitates the inter-thread communications because it is accomplished simply by sharing data in memory. Furthermore, the TILE processor implements inter e intra-core cache coherence techniques that leverage the user of worrying about correctness of data. Based on these considerations the multithreading approach was chosen for the encoder.

3.2.2. Data decomposition

Another issue has to do with the decomposition of data, i.e. how to partition and distribute the data space among the cores. The encoding problem has not a recursive nature and so it is clearly preferable a geometric decomposition ideally suited to the data dependencies of the algorithm. It seems that macroblock decomposition of data is the right choice. Additionally, this decomposition enables the use of a single-program multiple-data (SPMD) model that eases programming by means of parametrizing the input to the code.

3.2.3. Core processing threads

So far we have decided to use threads to process macroblocks; the question now is how to organize those threads. A digital cinema video frame is composed of 27456 macroblocks; if any single thread is responsible for a single macroblock we would need the same amount of threads. Even if spread into 60 cores the number far exceeds the Linux threading facilities. A better choice is to partition the data not into macroblocks but into rows of macroblocks; this yields just 108 threads, a more manageable figure that not compromises the SPMD model.

Such a thread assignment can be still improved. Programs typically let the threads die once they have finished their work, but thread creation and termination has some overhead that can be obviated by recycling them. This is a simple technique, usually seen in digital signal processing, in which a thread created at startup runs forever until explicitly killed. For this technique to be useful it requires two synchronization points, the first to ensure that data is available while the second to signal that the work is finished. The exchange between thread management overhead and synchronization is worthwhile.

Further improvement arises if we avoid thread scheduling and time sharing in any single core as it eliminates the operating system kernel overhead devoted to task switching. In such a scenario each thread dynamically selects the next row to be processed as soon it has finished with the current. This technique, known as thread pooling, is especially well suited to the TILE architecture since the pool can be spread among the cores, each one running a single thread. The MDE has a provision for exploiting this setting: the so called dataplanes in which the standard Linux kernel is substituted by a zero overhead kernel. A nice result of using a thread pool is a fair load balancing, which is not always easy to get.

It remains to determine the scope of the row processing, i.e. which algorithm tasks the row threads perform. Referring to the above list of macroblock level tasks it is worthwhile that the row threads be in charge of reading all the row input pixels, select the predictor, and so on, including the entropy encoding; the last task would be impossible if CABAC were used instead of CAVLC because CABAC, being recursive, needs data from the last macroblock of the previous row. In such a case the row threads could not proceed in parallel or, if they did, they should store all the information of the macroblocks and afterwards apply the entropy encoding. This would represent a severe waste of memory and an unmanageable hot spot. However, the CAVLC encoding so partitioned also has a drawback: since the bit alignment of the row data in the NAL unit is unknown, the whole unit must be realigned. Anyway, the entropy encoding in parallel is advantageous.

The preceding paragraphs have focused on analyzing the processing of macroblock rows but we have said anything about the issue described in Section 3.1.2 concerning Amdahl's Law: the re-alignment and byte stuffing of CAVLC encoded data are limiting factors of parallel performance that may saturate the speed up. The implementation dedicates a core, known as framer, running a single thread of manually optimized assembler code to address this problem.

Figure 4 shows a simplified time line of 12 row threads working in parallel. It can be seen that the whole process resembles a macroblock pipeline, although technically speaking it is not. Some details are worth being described:

- All time intervals are sketched alike; actually, times depend on input data.

- Each row, except the first, must start with a delay at least twice the handling time of a single macroblock to ensure that the boundary pixels are available.

- The total time spent at any frame is much greater than the time required for packetizing its encoded data into a NAL unit, as expected for a pipelined structure.

- The video input is not sorted in a natural way, e.g. row 0 at frame N (the green one) needs data before row 11 at frame N–1 (the blue one). The input procedure must take this fact into account and allow for a non ordered access to the data.

- At any moment in time multiple frames may be simultaneously being processed; the worst case arises at the frame boundaries (T0 in the figure). It is not difficult to prove that if we assume a constant, say mean, macroblock processing time value the number of simultaneous frames can be as great as:

$$n = \left\lceil \frac{2(C - 1) + R}{R} \right\rceil \tag{2}$$

being C the number of macroblocks in a row and R the number of rows. Furthermore, the row threads could start working with the frame N+1 before the framing thread has had an opportunity to evict the row data of frame N. A simple n-buffer strategy at the row thread output is enough to solve this trouble.

- The framing proceeds in bursts at the beginning of a frame because the encoded data is available but afterwards it must wait for the rows to terminate. This drawback puts the framer thread even more under pressure as there may be time intervals during which it cannot perform any work.

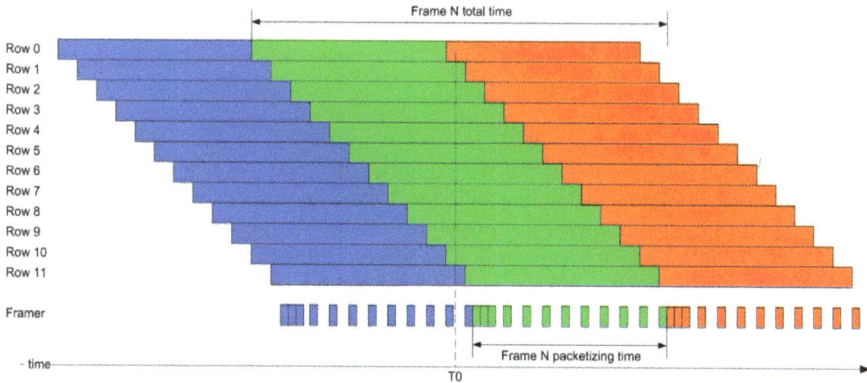

Figure 4. Time Line of Row Processing

3.2.4. Auxiliary threads

There are two tasks that still remain to be assigned to threads: the RTSP service and the RDO. The RTSP service can advantageously be implemented using two threads: the first one devoted to the service itself while the second in charge of the subsidiary real-time control protocol (RTCP). No one of these threads requires a great amount of CPU resources but the logical division facilitates software coding.

With regard to RDO, it could be as complex as desired in order to obtain accurate estimates of the bit rate and so select the best quantization parameter QP. But we have said repeatedly that optimizing the bit rate is not a priority of our implementation and a simple and fast PID (proportional integral derivative) controller algorithm is enough for our purposes. The only remarkable question is that adjusting the algorithm parameters should be made taking into account that input data are delayed due to the pipelined behavior of the processing. This RDO computation could be performed by the framer thread but we preferred to do it in a separate thread for the system to be more flexible in case of need.

3.2.5. Input and output

The best solution for input and output is that their functionality runs into two separate hypervisor drivers. Doing so, all I/O data could flow over the I/O Dynamic Network (IDN) that connects all tiles with the on-chip devices alleviating the burden of memory sharing at the user level. In addition, this schema obviates the intermediate level of buffering needed between the program and the Linux kernel drivers.

The output driver is just a packet based service tailored to the handling of the RTP payload over IP. A notable optimization feature is that the driver uses only fixed size buffers that fit into the Ethernet jumbo packets with two objectives: to reduce overhead and to avoid IP fragmentation.

On the other hand, the input driver is programmed as a server that handles the necessary buffering for extracting and reordering the camera data and delivers it at the pace enforced by the row threads requests.

3.3. Orchestration

The aim of the orchestration phase is to design the mechanisms that will ensure the proper synchronization among threads; i.e. that all control and data dependencies are met. Section 3.1.3 dealt with data dependencies; the control dependencies arise from the task assignment to threads, specifically to take advantage of the always live thread pattern.

The synchronization primitives used are those provided by the POSIX 1b and 1c extensions, available in any Linux box and in the Tilera's software stack. The use of these primitives is easier than programming custom ones, although its performance is not always the best; but having selected a coarse data grain for the implementation their impact is very limited.

In essence, we use semaphores for synchronizing threads and read-write locks to protect the macroblock boundary data. The use of the latter instead of the usual mutexes allows a higher degree of parallelism as it does not blocks any reading thread if the writer has not acquired the lock. Bearing in mind that we have designed the assignment of tasks to threads so that there is only one thread on each core, the read-write locks can have spin flavor to avoid putting the threads to sleep while waiting for the lock.

3.4. Distribution

So far we have always used 60 as the number of cores dedicated to encoding but the Tilera processor has 64; let see why. On the one hand, the framer thread, being the major potential bottleneck of the system, claims a core for itself; on the other hand, for the input and output hypervisor drivers to do their work overlapped with the algorithmic computation they each need a dedicated core. The remaining core hosts all the other auxiliary threads: RTSP server, RTCP, RDO and the C main thread that just waits for the program to exit.

The last issue to be addressed is how to distribute the threads, i.e. to determine in which physical cores they will run. The best way to make the distribution is keeping as far as possible the data locality since the latency for accessing an adjacent core's cache memory is much cheaper than accessing any other core's cache. This arrangement is easily attained for the row processing cores; unfortunately, there is no way for the framer core to be adjacent to all row processing cores. The selected distribution is shown in figure 5, in which row cores are shown in blue with the closed adjacency path in light blue. The framer core is shown in dark blue, input and output cores in orange and the auxiliary core in green. The position of input and output deserves a comment; they are physically very close to their corresponding hardware as Tilera recommends.

This distribution scales almost linearly for any video resolution with at least 60 rows of macroblocks, i.e. 960 pixels high, including high-definition (1080 pixels, 68 rows) and above. With lower resolutions the row cores will not all be active so there will be a degradation of performance.

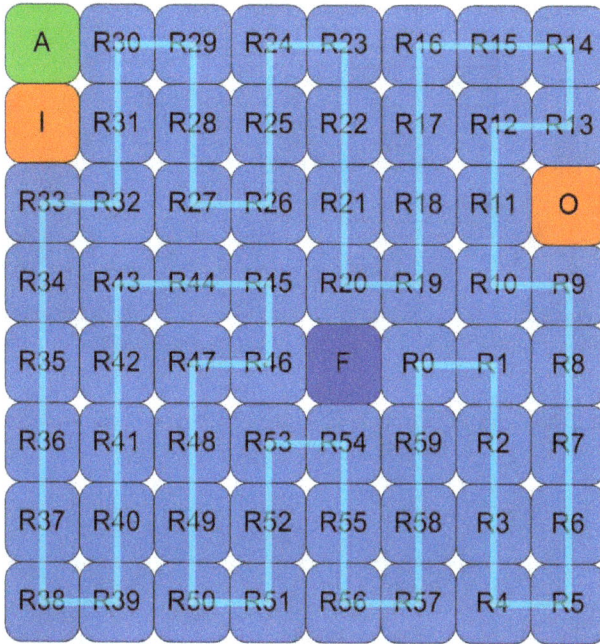

Figure 5. Distribution of Threads into Processor Cores

4. Results

In order to evaluate the results the freely distributable test video sequences Park-Joy has been encoded in different sizes using a constant quantization parameter QP = 18. This figure allows an encoding without noticeable visual degradation. Park-Joy contains small figures of running people; sometimes large objects - unfocused trees near the camera - move to the left as a result of a strictly horizontal camera movement, overlapping the entire scene. At the end of the sequence the camera slows the motion.

In sequential runs the mean time spent in macroblock encoding is 40.27 µs in Linux and 79.54 µs on the TILE processor @ 866 MHz. The differences are due to the operating clock and the architectural dissimilarities. It is easy to see that the optimizations undertaken in the Tilera side have been successful since the clock speed is reduced by a factor of 2.77 while the time ratio is only 1.98. Note that the Linux code has been optimized only at the C level and thus not using the SIMD instructions provided by the MMX or SSE instruction extensions.

The same run on the Tilera's simulator in functional mode in, which the cache hazards are not fully considered, yields 57.07 µs. It is apparent that the TILE core cache memory is not large enough to hold all code and data and thus incurring in a high rate of capacity misses.

In parallel runs the cache problem becomes more evident, as shows the following table:

Image Size	Simulator	Hardware
1280x720	239,82 fps	181,65 fps
1920x1080	155,98 fps	99,44 fps
3840x2160	37,37 fps	25,66 fps

Table 2. Encoded Frames per Second

It is worth to mention that the performance boosts around 32.5% in mean by avoiding the 8x8 block encoding of luma. This figure puts a little more spice to the controversy over the inclusion of this technique in the standard.

The following graph shows the throughput measured as time per macroblock (blue) and number of macroblocks per second (green) versus resolution.

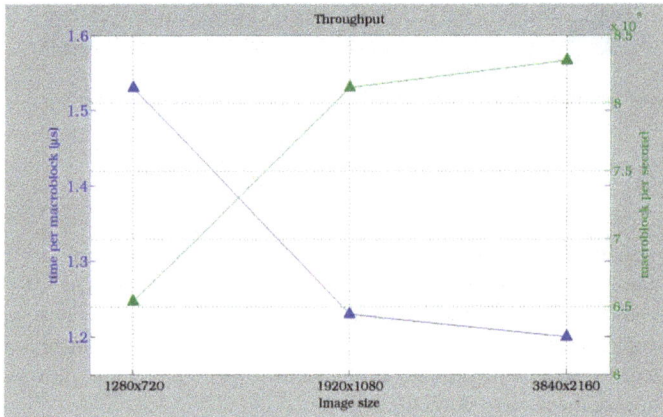

Figure 6. Throughput

It can be seen that the throughput degrades abruptly when the number of used row cores is less than available; the 1280x720 uses 45 cores, while the other uses all 60.

The next graph shows the speed up as a function of the number of row cores. The shape of the graph is quite linear but the slope is less than 1 as predicted by Amdahl's Law.

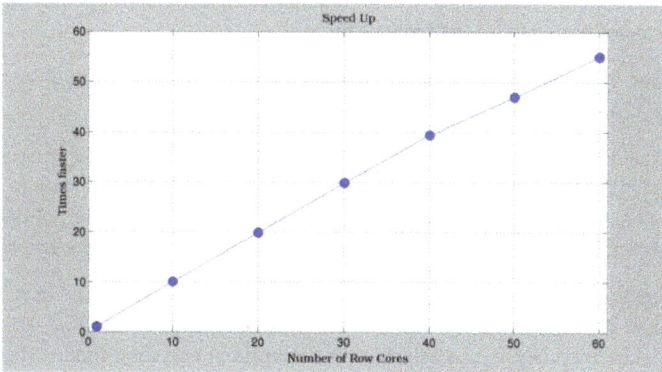

Figure 7. Speed Up

4.1. Some TILE*Pro*64™ troubles

Despite the enormous amount of silicon and functionality provided by the TILE*Pro*64™ processor, some flaws have been detected:

- It is quite hard to optimize the code using intrinsics or assembler; it would be nice if the documentation [13] contain examples, tips and tricks.

- The processor instruction set architecture contains basic instructions for bit and byte rearrangement at the register level; these include the "byte exchange", "byte/word interleave", and "masked merge word" instructions. However, it lacks bit-field extract and insert and byte/word shuffle instructions. These capabilities are incorporated in the new generation of Tilera processors, the TILE-Gx series.

- Correct use of branches is difficult, even for the compiler; branch mispredictions result in pipeline hazards that increase instruction latency. Fortunately, the feedback based optimization technique [14] alleviates this issue but it is cumbersome when optimizing source code.

- Finally, the most important limitation for video encoding is the amount of cache per core; 64 Kbytes of L2 is not enough for code and data, leading to many cache-capacity misses and therefore many stalled cycles. The TILE-Gx series has 256 Kbytes of L2 cache per core, without any doubt a must for achieving better video encoding performance.

5. The future of video coding

Video coding technology will not stop at H.264. A new draft standard known as HEVC (aka H.265 and MPEG-H Part 2) is still under development. It features important improvements over H.264 centered on achieving bit rate reductions of about 50% and supports a wider range of high definition resolutions. Computational complexity consequently increases by an estimated factor of two to ten, and maybe more.

The techniques by which those enhancements are realized should be analyzed from the point of view of our implementation.

As we mentioned before, the CABAC procedure as defined in H.264 is not amenable to parallelization. In HEVC special care has been taken to reduce data dependencies in the particular version of CABAC it implements. However, data dependencies have not disappeared, and this poses severe problems – albeit less so – for implementation on parallel processors.

In order to ameliorate spatial prediction, new modes have been defined in HEVC. In particular the number of modes is 35 versus 9 in H.264. We have already said that 60% computation time is taken by analysis and selection of the optimal encoding mode. Therefore one must expect a considerable increase in computation needs due simply to the number of prediction modes that must be explored; more even so, given the fact that the image is divided not in uniform macroblocks but in coding tree blocks (or coding units CU) with an inner structure of variable sizes of their own (64x64, 32x32, 16x16).

6. Conclusion

The overall performance of Tilera's TILE*Pro*64™ can be said to be outstanding for video coding applications. In the particular case of low-latency H.264 encoding the largest difficulties arises for the highest resolution values of the video stream. For these, large amounts of memory are required, exceeding what is readily available within the processor; the main limitation resulting from the relatively small cache memory. Notwithstanding this fact, we have been able to find memory management schemes and workarounds that make real time encoding possible even at the highest resolutions (4096x1716, 24 fps) contemplated in this work.

For other video codec applications, expectations are high. Inter frame prediction can probably be traded off by lower resolution values. The main conclusion being that the processor architecture is adequate for established coders, whose bases were laid a few years ago and which are still the subject of implementation research.

New developments such as HVEC hold enormous promise, but the difficulties surrounding real-time implementation are challenging, to say the least. It is likely that several years of research are needed to significantly advance in that direction. Of course this raises the question whether the architecture will be up to the coding schemes under development and/or what enhancements will be necessary.

The results of this work do not stop at video coding. Applications to novel fields such as virtual advertising and augmented reality in medicine are under study for current and future projects.

Acknowledgements

The. authors gratefully acknowledge the support provided by project IDI-20100823 of Spanish Government's *Ministerio de Economía y Competitividad,* under the leadership of Datatech SDA who also acknowledges that support. Project TEC2009-14219-C03-01 also provided support for this work.

The authors also acknowledge the continuing support and cooperation of Datatech SDA for ongoing developments of Tilera's processor capabilities: real-time video analysis, virtual advertising and augmented reality in medicine.

Author details

José Parera-Bermúdez, Javier Casajús-Quirós and Igor Arambasic

*Address all correspondence to: jose.parera@upm.es

Department of Signals, Systems and Radiocommunications, Polytechnic University of Madrid, Spain

References

[1] ITU-T Rec. H.264 I ISO/IEC 14496-10 version 16, *Advanced video coding for generic audiovisual services,* January 2012. http://www.itu.int/rec/dologin_pub.asp?lang=e&id=T-REC-H.264-201201-I!!PDF-E&type=items (accessed 9 September 2012)

[2] RFC2326, *Real Time Streaming Protocol (RTSP),* IETF, 1998. http://datatracker.ietf.org/doc/rfc2326/ (accessed 9 September 2012)

[3] RFC6184, *RTP Payload Format for H.264 Video,* IETF, 2011. http://datatracker.ietf.org/doc/rfc6184/ (accessed 9 September 2012)

[4] Tilera Corporation, *TILE Processor Architecture Overview for the TILEPro Series,* 2009.

[5] Tilera Corporation, *Multicore Development Environment: Programming the TILE Processor,* 2009.

[6] Takeuchi, Y., Nakata, Y., Kawaguchi, H. & Yoshimoto, M. *Scalable parallel processing for H.264 encoding application to multi/many-core processor.* International Conference on Intelligent Control and Information Processing (ICICIP), August 13-15, 2010, Dalian, China. doi: 10.1109/ICICIP.2010.5565292

[7] Gove D. *Multicore Application Programming: for Windows, Linux and Oracle Solaris,* Addison-Wesley, 2011.

[8] Richardson I. *The H.264 Advanced Video Compression Standard, Second Edition*, John Wiley & Sons, 2010.

[9] Wiegand T. & Sullivan G.J. *Overview of the H.264/AVC Video Coding Standard*, IEEE Transactions on Circuits and Systems for Video Technology 2003;13(7):560-576. doi: 10.1109/TCSVT.2003.815165

[10] Sullivan G.J., Topiwala P. & Luthra A. *The H.264/AVC Advanced Video Coding Standard: Overview and Introduction to the Fidelity Range Extensions*, SPIE Conference on Applications of Digital Image Processing XXVII, Special Session on Advances in the New Emerging Standard: H.264/AVC, 2004. doi: 10.1117/12.564457

[11] Milani S. *Spatial prediction in the H.264/AVC FREx coder and its optimization*, In: Miron S. (ed.) *Signal Processing*, Rijeka: InTech; 2010. http://www.intechopen.com/books/signal-processing/spatial-prediction-in-the-h-264-avc-frext-coder-and-its-optimization (accessed 9 September 2012)

[12] Park J.S. & Song, H.J. *Selective Intra Prediction Mode Decision for H.264/AVC Encoders*, World Academy of Science, Engineering and Technology 13, 2008. http://www.waset.org/journals/waset/v13/v13-104.pdf (accessed 9 September 2012)

[13] Tilera Corporation, *User Architecture Manual*, 2010.

[14] Tilera Corporation, *Optimization Guide*, 2010.

Chapter 6

A Real-Time Video Encoding Scheme Based on the Contourlet Transform

Stamos Katsigiannis, Georgios Papaioannou and
Dimitris Maroulis

Additional information is available at the end of the chapter

1. Introduction

Real-time video communication over the internet and other heterogeneous IP networks has become a significant part of modern communications, underlining the need for highly efficient video coding algorithms. The most desirable characteristic of such an algorithm would be the ability to maintain satisfactory visual quality while achieving good compression. Additional advantageous characteristics would be low computational complexity and real-time performance, allowing the algorithm to be used in a wide variety of less powerful computers. Transmission of video over the network would benefit by the ability to adapt to the network's end-to-end bandwidth and transmitter/receiver resources, as well as by resistance to packet losses that might occur. Additionally, scalability and resistance to noise would be highly advantageous characteristics for a modern video compression algorithm. Most state of the art video compression techniques like the H.264, DivX/Xvid, MPEG2 fail to achieve real time performance without the use of dedicated hardware due to their high computational complexity. Moreover, in order to achieve optimal compression and quality they depend on multipass statistical and structural analysis of the whole video content, which cannot happen in cases of live video stream generation as in the case of video-conferencing.

In this chapter, a more elaborate analysis of a novel algorithm for high-quality real-time video encoding, originally proposed in [1], is presented. The algorithm is designed for content obtained from low resolution sources like web cameras, surveillance cameras, etc. Critical to the efficiency of video encoding algorithm design is the selection of a suitable image representation method. Texture representation methods proposed in the literature that utilize the Fourier transform, the Discrete Cosine transform, the Wavelet transform as well as other frequency domain methods have been extensively used for image and video encoding. Nevertheless, these methods have some limitations that have been partially addressed by the Contourlet Transform (CT) [2], which our video encoding algorithm is based on. The Contourlet Transform offers multiscale and directional decomposition, providing anisotropy and directionality, features missing from traditional transforms like the Discrete Wavelet Transform [2]. In recent years, the Contourlet Transform has been successfully utilised in a variety of texture analysis applications, including synthetic aperture radar (SAR) [3], medical and natural image classification [4], image denoising [5], despeckling of images, image compression, etc. By harnessing the computational power offered by modern graphics processing units (GPUs), a gpu-based contourlet transform is able to provide an image representation method with advantageous characteristics, while maintaining a fast performance.

The rest of this chapter is organised in four sections. First, some background knowledge and information needed for better understanding the algorithm is presented in section 2. Then, the aforementioned video encoding algorithm is presented in section 3, whereas an experimental study for the evaluation of the algorithm is provided in section 4. Conclusions and future perspectives of this work are presented in section 5.

2. Background

2.1. The Contourlet Transform

The Contourlet Transform (CT) is a directional multiscale image representation scheme proposed by Do and Vetterli, which is effective in representing smooth contours in different directions of an image, thus providing directionality and anisotropy [2]. The method utilizes a double filter bank in which, first the Laplacian Pyramid (LP) [6] detects the point discontinuities of the image and then the Directional Filter Bank (DFB) [7] links those point discontinuities into linear structures. The LP provides a way to obtain multiscale decomposition. In each LP level, a downsampled lowpass version of the original image and a more detailed image with the supplementary high frequencies containing the point discontinuities are obtained. This scheme can be iterated continuously in the lowpass image and is restricted only by the size of the original image due to the downsampling. The DFB is a 2D directional filter bank that can achieve perfect reconstruction, which is an important characteristic for image and video encoding applications. The simplified DFB used for the contourlet transform consists of two stages and leads to 2^l subbands with wedge-shaped frequency partitioning [8], with l being the level of decomposition. The first stage of the DFB is a two-channel quincunx filter bank [9] with fan filters that divides the 2D spectrum into vertical and horizontal directions, while the second stage is a shearing operator that just reorders the samples. By adding a 45 degrees shearing operator and its inverse before and after a two-channel filter bank, a different directional frequency partition is obtained (diagonal directions), while maintaining the ability to perfectly reconstruct the original image, since the sampling locations coincide with the (integer) pixel grid.

The combination of the LP and the DFB is a double filter bank named Pyramidal Directional Filter Bank (PDFB). In order to capture the directional information, bandpass images from the LP decomposition are fed into a DFB. This scheme can be repeated on the coarser image levels. The combined result is the contourlet filter bank, which is a double iterated filter bank that decomposes images into directional subbands at multiple scales. The contourlet coefficients have a similarity with wavelet coefficients since most of them are almost zero and only few of them, located near the edge of the objects, have large magnitudes [10]. In the presented algorithm, the Cohen and Daubechies 9-7 filters [11] have been utilized for the Laplacian Pyramid. For the Directional Filter Bank, these filters were mapped into their corresponding 2D filters using the McClellan transform as proposed by Do and Vetterli in [2]. It must be noted that these filters are not considered as optimal. The creation of optimal filters for the contourlet filter bank remains an open research topic. An outline of the Contourlet Transform is presented on Figure 1, while an example of decomposition is shown on Figure 2.

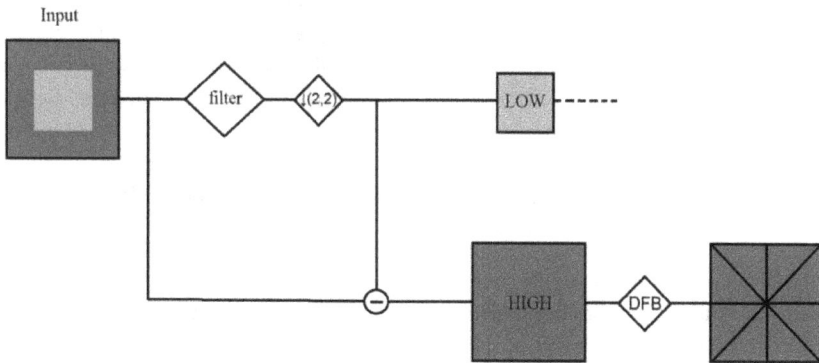

Figure 1. The Contourlet Filter Bank.

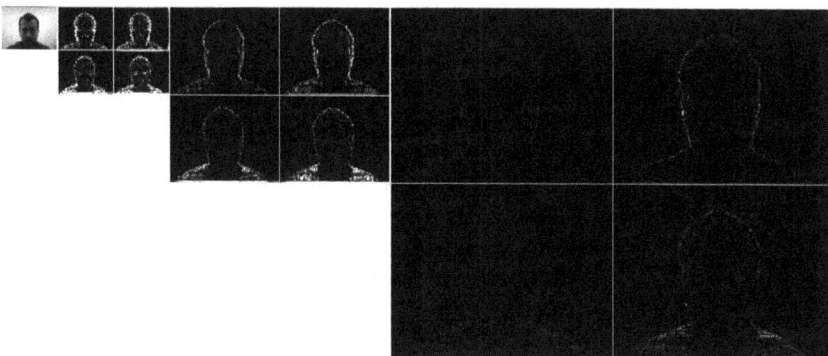

Figure 2. Example of contourlet transform decomposition of a greyscale image. Three levels of decomposition with the Laplacian Pyramid were applied, each then decomposed into four directional subbands using the Directional Filter Bank.

2.2. GPU-based contourlet transform

By analysing the structure of the contourlet transform, it is evident that its most computational-ly intensive part is the calculation of all the 2D convolutions needed for complete decomposi-tion or reconstruction. Calculating the convolutions on the CPU using the 2D convolution definition is not feasible for real-time applications since performance suffers significantly due to the computational complexity. Utilizing the DFT or the FFT in order to achieve better per-formance provides significantly faster implementations but still fails to achieve satisfactory re-al-time performance, especially in mobile platforms such as laptops and tablet PCs. The benefits of the FFT for the calculation of 2D convolution can only be fully exploited by an archi-tecture supporting parallel computations. Modern personal computers are commonly equip-ped with powerful graphics processors (GPUs), which in the case of live video capture from web or surveillance cameras are underutilized. Intensive, repetitive computations that can be computed in parallel can be accelerated by harnessing this "dormant" computational power. General purpose computing on graphics processing units (GPGPU) is the set of techniques that use a GPU, which is otherwise specialized in handling computations for the display of comput-er graphics, in order to perform computations traditionally handled by a CPU. The highly par-allel structure of GPUs makes them more effective than general-purpose CPUs for algorithms where processing of large blocks of data can be done in parallel.

For the GPU implementation of the contourlet transform, the NVIDIA Compute Unified De-vice Architecture (CUDA) has been selected due to the extensive capabilities and specialized API it offers. CUDA is a general purpose parallel computing architecture that allows the parallel compute engine in NVIDIA GPUs to be used in order to solve complex computa-tional problems that are outside the scope of graphics algorithms. In order to compute the contourlet transform, first the image and the filters are transferred from the main memory to the GPU dedicated memory. Then, the contourlet transform of the image is calculated by migrating all the calculations on the GPU in order to reduce the unnecessary transfers to and from the main memory that introduce delay to the computations. The 2D convolutions required are calculated by means of the FFT. After calculating the contourlet transform of the image, the output is transferred back to the main memory and the GPU memory is freed. Considering that this implementation will be used for video encoding, the filters are loaded once at the GPU memory since they will not change from frame to frame. In order to evalu-ate the performance of this approach, various implementations of the contourlet transform were developed, both for the CPU and the GPU. These implementations were based on the FFT (frequency domain) and the 2D convolution definition (spatial domain). Except for the basic GPU implementation using spatial domain convolution, other out-of-core implementa-tions were developed, based on the 2D convolution definition and utilizing memory man-agement schemes in order to support larger frames when the GPU memory is not sufficient [12]. The GPU implementation based on the FFT outperformed all the aforementioned im-plementations in our tests and was therefore the method of choice for our video encoder.

2.3. The YCoCg colour space

Inspired by recent work on real-time RGB frame buffer compression using chrominance sub-sampling based on the YCoCg colour transform [13], we investigated the use of these techniques in conjunction with the contourlet transform to efficiently encode colour video frames.

The human visual system is significantly more sensitive to variations of luminance compared to variations of chrominance. Encoding the luminance channel of an image with higher accuracy than the chrominance channels provides a simple low complexity compression scheme, while maintaining satisfactory visual quality. Various image and video compression algorithms take advantage of this fact in order to achieve increased efficiency. First introduced in H.264 compression, the RGB to YCoCg transform decomposes a colour image into luminance (Y), orange chrominance (Co) and green chrominance (Cg) components and has been shown to exhibit better decorrelation properties than YCbCr and similar transforms [14]. It was developed primarily to address some limitations of the different YCbCr colour spaces [15]. The transform and its reverse are calculated by the following equations:

$$Y = R/4 + G/2 + B/4 \tag{1}$$

$$Co = R/2 - B/2 \tag{2}$$

$$Cg = -R/4 + G/2 - B/4 \tag{3}$$

$$R = Y + Co - Cg \tag{4}$$

$$G = Y + Cg \tag{5}$$

$$B = Y - Co - Cg \tag{6}$$

Image set	Number of images	Average PSNR (dB)
Kodak	23	59.27
Canon	18	59.05
Outdoor scene images	963	58.87

Table 1. Average PSNR obtained for each image set after transforming from RGB to YCoCg and back using the same precision for the RGB and YCoCg components.

In order for the reverse transform to be perfect and to avoid rounding errors, the Co and Cg components should be stored with higher precision than the RGB components. Experiments using 23 images from the Kodak image set and 18 images from the Canon image set, all obtained from [16], as well as 963 outdoor scene images obtained from [17], showed that using the same precision for the YCoCg and RGB data when transforming from RGB to YCoCg and back results in an average PSNR of more than 58.87 dB for all the image sets, as shown in Table 1. This loss of quality cannot be perceived by the human visual system, resulting to no visible altera-tion of the image. Nevertheless, it indicates the highest quality possible when used for image compression.

3. The presented algorithm

Listings 1 and 2 depict the presented algorithm for encoding and decoding respectively. Input frames are considered to be in the RGB format. The first step of the algorithm is the conversion from RGB to YCoCg colour space for further manipulation of the luminance and chrominance channels. The luminance channel is decomposed using the contourlet transform, while chromi-nance channels are subsampled by a user-defined factor N. The levels and filters for contourlet transform decomposition are also defined by the user. From the contourlet coefficients ob-tained by decomposing the luminance channel, only a user-specified percentage of the most significant ones are retained. Then, the precision allocated for storing the contourlet coefficients is reduced. All computations up to this stage are performed on the GPU, avoiding unnecessary memory transfers from the main memory to the GPU memory and vice versa. After reducing the precision of the retained contourlet coefficients of the luminance channel, the directional subbands are encoded using a run length encoding scheme that encodes only zero valued ele-ments. The large sequences of zero-valued contourlet coefficients that occur after the insignifi-cant coefficient truncation make run length encoding ideal for their encoding.

The encoding algorithm
1: Start
2: Input RGB frame
3: Convert to YCoCg
4: Downsample Co and Cg by N
5: Decompose Y with the Contourlet Transform
6: Keep the M% most significant CT coefficients
7: Round the CT coefficients to the n-th decimal
8: IF the frame is an internal frame
9:   Calculate the frame as the difference between the frame and the previous keyframe
10:   Run-length encoding of Co, Cg and the lowpass CT component of Y
11: END IF
12: Run-length encoding of the directional subbands of Y
13: Adjust precision of all components
14: IF frame is NOT the last frame
15:    GOTO Start
16: END IF
17: Finish

Listing 1. Steps of the encoding algorithm. Highlighted steps refer to calculations performed on the GPU, while the other steps refer to calculations performed on the CPU.

The algorithm divides the video frames into two categories; keyframes and internal frames. Keyframes are frames that are encoded using the steps described in the previous paragraph and internal frames are the frames between two key frames. The interval between two keyframes is a user defined parameter. At the step before the run-length encoding, when a frame is identified as an internal frame all its components are calculated as the difference between the respective components of the frame and those of the previous key frame. This step is processed on the GPU while all the remaining steps of the algorithm are performed on the CPU unless otherwise stated. Then, run length encoding is applied to the chromatic channels, the low frequency contourlet component of the luminance channel, as well as the directional subbands of the luminance channel. It must be noted that steps that are executed on the CPU are inherently serial and cannot be efficiently mapped to a GPU.

The last stage of the algorithm consists of the selection of the optimal precision for each video component. The user can select between lossless or lossy change of precision, directly affecting the output's visual quality.

The decoding algorithm
1: Start
2: Input encoded frame
3: IF the frame is a keyframe
4:    Decode the run-length encoded directional subbands of Y
5:   Keep keyframe in memory and discard old keyframe
6: ELSE IF the frame is an internal frame
7:    Decode the run-length encoded Co, Cg, lowpass CT component of Y
8:    Decode the run-length encoded directional subbands of Y
9:   Calculate the frame as the sum of the frame and the previous keyframe
10: END IF
11: Upsample Co and Cg by N
12: Reconstruct Y
13: Convert to RGB
14: IF frame is NOT the last frame
15:    GOTO Start
16: END IF
17: Finish

Listing 2. Steps of the decoding algorithm. Highlighted steps refer to calculations performed on the GPU, while the other steps refer to calculations performed on the CPU.

3.1. Chrominance channel subsampling

Exploiting the fact that the human visual system is relatively insensitive to chrominance variations, in order to achieve compression, the chrominance channels Co and Cg are subsampled by a user-defined factor N that directly affects the output's visual quality and the compression achieved. The chrominance channels are stored in lower resolution, thus providing compression. For the reconstruction of the chrominance channels at the decoding stage, the missing chrominance values are replaced with the nearest available subsampled chrominance values. This approach is simple and naïve but has been selected due to the significantly smaller number of (costly) memory fetches and minimal computation cost, compared to other methods like bilinear interpolation. Utilizing the nearest neighbour reconstruction approach can introduce artifacts in the form of mosaic patterns in regions with strong chrominance transitions depending on the subsampling factor. In order to address this problem, given adequate computational resources, the receiver can choose to use the bilinear interpolation approach. Figure 3 shows an example of subsampling the Co and Cg chrominance channels by various factors, while using the nearest neighbour and the bilinear interpolation approach for reconstruction. Only a small, magnified part of the "baboon" image used is shown for clarity. As demonstrated in Figure 3, subsampling by a factor of 2 or 4 does not have a drastic effect on visual quality. Further subsampling leads to visible artifacts indicating the need for an optimal trade-off between quality and compression.

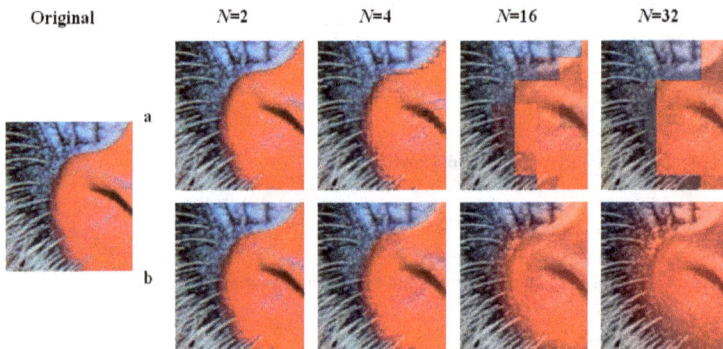

Figure 3. Example of chroma subsampling by factor N of the Co and Cg channels of the "baboon" image. Row (a) depicts images reconstructed using the nearest neighbour method, while (b) those reconstructed using bilinear interpolation.

3.2. Contourlet Transform decomposition of luminance channel and quality selection

The luminance channel of the frame is decomposed using the contourlet transform. Decomposition levels, as well as the filters used are user-defined and directly affect the quality of the output. Decomposition at multiple scales offers better compression while providing scalability, i.e. multiple resolutions inside the same video stream. This characteristic allows video coding algorithms to adapt to the network's end-to-end bandwidth and transmitter/receiver resources. The quality for each receiver can be adjusted without re-encoding the video frames at the source, by just dropping the encoded information referring to higher resolution than needed.

In order to achieve compression, after the decomposition of the luminance channel with the contourlet transform, a user-defined amount of the contourlet coefficients from the directional subbands are dropped by means of keeping only the most significant coefficients. The amount of coefficients dropped drastically affects the output's visual quality as well as the compression ratio. Contourlet coefficients with large magnitudes are considered more significant than coefficients with smaller magnitudes. Exploiting this fact, a common method for selecting the most significant contourlet coefficients is to keep the M most significant coefficients, or respective percentage, while dropping all the others [2] (coefficient truncation). This procedure leads to a large number of zero-valued sequences inside the elements of the directional subbands, a fact exploited by using run length encoding in order to achieve even higher compression. Considering the values and the distribution of contourlet coefficients at the directional subbands, only the zero-valued coefficients are run length encoded along the horizontal direction. Compression gained by run length encoding of all the different values is minimum and does not justify the increased computational cost. It is worth mentioning that dropping all the contourlet coefficients is similar to lowering the luminance channel's resolution while applying a lowpass filter and then, at the decoding stage, upscaling it without reincorporating the high frequency content.

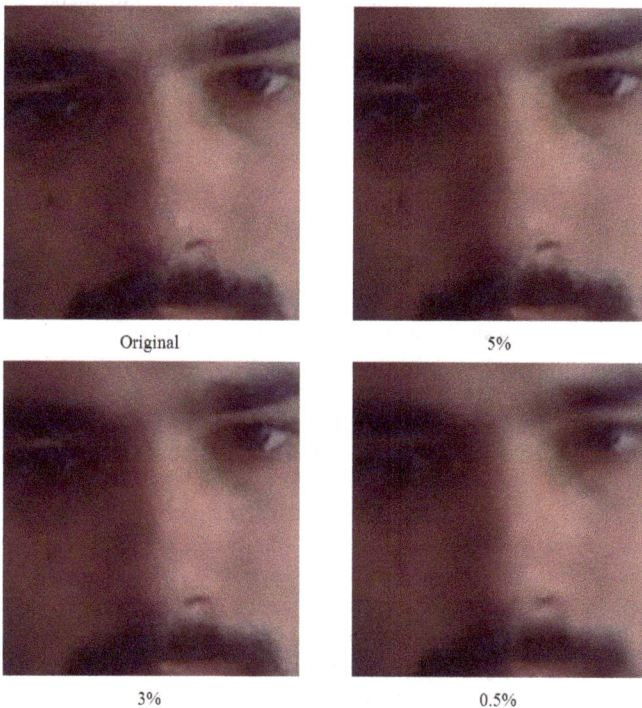

Figure 4. Example of smoothing due to the dropping of contourlet coefficients. The caption indicates the percentage of the contourlet coefficients retained. Images are cropped and scaled to 200% of their original size.

Keeping only the most significant contourlet coefficients also provides a means to suppress the noise induced by low-quality sensors usually encountered in web-cameras. Random noise is largely unstructured and therefore not likely to generate significant contourlet coefficients [2]. As a result, keeping only the most significant contourlet coefficients provides enhanced visual quality, which is a highly desirable characteristic since no additional filtering of the video stream is required in order to reduce the noise level. On Figure 4, an example of smoothing due to the dropping of contourlet coefficients is shown. Mosaicing artifacts and noise introduced due to the low quality of the web camera's sensor are suppressed and replaced by a fuzzier texture, resulting in a smoother and more perceptually acceptable image.

At the contourlet transform decomposition stage, 32 bit single precision floating point elements are used in order to avoid rounding errors and precision loss. Experiments with the precision allocated for the contourlet coefficients showed that the contourlet transform exhibits resistance to quality loss due to arithmetic precision reduction. This fact is exploited in order to achieve better compression by reducing the precision of the contourlet coefficients through rounding to a specific decimal point. Visual quality is not affected at all when only one decimal or more are kept. Rounding to the integer provides a PSNR of more than 60 dB when only the directional subbands' coefficients are rounded. Additionally, also rounding the low pass content provides a PSNR of more than 55 dB. In both cases, the loss of quality cannot be perceived by the human visual system and is considered as insignificant. For these experiments, the images were first transformed into the YCoCg colour space. Then the luminance channel was decomposed using the contourlet transform and the contourlet coefficients were rounded. No alteration was done to the chrominance channels. After the manipulation of the contourlet coefficients, the luminance channel was reconstructed and the image was transformed back into the RGB colour space.

3.3. Frame types

As mentioned before, frames are divided into keyframes and internal frames, with an internal frame being the difference between the current frame and the respective keyframe. Consecutive frames tend to have small variations, with many identical regions. This fact can be exploited by calculating the difference between a frame and the keyframe. This procedure provides components with large sequences of zero values leading to improved compression through the run length encoding stage. Especially in the case of video-conferencing or surveillance video, the background tends to be static, with slight or no variations at all. The occurrence of static background leads to many parts of the consecutive frames to be identical. As a result, calculating the difference of each frame from its respective keyframe provides large sequences of zero values leading to improved compression when run length encoding is applied. Run length encoding of the difference of contourlet-transformed images is even more efficient, since static noise is drastically suppressed by the coefficient truncation. Experiments showed that the optimal compression is achieved for a relatively small interval between keyframes, in the region of 5-7 internal frames, providing small groups of pictures (GOP) that depend to a keyframe. This characteristic makes the algorithm more resistant to packet loses when transmitting over a network. In the case of a scene change, consecutive frames drastically differ from each other and the compression achieved for the internal frames until the next keyframe is similar to that of a keyframe. If this scenario occurs, having small intervals between consecutive keyframes reduces the number of non optimally encoded frames. Nevertheless, in cases where the video is expected to be mostly static, like surveillance video for example, a larger interval between keyframes will provide considerably better compression.

3.4. Other supported colour spaces

Except for the YCoCg colour space, the algorithm supports the YCbCr and Greyscale colour spaces without the need to alter its core functionality. The process of encoding greyscale videos consists of handling the video as a colour video with only the luminance channel. All the steps of the algorithm are calculated except for those referring to the chromatic channels. On the other hand, due to the similarity of the YCbCr colour space with the YCoCg colour space the algorithm remains the same. The only difference is the RGB-to-YCbCr conversion at the encoder and the YCbCr-to-RGB conversion at the decoder. The luminance channel is identically handled as in the YCoCg-based algorithm, and the same holds for the replacement of CoCg channels with the CbCr ones. Nevertheless, the CbCr channels have a different range of values compared to CoCg channels. As a consequence, the optimal precision for the CbCr channels differs from that of the CoCg channels and has to be taken into consideration.

4. Quality and performance analysis

For evaluating the presented algorithm, two videos were captured using a VGA web camera that supported a maximum resolution of 640x480 pixels. Low resolution web cameras are very common on everyday personal computer systems showcasing the need to design video encoding algorithms that take into consideration the problems arising due to low-quality sensors. The videos captured were a typical video-conference sequence with static background showing the upper part of the human body and containing some motion, and a surveillance video with almost no motion depicting the entrance of a building.

The captured videos were encoded using the YCoCg, YCbCr and Greyscale colour spaces. The chrominance channels of the colour videos were subsampled by a factor of 4 and the video stream contained two resolutions: the original VGA (640x480) as well as the lower QVGA (320x240). The method utilized for the reconstruction of the chrominance channels was the nearest neighbour method. The percentage of the most significant contourlet coefficients of the luminance channel retained was adjusted for each encoded video, providing results of various quality and compression levels. Furthermore, at each scale, the luminance channel's high frequency content was decomposed into four directional subbands. In order to test the algorithm using the YCbCr colour space, the RGB to YCbCr conversion formula for SDTV found in [18] was utilised. For the Greyscale colour space, the aforementioned videos were converted from RGB to greyscale using the standard NTSC conversion formula [18] that is used for calculating the effective luminance of a pixel:

$$Y(i, j) = 0.2989 \cdot R(i, j) + 0.5870 \cdot G(i, j) + 0.1140 \cdot B(i, j) \tag{7}$$

The sample videos were encoded using a variety of parameters. The mean PSNR value for each video was calculated based on a set of different percentages of contourlet coefficients to be retained. The compression ratios achieved when using the scheme that incorporates both key frames and internal frames and when compressing all the frames as keyframes were also calculated. The interval between the key frames was set to five frames for the video-conference sample video and to twenty frames for the surveillance video. Detailed results are shown on Tables 2, 3 and 4 while sample frames of the encoded videos utilizing the YCoCg, YCbCr and Greyscale colour space for a set of settings are shown on Figures 5-10.

Examining the compression ratios achieved, it is shown that utilizing the keyframe and internal frame scheme outperforms the naive method of encoding all the frames the same way, as expected. However, the selection of an efficient entropy encoding algorithm that will further enhance the compression ability of our algorithm is still an open issue. Another interesting observation is that the contourlet transform exhibits substantial resistance to the loss of contourlet coefficients. Even when only 5% of its original coefficients are retained, the visual quality of the image is not seriously affected. This fact underlines the efficiency of the contourlet transform in approximating natural images using a small number of descriptors and justifies its utilization in this algorithm. The slightly lower PSNR achieved for the surveillance video sample can be explained due to the higher complexity of the scene compared to the video conference sample. More complex scenes contain higher frequency content, a portion of which is then discarded by dropping the contourlet coefficients.

(a) Video conference sample				(b) Video surveillance sample			
PSNR (dB)				PSNR (dB)			
Contourlet coefficients retained (%)	YCoCg	YCbCr	Greyscale	Contourlet coefficients retained (%)	YCoCg	YCbCr	Greyscale
10	45.11	44.77	52.04	10	44.18	44.03	50.11
5	44.53	44.29	49.71	5	43.54	43.45	47.88
3	43.88	43.70	47.72	3	42.96	42.89	46.33
1	42.30	42.23	44.28	1	41.57	41.50	43.45
0.5	41.62	41.56	43.10	0.5	40.80	40.76	42.17
0.2	41.30	41.25	42.60	0.2	40.17	40.14	41.21
0	39.15	39.13	39.82	0	39.59	39.55	40.39

Table 2. PSNRs achieved for the (a) video conference and (b) video surveillance samples, retaining various percentages of contourlet coefficients and utilizing the YCoCg. YCbCr and Greyscale colour spaces.

Video conference sample						
	Compression ratio					
Contourlet coefficients retained (%)	YCoCg		YCbCr		Greyscale	
	Only keyframes	Keyframes & Internal frames	Only keyframes	Keyframes & Internal frames	Only keyframes	Keyframes & Internal frames
10	4.96:1	11.06:1	4.93:1	12.05:1	2.09:1	3.49:1
5	6.44:1	14.39:1	6.44:1	16.31:1	2.94:1	4.44:1
3	7.36:1	16.39:1	7.37:1	18.98:1	3.56:1	5.02:1
1	8.71:1	19.46:1	8.70:1	23.15:1	4.57:1	5.85:1
0.5	9.07:1	20.24:1	9.06:1	24.33:1	4.87:1	6.07:1
0.2	9.22:1	20.62:1	9.21:1	24.81:1	5.00:1	6.17:1
0	11.71:1	25.84:1	11.71:1	32.89:1	7.65:1	7.53:1

Table 3. Compression ratios achieved for the video conference sample, retaining various percentages of contourlet coefficients and utilizing the YCoCg. YCbCr and Greyscale colour spaces.

Video surveillance sample						
	Compression ratio					
Contourlet coefficients retained (%)	YCoCg		YCbCr		Greyscale	
	Only keyframes	Keyframes & Internal frames	Only keyframes	Keyframes & Internal frames	Only keyframes	Keyframes & Internal frames
10	4.35:1	21.55:1	4.35:1	22.73:1	1.78:1	7.26:1
5	5.89:1	28.74:1	5.92:1	31.06:1	2.63:1	9.68:1
3	6.85:1	32.89:1	6.89:1	35.97:1	3.23:1	11.09:1
1	8.14:1	38.02:1	8.18:1	42.19:1	4.15:1	12.79:1
0.5	8.58:1	39.53:1	8.61:1	44.05:1	4.48:1	13.30:1
0.2	8.89:1	40.65:1	8.92:1	45.45:1	4.74:1	13.70:1
0	11.71:1	49.26:1	11.71:1	56.50:1	7.65:1	16.58:1

Table 4. Compression ratios achieved for the video surveillance sample, retaining various percentages of contourlet coefficients and utilizing the YCoCg. YCbCr and Greyscale colour spaces.

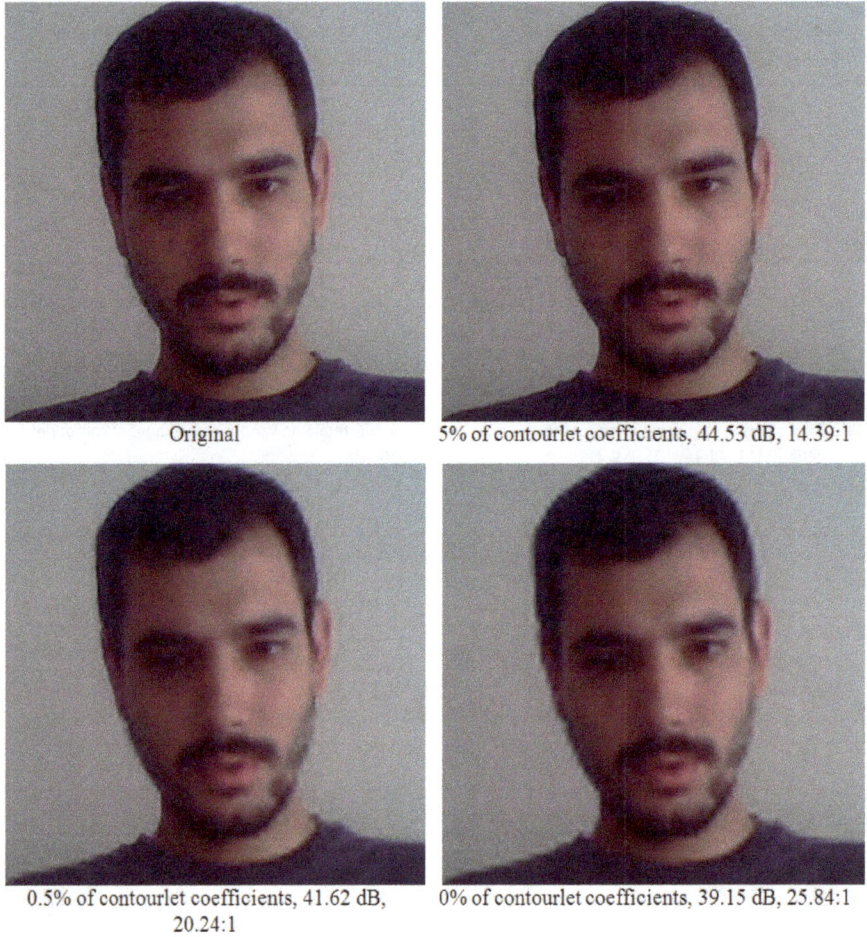

Figure 5. Sample frame of the encoded video-conference video for each setting using the *YCoCg* colour space. The frame has been resized and cropped to fit the figure.

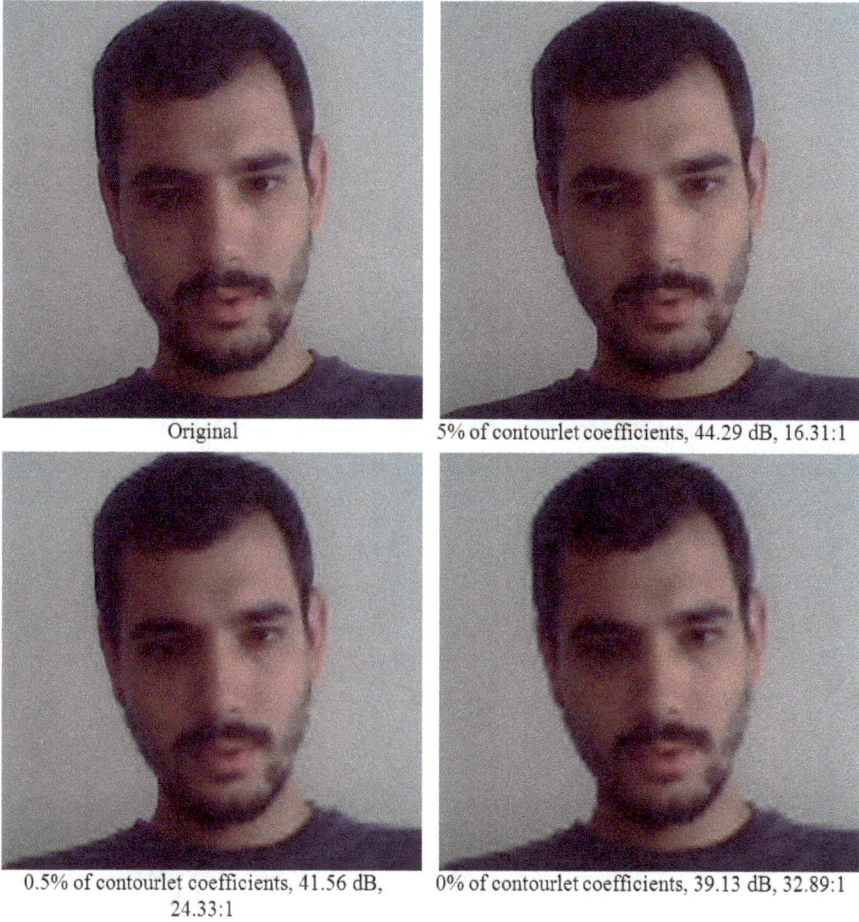

Figure 6. Sample frame of the encoded video-conference video for each setting using the *YCbCr* colour space. The frame has been resized and cropped to fit the figure.

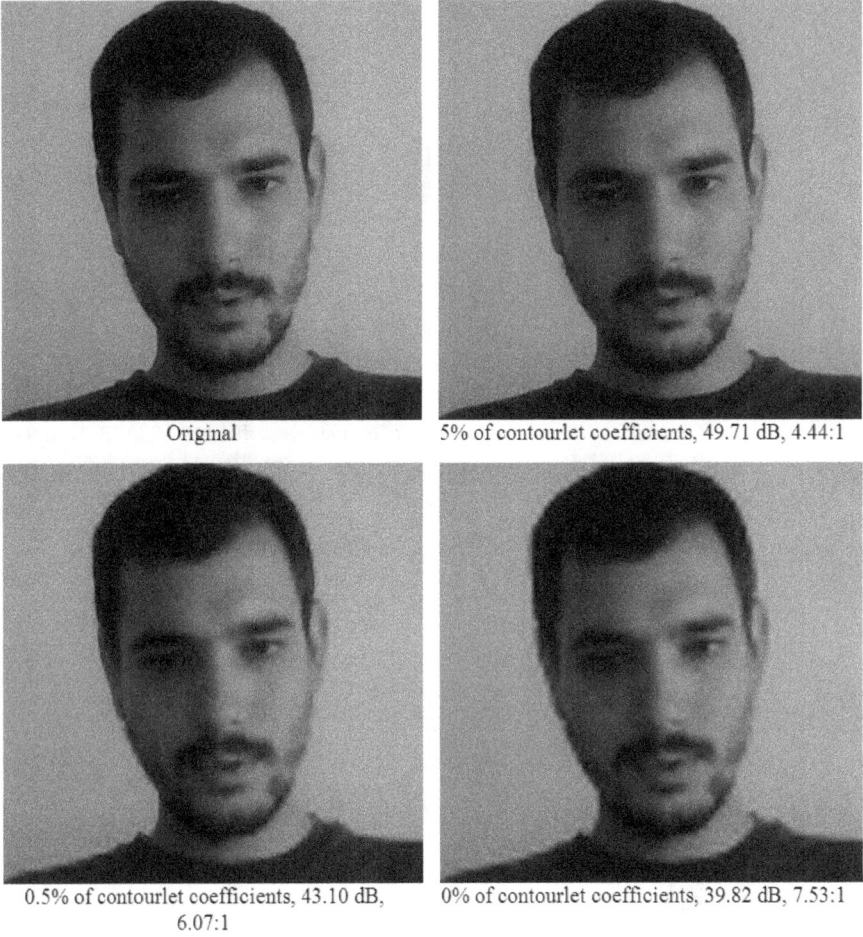

Figure 7. Sample frame of the encoded video-conference video for each setting using the *Greyscale* colour space. The frame has been resized and cropped to fit the figure.

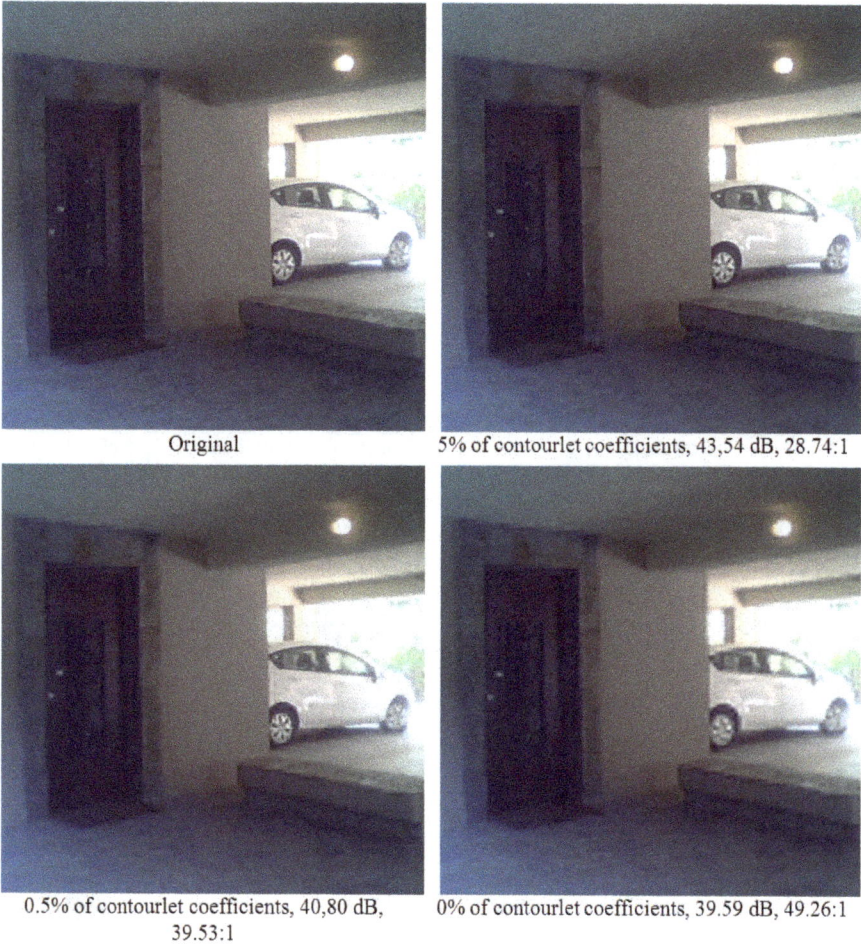

Original

5% of contourlet coefficients, 43,54 dB, 28.74:1

0.5% of contourlet coefficients, 40,80 dB, 39.53:1

0% of contourlet coefficients, 39.59 dB, 49.26:1

Figure 8. Sample frame of the encoded video surveillance video for each setting using the *YCoCg* colour space. The frame has been resized and cropped to fit the figure.

Original

5% of contourlet coefficients, 43.45 dB, 31.06:1

0.5% of contourlet coefficients, 40.76 dB, 44.05:1

0% of contourlet coefficients, 39.55 dB, 56.50:1

Figure 9. Sample frame of the encoded video surveillance video for each setting using the *YCbCr* colour space. The frame has been resized and cropped to fit the figure.

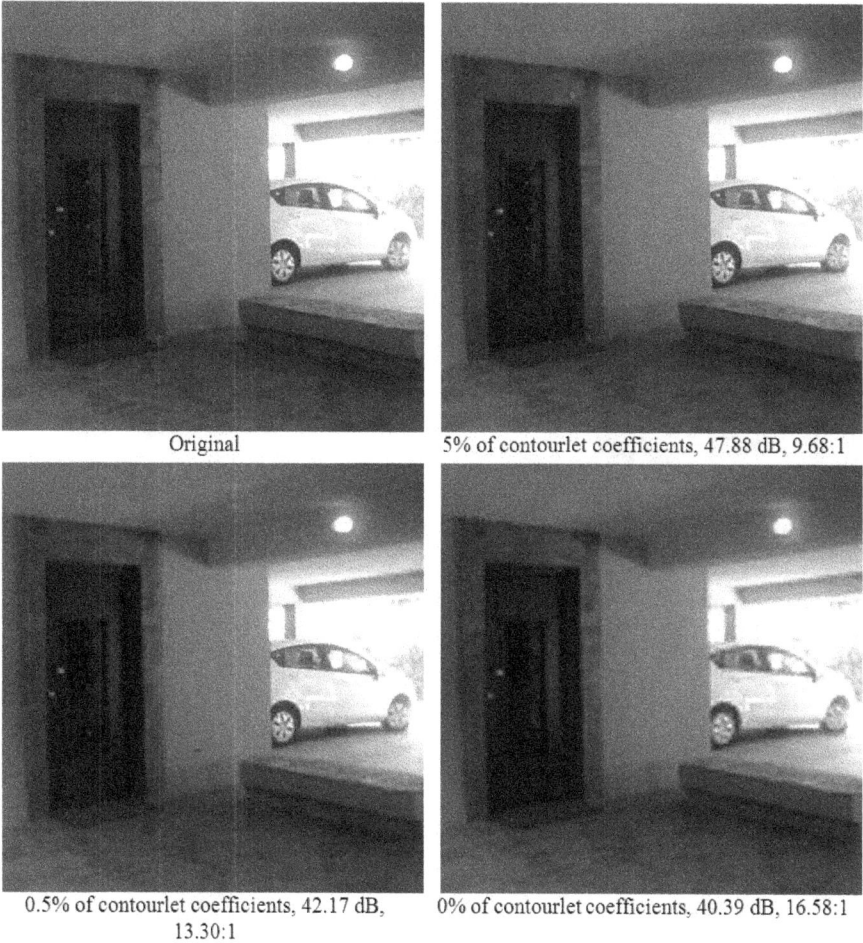

Original

5% of contourlet coefficients, 47.88 dB, 9.68:1

0.5% of contourlet coefficients, 42.17 dB, 13.30:1

0% of contourlet coefficients, 40.39 dB, 16.58:1

Figure 10. Sample frame of the encoded video surveillance video for each setting using the *Greyscale* colour space. The frame has been resized and cropped to fit the figure.

Considering the YCoCg and the YCbCr colour spaces, for the two video samples tested, it is shown on Tables 2-4 and Figures 11 and 12 that the YCoCg colour space achieves slightly better visual quality (higher PSNR), while the YCbCr colour space provides better compression (higher compression ratio). The Greyscale examples cannot be directly compared to the colour samples since the calculated PSNR characterizes the original and encoded greyscale samples. Nevertheless, it is clear that in the case of Greyscale colour space, compression suffers greatly compared to the other colour spaces.

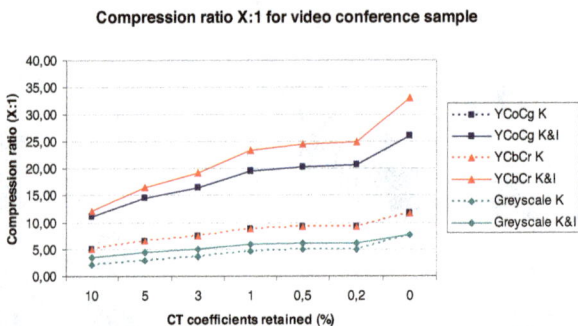

Figure 11. Compression ratio vs percentage of contourlet coefficients retained diagram, for the video conference sample, utilizing the YCoCg. YCbCr and Greyscale colour spaces. K refers to using only keyframes, while K&I refers to using both keyframes and internal frames.

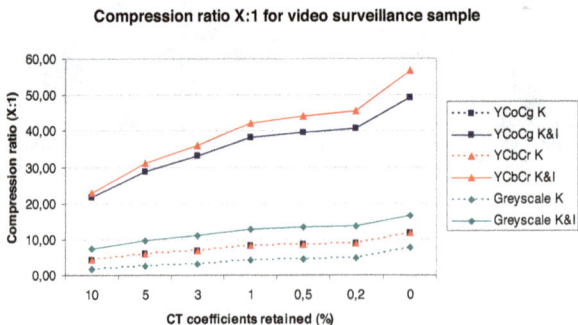

Figure 12. Compression ratio vs percentage of contourlet coefficients retained diagram, for the video surveillance sample, utilizing the YCoCg. YCbCr and Greyscale colour spaces. K refers to using only keyframes, while K&I refers to using both keyframes and internal frames.

Average execution times for the basic operations of the encoding and decoding algorithm for a frame of the video conference sample are presented on Table 5. Parameters were kept the same as in the previous examples and the computer utilised for the performance tests was equipped with an Intel Core i3 CPU, 4 GB of memory and a NVIDIA GeForce 430 graphics card with 1 GB of memory.

Operation	Time (ms)
Transfer of RGB frame to GPU memory	1.385
Transfer of encoded frame to main memory	1.050
Conversion from RGB to YCoCg	1.067
Conversion from YCoCg to RGB	0.402
Contourlet transform decomposition	59.040
Contourlet transform reconstruction	57.102
Run-length encoding of directional subbands	2.424
Run-length decoding of directional subbands	7.008
Contourlet coefficients dropping	0.492

Table 5. Average execution times (in milliseconds) for the basic operations of the algorithm for a 640x480 video frame. The chrominance channels were subsampled by a factor of 4 and the video stream contained the original VGA (640x480) as well as the lower QVGA (320x240) resolution.

5. Conclusions

In this chapter, a low complexity algorithm for real-time video encoding based on the contourlet transform and optimized for video conferencing applications and surveillance cameras has been presented and evaluated. The algorithm provides a scalable video compression scheme ideal for video conferencing content as it achieves high quality encoding and increased compression efficiency for static regions of the image, while maintaining low complexity and adaptability to the receivers resources. A video stream can contain various resolutions avoiding the need for reencoding at the source. The receiver can select the desired quality by dropping the components referring to higher quality than needed. Furthermore, the algorithm has the inherent ability to suppress the noise induced by low-quality sensors, without the need of an extra denoising or image enhancement stage, due to the manipulation of the structural characteristics of the video through the rejection of insignificant contourlet transform coefficients. In the case of long recordings for surveillance systems, where higher compression is needed, the visual quality degradation is much more eye-friendly than with other well established video compression methods, as it introduces fuzziness and blurring instead of artificial block artifacts, providing smoother images and facilitating image rectification/recognition procedures. Additionally, due to the relatively small GOPs, the algorithm is more resistant to frame losses that can occur during transmission over IP networks. Another advantageous characteristic of the presented algorithm is that its most computationally intensive parts are calculated on the GPU. The utilization of the usually "dormant" GPU computational power lets the CPU to be utilized for other tasks, further enhancing the multitasking capacity of the system and enabling the users to take full advantage of their computational capabilities. The experimental evaluation of the presented algorithm provided promising results. Nevertheless, in order to compete for compression efficiency with state of the art video compression algorithms, a highly efficient entropy encoding scheme has to be incorporated to the algorithm. Modern entropy encoding methods tend to be complex and computationally intensive. As a result, the optimal trade-off between compression rates and complexity has to be decided in order to retain the low complexity and real time characteristics of our algorithm.

Acknowledgements

The authors would like to thank Pavlos Mavridis for his fruitful advice concerning the YCoCg colour space.

Author details

Stamos Katsigiannis[1*], Georgios Papaioannou[2] and Dimitris Maroulis[1]

*Address all correspondence to: stamos@di.uoa.gr

1 Department of Informatics and Telecommunications, National and Kapodistrian University of Athens, Athens, Greece

2 Department of Informatics, Athens University of Economics and Business, Athens, Greece

References

[1] Katsigiannis, S., Papaioannou, G., & Maroulis, D. (2012). A contourlet transform based algorithm for real-time video encoding. *Proceedings of SPIE*, 8437, 843704, doi: 10.1117/12.924327.

[2] , M. N., & Vetterli, M. (2005). The contourlet transform: an efficient directional multi-resolution image representation. *IEEE Transactions on Image Processing*, 14(12), 2091-2106, doi: 10.1109/TIP.2005.859376.

[3] Liu, Z. (2008). Minimum Distance Texture Classification of SAR Images in Contourlet Domain. *Proceedings of the 2008 International Conference on Computer Science and Software Engineering, CSSE*, doi: 10.1109/IASP.2010.5476106.

[4] Katsigiannis, S., Keramidas, E., & Maroulis, D. (2010). A Contourlet Transform Feature Extraction Scheme for Ultrasound Thyroid Texture Classification. *Engineering Intelligent Systems*, 18, 3/4.

[5] Liu, Z., & Xu, H. (2010, 9-11 April 2010). Image denoising using Contourlet and two-dimensional Principle Component Analysis. Xi'an, China. *Proceedings of 2010 International Conference on Image Analysis and Signal Processing, IASP*, 309-313, doi: 10.1109/IASP.2010.5476106.

[6] Burt, P. J., & Adelson, E. H. (1983). The Laplacian Pyramid as a compact image code. *IEEE Transactions on Communications*, 31(4), 532-540, doi: 10.1109/TCOM.1983.1095851.

[7] Bamberger, R. H., & Smith, M. J. T. (1992). A filter bank for the directional decomposition of images: Theory and design. *IEEE Transactions on Signal Processing*, 40(4), 882-893, doi: 10.1109/78.258085.

[8] Shapiro, J. M. (1993). Embedded image coding using zerotrees of wavelet coefficients. *IEEE Transactions on Signal Processing*, 41(12), 3445-3462, doi: 10.1109/78.258085.

[9] Vetterli, M. (1984). Multidimensional subband coding: Some theory and algorithms. *Signal Processing 1984*, 6(2), 97-112, doi: 10.1016/0165-1684(84)90012-4.

[10] Yifan, Z., & Liangzheng, X. (2008). Contourlet-based feature extraction on texture images. *Proceedings of the 2008 International Conference on Computer Science and Software Engineering, CSSE*, 221-224, doi: 10.1109/CSSE.

[11] Cohen, A., Daubechies, I., & Feauveau, J. C. (1992). Biorthogonal bases of compactly supported wavelets. *Communications on Pure and Applied Mathematics*, 45(5), 485-560, doi: 10.1002/cpa.3160450502.

[12] Katsigiannis, S. (2011). Acceleration of the Contourlet Transform. *M.Sc. thesis*, Athens University of Economics and Business.

[13] Mavridis, P., & Papaioannou, G. (2013). The Compact YCoCg Frame Buffer. *GPU Pro 4*, CRC Press.

[14] Malvar, H., & Sullivan, G. (2003). YCoCg-R: A Color Space with RGB Reversibility and Low Dynamic Range. *Joint Video Team (JVT) of ISO/IEC MPEG & ITU-T VCEG*, Document No. JVTI014r3.

[15] Van Rijsselbergen, D. (2005). YCoCg(-R) Color space conversion on the GPU. *6th FirW PhD Symposium*, Faculty of Engineering, Ghent University, paper no. 102.

[16] *Center for Image Processing Research (CIPR), Rensselaer Polytechnic Institute*, http://www.cipr.rpi.edu/resource/stills/index.html, accessed 1 July 2012.

[17] *Object and Concept Recognition for Content-Based Image Retrieval Groundtruth Database*, University of Washington, http://www.cs.washington.edu/research/imagedatabase/groundtruth/, accessed 1 July 2012.

[18] International Telecommunication Union. (2011). Studio encoding parameters of digital television for standard 4:3 and wide screen 16:9 aspect ratios. *Recommendation BT. 601-7 (03/11).*

Chapter 7

Low Complexity Interpolation Filters for Motion Estimation and Application to the H.264 Encoders

Georgios Georgis, George Lentaris and
Dionysios Reisis

Additional information is available at the end of the chapter

1. Introduction

Techniques for image super-resolution play an important role in a plethora of applications, which include video compression and motion estimation. The detection of the fractional displacements among frames facilitates the removal of temporal redundancy and improves the video quality by 2-4 dB PSNR [12], [2]. However, the increased complexity of the Fractional Motion Estimation (FME) process adds a significant computational load to the encoder and sets constraints to real-time designs. Researchers have performed timing analysis for the motion estimation process and they reported that FME accounts for almost half of the entire motion estimation period, which in turn accounts for 60-90% of the total encoding time depending on the design configuration [12].

The FME bases on an interpolation procedure to increase the resolution of any frame region by generating sub-pixels between the original pixels. In mathematics, interpolation refers to the construction of an interpolant function whose plot covers (i.e. passes through) all required points. Known points of a sample area are referred to as having integer interval or displacement, depending on whether they are time or frequency-domain (TD or FD) samples respectively. Similarly, unknown samples which have to be approximated through an interpolant function, are said to have fractional interval or displacement respectively. In images, the interpolation takes place in a two-dimensional frequency-domain grid, where the problem of calculating fractional displacements can be facilitated by focusing on an area of four initially known pixels which reside on the corners of a unit square (Fig. 1). Hence, regardless of the interpolation factor, it is adequate to calculate pixels with arbitrary displacements in the unit square and extend the calculation for every unit square, which belongs to the frame.

Most of the non-adaptive techniques presented in the bibliography, base on solving piece-wise polynomial functions of varying degrees in order to calculate the interpolated signal. The resulting polynomial solution leads to sets of coefficients to be applied on consecutive sample points in the grid, which most often extend beyond the unit square. Examples of the above approach are first, the Bilinear interpolation [8] with first order polynomials and us-ing two pixels in each dimension and second, the Bicubic interpolation [9] which is derived from third order polynomials and uses four pixels in each dimension. On the other hand, Lanczos interpolation coefficients [10] stem from windowing a *sinc* function. Therefore, the number of pixels required by the Lanczos approach depends on the choice of the order of the interpolation function. More complex techniques applied to video encoding, employ edge-detection, error function minimization, or super-resolution (SR) procedures originating from theoretical signal processing methods. Among these techniques, the most commonly used is the edge-detection, which characterizes pixels or areas in an image belonging to an edge (luminance inconsistency). Edge-detection is also utilized for preventing aliasing fre-quency components to be encoded and transmitted.

Modern compression standards specify the exact filter to use in the Motion Compensation module, a fact allowing the encoder and the decoder to create and use identical reference frames. In particular, H.264/AVC specifies a 6-tap filter for generating sub-pixels between the pixels of the original image, which are called half-pixels with accuracy $\frac{1}{2}$ [3]. Also, it de-fines a low cost 2-tap interpolation filter for generating sub-pixels between half- and original pixels, which are defined as quarter-pixels with accuracy $\frac{1}{4}$. Even though it is a common practice among the encoder designers to integrate the standard 6-tap filter also in the Esti-mation module (before Compensation), the fact is that the interpolation technique used for detecting the displacements (not computing their residual) is an open choice following cer-tain performance trade-offs.

Aiming at speeding up the Estimation, a process of considerably higher computational de-mand than the Compensation, this chapter builds on the potential to implement a lower complexity interpolation technique instead of using the costly H.264 6-tap filter. For this purpose, we show the results of integrating in the Estimation module several distinct interpo-lation techniques not included in the H.264 standard. We keep the standard H.264/AVC Compensation and we measure the impact of the above techniques first on the time re-quired to process the up-sampling and second on the video quality achieved by the predic-tion engine.

Related results in the bibliography include techniques, which lead to avoid or replace the standard computations [4] [5] [13], or minimize the search area [14]. Researchers in [4] calcu-late the number of operations required for each pixel in cases where 8-to-2-tap filters and the Sum of Absolute Differences (SAD) metric is utilized. Then, they perform statistical analysis in CIF sequences encoded using bitrates from 0.5 to 1Mbps, to determine the recurrence of a motion vector when the aforementioned filter lengths are applied. The authors of [5] and [13] initially focus on reducing the number of taps and the multiplication operations, by pro-posing a filter which requires only shifts and additions. Then they propose adaptive thresholds to bypass the interpolation process based on the computed SAD value. Recent developments towards replacing the H.264 / AVC (High Efficiency Video Coding or H.265 or MPEG-H part 2) [16], combine Rate-Distortion minimization and adjustments to local image characteristics [15], [17], [18], [19]. Effectively, these techniques switch between standard and directionally adaptive interpolation kernels and they take this decision by examining each frame either on a pixel or macroblock basis.

Conventional Super-resolution (SR) techniques are generally considered to be prohibitively expensive when encoding video sequences. However, in many cases the learning-based super-resolution techniques are considered to be valid [20]. Consisting of a training phase, where low and high-resolution image patches are matched and a synthesis phase, where low resolution patches kept in the dictionary are used to oversample, learning-based SR provides increased PSNR whilst expanding storage and memory access requirements. Researchers and engineers have also focused on methodologies for designing the H.264 6-tap filter, which are able to efficiently support its increased memory requirements [2] [6] [7]. The H.264 filter needs quite a number of data to be stored for its operation because its specifications include a kernel with coefficients $\langle 1, -5, 20, 20, -5, 1 \rangle$, which are multiplied with six consecutive pixels of the frame either in column or row format. The resulting six products are accumulated and normalized for the generation of a single half-pixel, which is produced between the 3rd and the 4th tap. The operation described above must be repeated for producing each "horizontal" and "vertical" half-pixel by sliding the kernel on the frame, both in row and column order. Moreover, there exist as many "diagonal" half-pixels to be generated by applying the kernel on previously computed horizontal or vertical half-pixels. That is to say, depending on its position, we must process 6 or 36 frame pixels to compute a single half-pixel. To avoid the cost of implementing the H.264 filter in the Estimation module, the current chapter studies a set of interpolation techniques and compares their performance. The techniques presented here are similar to the standard filter but they use less than 6 taps [8] [9] [10]. Moreover, a subset of these techniques features the exploitation of gradients in the image [11].

The chapter is organized as follows: Section 2 shows three commonly used interpolation techniques, proposes three novel techniques and describes the differences among those commonly used and the proposed. Section III reports the performance results achieved by the interpolation techniques and by comparing these shows the gains of using the proposed. Finally, Section IV concludes the chapter.

2. Interpolation techniques

The current section presents six interpolation techniques. The first three (3) are known in the literature and are commonly used techniques. The other three (3) have been recently introduced [13] and their design targets the improvement of the interpolation process.

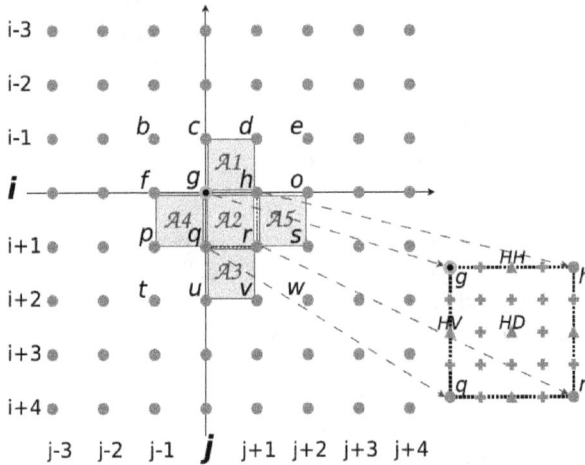

Figure 1. Pixels on the image grid and magnification of a 1×1 area showing sub-pixel positions (right). The symbols facilitate the description of filters.

Each video frame consists of pixels and we consider each pixel of the original image located at a distinct position (i, j) of a two dimensional (2D) grid with $i, j \in N$ denoting the vertical and horizontal coordinates of the pixel, respectively. The sub-pixels can be generated next to any pixel (i, j) at the positions $(i+k, j+l)$ with $k,l \in \left\{0, \frac{1}{4}, \frac{1}{2}, \frac{3}{4}\right\}$.

We distinguish between quarter-pixels and half-pixels, for which $k,l \notin \left\{\frac{1}{4}, \frac{3}{4}\right\}$. The half-pixels are further categorized as half-horizontal, half-vertical, or half-diagonal (those located at the positions given by $\left(i + \frac{1}{2}, j + \frac{1}{2}\right)$). Fig. 1 depicts part of the original image grid and magnifies a small area while the right-hand side magnifies an interior square region to show all sub-pixel positions (according to H.264/AVC). Moreover, Fig. 1 marks pixels and regions on the grid to be used as references with designated letters as a notation to be followed for the remaining of the paper.

A half-pixel is generated by an interpolation procedure operating on a set of neighboring, integer-position pixels located around the position of interest. We study the following interpolations:

2.1 Bilinear

This technique is actually the simplest of all the techniques presented in this chapter. In practice, this technique consists of a simple averaging of the two original pixels, which are adjacent to the half-horizontal or the half-vertical pixel to be generated (i.e., 2-tap FIR filter) [8]. For the half-diagonal (HD), the technique computes the average of the four (4) pixels $\{g, h, q, r\}$ surrounding the half-diagonal position as shown in (Fig. 1).

2.2 Bicubic

The Bicubic technique uses as a base the solution of third order polynomials [9]. In this chapter we examine the parameterized form of the underlying equation using $a \in [-1, 0]$ to provide sharpness variance in the interpolated image. We focus on the following values: $a= -1$, $a= -0.75$, and $a= -0.5$. These values result in three distinct kernels, which are characterized by the convolution coefficients $\langle -1, 5, 5, -1 \rangle$, $\langle -3, 19, 19, -3 \rangle$ and $\langle -1, 9, 9, -1 \rangle$, respectively. Such a quadruplet is multiplied with four (4) consecutive image pixels to generate their intermediate half-pixel. To compute the half-diagonal pixel, the Bicubic technique requires the calculations of the corresponding four half-horizontal (a total of 16 multiplications) and then apply the coefficients on the resulting pixels to produce the target half-diagonal. Hence, overall it uses 16 image pixels with the requirement of 20 multiplications.

2.3 Lanczos

This technique is similar to the H.264/AVC interpolation and with a third order Lanczos equation, it uses a 6-tap FIR filter. Overall, the technique bases on the Sinc function [10]. In this chapter we examine the kernel with coefficients given by $\left\langle \frac{12}{50\pi^2}, -\frac{12}{9\pi^2}, \frac{6}{\pi^2}, \frac{6}{\pi^2}, -\frac{12}{9\pi^2}, \frac{12}{50\pi^2} \right\rangle$. Lanczos half-pixels are generated by a trivial convolution procedure, as in the case of the H.264/AVC filter (a single half-diagonal pixel depends on 36 integer pixels). Note here that, the H.264/AVC standard defines a 6-tap filter for use in motion compensation with coefficients $\langle 1, -5, 20, 20, -5, 1 \rangle$.

2.4 Data-Dependent Triangulation

The first of the recently introduced techniques in [13] is actually a modification of the approach, which was presented in [11]. The authors in [11] use an edge-detection technique for determining the exact set of integer pixels, which will be given as input to the interpolation function. We study here a special case of Data-Dependent Triangulation (DDT), which examines only 4 pixels. To describe the technique, we consider the generation of the half-horizontal (HH) pixel Y_D^{HH} at $(i, j+\frac{1}{2})$ and the half-vertical (HV) pixel Y_D^{HV} at $(i+\frac{1}{2}, j)$ as shown in Fig. 1.

We examine the luma differences of pixels $\{g, h, q, r\}$ to determine whether an edge crosses their enclosed region: if it holds that $|Y_g - Y_r| > |Y_h - Y_q|$, then we will detect an edge at hq, else we will detect an edge at rg. In the first case, that is there is an edge at hq which is denoted as $_h^q E_D$, we assume that pixels $\{g, h, q\}$ form a homogeneous triangular and we compute:

$$Y_D^{HH} = Clip_{div_D}^R (w_1 Y_g + w_1 Y_h + w_2 Y_q)$$
$$Y_D^{HV} = Clip_{div_D}^R (w_1 Y_g + w_1 Y_q + w_2 Y_h)$$

$$(1)$$

Where $Clip_{div_D}^R$ is a normalization function (divides by $div_D = 2w_1 + w_2$, clips value in $[0, 255]$). Factors $w_1 > w_2$ are used to increase the luma weights of the neighbors residing next to the generated sub-pixel. The examination of a large number of factors has resulted in highest PSNR for $w_1 = 7$ and $w_2 = 2$ (given that $div_D = 2^4$). The second case refers to the detection of an edge at rg (when there is the edge ${}_r^g E_D$). In this case, we use the same idea as above (orientation and weights) but we modify accordingly the luma inputs of (1). In the case of a homogeneous square $ghqr$ the technique degenerates to a simple bilinear filter (i.e. $w_1 = 1$, $w_2 = 0$).

The technique generates the half-diagonal pixel by including a second gradient check, which follows the detection of the edge ${}_h^q E_D$, or the edge ${}_r^g E_D$. The idea is to identify the most homogeneous triangle in the enclosed area $A2$ shown in the Fig. 1. Thereby, in the case of ${}_h^q E_D$, we check $|Y_g - Y_q| + |Y_g - Y_h| < |Y_r - Y_q| + |Y_r - Y_h|$, otherwise we check $|Y_h - Y_g| + |Y_h - Y_r| < |Y_q - Y_g| + |Y_g - Y_r|$ to decide if the HD pixel resides *above* (<) or *below* (>) the edge. Extending our notation with abv and blw superscripts, we describe the modified DDT (mDDT) computation as:

$$
Y_D^{HD} = \begin{cases}
Clip_{div_D}^R \left(w_1 Y_h + w_1 Y_q + w_2 Y_g \right) & if_h^q \ E_D^{abv} \\
Clip_{div_D}^R \left(w_1 Y_h + w_1 Y_q + w_2 Y_r \right) & if_h^q \ E_D^{blw} \\
Clip_{div_D}^R \left(w_1 Y_h + w_1 Y_q + w_2 Y_h \right) & if_r^q \ E_D^{abv} \\
Clip_{div_D}^R \left(w_1 Y_h + w_1 Y_q + w_2 Y_g \right) & if_h^q \ E_D^{abv} \\
Clip_4^R \left(Y_q + Y_h + Y_r \right) & if_r^g \ unif
\end{cases}
\tag{2}
$$

Where the values of the w_1, w_2 and $Clip_{div_D}^R$ are as described in (1).

An alternative approach uses the equation 1 to develop a simpler HD generation technique, we call this technique $mDDT'$, which relies directly on the first DDT check and performs a bilinear operation on the two pixels of the detected edge, i.e., $Y_{D'}^{HD} = Clip_2^R(Y_h + Y_q)$ if ${}_h^q E_D$.

We further improve the $mDDT'$ and produce the ($mDDT'$) technique by modifying the final operation to subtract the remaining two off-diagonal pixels (as a high-pass FIR), i.e., $Y_{D''}^{HD} = Clip_{D''}^R(w_1 Y_h + w_1 Y_q - w_2 Y_g - w_2 Y_r)$ if ${}_h^q E_D$. The latter operation although it increases the amount of calculations, it results in better PSNR compared to the $mDDT'$.

2.5 CrossHD

The second approach is called CrossHD [13] and bases on an edge-oriented technique. The advantages of CrossHD compared to the DDT mentioned above, is that it improves on the locality of the aforementioned DDT detections by comparing the luminance difference of areas –instead of single pixels. This technique computes the luma of a small square area by adding the pixels, which are located at its four corners. For instance, for the example given in Fig. 1, we get that: $Y_{A1} = Y_c + Y_D + Y_g + Y_h$. The technique examines the outcome of the $|Y_{A4} - Y_{A5}| > |Y_{A1} - Y_{A3}|$ operation to decide if there exists a vertical (>) or horizontal (<) edge crossing the area A_2 . In the case of a vertical edge crossing the area A_2 , we examine independently the areas A_1 , A_2 , and A_3 by using the simple DDT check to identify the directions of the edges crossing each of these three (3) areas. The majority of the edge directions found within A_1 , A_2 , and A_3 refines the assumed edge direction within A_2 , i.e., we conclude if $_h^q E_\chi$ or $_r^g E_\chi$. Note that, in the case of examining whether there exists a horizontal edge, the technique will examine the areas A_4 , A_2 , and A_5 . Finally, the HD pixel is generated by averaging the pixels, which reside on the detected edge: $Y_\chi^{HD} = Clip_2^R(Y_h + Y_q)$ if $_h^q E_\chi$, or $Y_\chi^{HD} = Clip_2^R(Y_r + Y_g)$ if $_r^g E_\chi$. If the technique does not detect any edge (i.e., in the homogeneous square A_2), it will average the pixels $\{g, h, q, r\}$.

2.6 CxScale

The third approach extends the aforementioned ideas to develop a technique called CxScale [13], which improves both the edge detection and the subsequent kernel selection. Here, the edge detection mechanism examines the luma gradients over an area of 8 neighboring integer pixels and the half-pixels are generated afterwards via a conditional use of bilinear and bicubic interpolators. The technique includes three steps:

1. The detection of a horizontal or vertical edge.

2. The possible refinement of its direction to an assumed diagonal.

3. The selection of inputs to a bicubic or a bilinear function.

The specifics of these steps depend on the position of the half-pixel to be generated. Beginning with the HH pixel, we examine $|(Y_f + Y_g) - (Y_h + Y_o)| < |(Y_c + Y_d) - (Y_q + Y_r)|$ to detect a horizontal edge, i.e. $_g^h E_c^{HH}$. When we detect a vertical edge (when ">"), we refine its direction by checking:

assume $_q^d E_c^{HH}$ if $|Yc - Yr| > |Yd - Yq|$ (from q to d)

assume $_q^d E_c^{HH}$ if $|Yc - Yr| < |Yd - Yq|$ (from r to c)

assume $_{A1}^{A3}E_c^{HH}$ if $|Yc - Yr| = |Yd - Yq|$ (strictly vertical)

Else, we assume a homogeneous area. Finally, we compute

$$
Y_C^{HD} = \begin{cases}
Clip_{32}^R(-3Y_f + 19Y_g + 19Y_h - 3Y_o) & if_g^h E_c^{HH} \\
Clip_{32}^R(-3Y_c + 19Y_d + 19Y_q - 3Y_r) & if_q^d E_c^{HH} \\
Clip_{32}^R(-3Y_d + 19Y_c + 19Y_r - 3Y_q) & if_r^c E_c^{HH} \\
Clip_2^R(Y_g + Y_h) & otherwise
\end{cases}
\tag{3}
$$

Similarly, the generation of the HV pixel begins by examining $|(Y_c + Y_g) - (Y_q + Y_u)| < |(Y_f + Y_p) - (Y_h + Y_r)|$ to detect a vertical edge $_q^g E_c^{HV}$. If we detect a horizontal edge (>), we refine its direction and we compute the pixel Y_c^{HV} as follows:

assume $_p^h E_c^{HH}$ if $|Y_f - Y_r| > |Y_p - Y_h|$ (from p to h)

assume $_f^r E_c^{HH}$ if $|Y_f - Y_r| < |Y_p - Y_h|$ (from f to r)

assume $_{A4}^{A5}E_c^{HH}$ if $|Y_f - Y_r| = |Y_p - Y_h|$ (strictly horizontal)

$$
Y_C^{HV} = \begin{cases}
Clip_{32}^R(-3Y_c + 19Y_g + 19Y_q - 3Y_u) & if_g^q E_c^{HV} \\
Clip_{32}^R(-3Y_f + 19Y_h + 19Y_p - 3Y_r) & if_p^h E_c^{HV} \\
Clip_{32}^R(-3Y_h + 19Y_f + 19Y_r - 3Y_p) & if_f^r E_c^{HV} \\
Clip_2^R(Y_g + Y_q) & otherwise
\end{cases}
\tag{4}
$$

To conclude the CxScale description, we refer to the HD pixel generation, which begins by examining $|(Y_b + Y_g) - (Y_r + Y_w)| > |(Y_e + Y_h) - (Y_q + Y_t)|$ to detect an edge at $_q^h E_c^{HD}$. Otherwise, we assume $_r^g E_c^{HD}$. Then

$$
Y_C^{HD} = \begin{cases}
Clip_{32}^R(-3Y_e + 19Y_h + 19Y_q - 3Y_t) & if_q^h E_c^{HD} \\
Clip_{32}^R(-3Y_b + 19Y_g + 19Y_r - 3Y_w) & if_r^g E_c^{HD}
\end{cases}
\tag{5}
$$

3. Performance Evaluation

To evaluate the performance of the interpolation techniques in the considered application, we execute multiple motion estimation procedures and the entire application is completed by including the standard H.264/AVC motion compensation. For the realization of each test, we let the estimation procedure to employ one of the six interpolation techniques described in the previous Section, which will detect the fractional motion. The compensation procedure bases solely on the resulting motion vectors for constructing the frame-predictors according to the standard 6-tap filter. Hence, we use a setup, which ensures that the encoder and the decoder will still be able to use identical reference frames for their predictions, i.e., we avoid the accumulation of errors introduced to the coding process due to the encoder and the decoder. More specifically, the estimation algorithm computes the Sum of Absolute Differences (SAD) for comparing 4×4 pixel candidates and it operates in two phases:

1. A "Diamond Search" matches the block to the best integer position candidate,

2. An exhaustive search in the vicinity of the integer match detects fractional motion by examining 8 candidate blocks located at distance $\pm \frac{1}{2}$ pixels.

Overall, the only parameter varying in this scheme is the interpolation technique used in the second phase of the algorithm, and thus, the quality variations among the output sequences (predictor frames) depend only on the efficiency of the interpolation. The results are shown in the following test reports, which display the PSNR of the output sequences and in particular, the DPSNR for each interpolation technique.

We have performed the simulations to measure the quality and the processing time by testing a variety of well-known videos and up to five (5) frame resolutions for each. The simulations setup with *videos*, number of frames and *resolution* has been: The *car-phone* with 90 frames, the *foreman* with 400 frames, the *container* with 300 frames in *QCIF*, the *coastguard, foreman, news* with 300 frames each in *CIF* and finally, the *blue sky, pedestrian, riverbed, rush-hour* with 100 each in *SD1, 720p* and *1080p*. Our prediction engine is written in C, it uses 1 reference

Filter:	H.264	Nearest Neighbor	Bicubic			Lanczos	CxScale	DDT
Resolution			a=-1	a=-0.5	a=-0.75			
QCIF	35.0379	-2.2069	-0.0142	-0.0214	-0.0105	0.0009	-0.3359	-0.2265
CIF	34.2930	-1.4229	-0.0150	-0.0340	-0.0166	-0.0042	-0.3697	-0.1994
SD1	33.1775	-0.5483	-0.0170	-0.0192	-0.0118	-0.0030	-0.2071	-0.1249
720p	32.3743	-0.3316	-0.0130	-0.0151	-0.0096	-0.0029	-0.1021	-0.0866
1080p	33.0837	-0.2084	-0.0122	-0.0172	-0.0123	-0.0042	-0.0810	-0.0697
total	33.4971	-0.8843	-0.0144	-0.0209	-0.0120	-0.0027	-0.2116	-0.1372

Table 1. PSNR of the H.264/AVC filter and DPSNR of other techniques when estimating in HH+HV positions (with H. 264 compensation).

frame and it is designed to efficiently substitute any filter. We begin by distinguishing between horizontal/vertical and diagonal interpolation. Table 1 reports the PSNR results of the algorithm examining fractional displacements only at the horizontal and vertical directions (4 candidates). The table shows the results of two 6-tap filters (H.264/AVC, Lanczos), three

4-tap filters (Bicubic), and two edge-detection based techniques (DDT, CxScale). Moreover, for sake of comparison, we include the PSNR results achieved by the Nearest Neighbor (NN) technique [8]. The table 1 shows the low PSNR results of the Nearest Neighbor (NN) technique [8], which evades interpolation computations by simply forwarding the value of the integer pixel next to the HH/HV position. This technique practically, does not involve fractional motion detection.

The NN results point out that, even with only 4 HH/HV candidates, the algorithm improves its prediction quality by up to 2 dB at low frame resolutions. Using another technique, the Lanczos 6-tap filter, results in almost equivalent quality with the standard H.264 filter. We approximated the Lanczos coefficients by integer values to achieve low complexity operations.

The exact values of the coefficients were set after extensive testing to $\langle 3, -17,78,78, -17,3 \rangle$. The performance of the remaining filters lies between the above two extremes of six taps (Lanczos) and zero taps (NN). More precisely, the best quality was achieved with the Bicubic filters. We have examined the performance of several Bicubic kernels, with parameters $-a \in \left\{ \frac{7}{8}, \frac{6}{8}, \frac{5}{8}, \frac{4}{8}, \frac{3}{8}, \frac{2}{8} \right\}$ and we report the most prominent of these in Table 1. As it is shown, for most frame resolutions the kernel with coefficients $\langle -3,19,19, -3 \rangle$ maximizes the quality and limits the expected PSNR degradation to almost 0.01 dB compared to the H.264 filter. That is, although the kernel with coefficients $\langle -1,5, 5, -1 \rangle$ seems –intuitively– a better approximation of the $\langle 1, -5,20,20, -5,1 \rangle$ kernel of H.264 (approximation achieved by merging the marginal taps, i.e., by assuming equal values for the corresponding pixels), the experimental results are in favor of $a=-0.75$. For this reason, CxScale adopts the kernel with coefficients $\langle -3,19,19, -3 \rangle$ for its Bicubic filtering. Edge-detection based techniques degrade the quality by 0.1 dB, a fact indicating that their induced error surface deviates from the 6-tap filters error surface. However, we note that if we omit the H.264 compensation, these edge-detection based techniques prevail in terms of PSNR, as well as subjective criteria, up to 0.1 dB even when they are compared to 6-taps filters and especially in high-definition videos. Table 1 shows

Filter:	H.264	Nearest Neighbor	Bicubic			Lanczos
Resolution			a=-1	a=-0.5	a=-0.75	
QCIF	34.7318	-1.8864	-0.0288	-0.0436	-0.0143	0.0004
CIF	33.9850	-1.1102	-0.0145	-0.0423	-0.0148	-0.0016
SD1	33.1292	-0.4790	-0.0247	-0.0241	-0.0117	-0.0032
720p	32.3766	-0.3178	-0.0176	-0.0188	-0.0092	-0.0031
1080p	33.0869	-0.1979	-0.0146	-0.0223	-0.0119	-0.0045
total	33.3785	-0.7512	-0.0202	-0.0292	-0.0121	-0.0004

Table 2. PSNR of the H.264/AVC filter and DPSNR of Nearest Neighbor, Bicubic and Lanczos when estimating in HD positions (with H.264 compensation).

Filter:	CxScale	mDDT	[11]	CrossHD	mDDT'
Resolution					
QCIF	-0.1010	-0.1728	-0.1095	-0.1299	-0.1595
CIF	-0.0740	-0.1731	-0.1219	-0.1414	-0.1595
SD1	-0.0396	-0.1217	-0.0860	-0.0972	-0.1070
720p	-0.0294	-0.0816	-0.0636	-0.0696	-0.0760
1080p	-0.0420	-0.0746	-0.05889	-0.0637	-0.0725
total	**-0.0495**	**-0.1220**	**-0.0864**	**-0.0984**	**-0.1121**

Table 3. DPSNR of CxScale, mDDT, [11], CrossHD and *mDDT'* when estimating in HD positions (with H.264 compensation).

that the performance of the DDT and the CxScale techniques improves as the frame resolution increases.

Next, we consider the report of results regarding the efficiency of the techniques interpolating half-diagonal pixels, which are more computationally demanding than the interpolation of HH/HV pixels. We program the search procedure to examine only 4 HD candidates. Tables 2 and 3 present the resulting PSNR for the techniques of Table 1, plus four edge-detection based techniques: CrossHD, the proposed HD generation based on DDT (mDDT), its first alternative (*mDDT'*), and the technique of [11] using bilinear filtering at its last stage. We can mention here that, when compared to the HH/HV candidates, the HD candidates add slightly less quality to the algorithm, especially in low resolution videos (e.g., as reported in the NN results). Qualitatively, we draw similar conclusions with Table 1 verifying that the Bicubic filtering, especially the kernel with values $\langle -3,19,19, -3 \rangle$ prevails the over edge-detection based techniques. However, the latter show different behavior when compared to the HH/HV case. More precisely, we deduce that the HD part of CxScale employs an effective

Interpolation technique	PSNR(dB)			Time (µsec)per MB
	QCIF	SD1	1080p	
H.264/AVC	35.4263	33.3687	33.1513	46.0
Lanczos	-0.0032	-0.0050	-0.0076	46.0
Bicubic, a=-0.75	-0.0215	-0.0177	-0.0202	30.6
DDT⊕ mDDT	-0.3513	-0.1782	-0.1018	21.3
DDT⊕ mDDT'	-0.3341	-0.1642	-0.0980	16.0
CxScale	-0.3801	-0.1798	-0.0904	45.6
DDT⊕ CrossHD	-0.3192	-0.1618	-0.0932	32.4
DDT⊕ CxSc(HD)	-0.3061	-0.1302	-0.0839	28.3
DDT⊕ [11](HD)	-0.2913	-0.1492	-0.0889	53.6

Table 4. Quality vs. Time when estimating in HH+HV+HD positions.

gradient check, which is combined with the Bicubic kernel to improve the quality of CxScale. Table 3 shows that it is the prevailing edge-detection based technique among these in the paper. In cases where the filters are using less taps, the CrossHD technique performs better than the DDT techniques.

We complete the evaluation by examining all 8 candidates and taking into account the examination of all pixels at HH, HV, and HD positions. For each technique, Table 4 reports the PSNR results and the time required (as a complexity measure) for generating 16×16 arbitrary half-pixels (averaging over HH, HV, and HD positions) as measured on a Core 2 x86-64 GPP architecture at 3GHz. Furthermore, we combine distinct HH/HV techniques and HD techniques by adopting the prevailing edge-detection mechanisms given in Tables 1, 2 and 3 (in Table 4, "$A \oplus B$" stands for "use technique A in HH/HV interpolation and technique B in HD"). Overall, Bicubic reduces the 6-tap filtering time by 33% and keeps the PSNR level as close as 0.02 dB to the maximum. DDT techniques reduce time by 65% (primarily due to the fast HD generation) with a cost of 0.1 dB. CxScale and [11], involve the time consuming gradient checks. However, the HD part of CxScale combined with DDT (for HH/HV) results in a hybrid technique featuring best PSNR among the edge-detection based techniques with almost 40% time improvement.

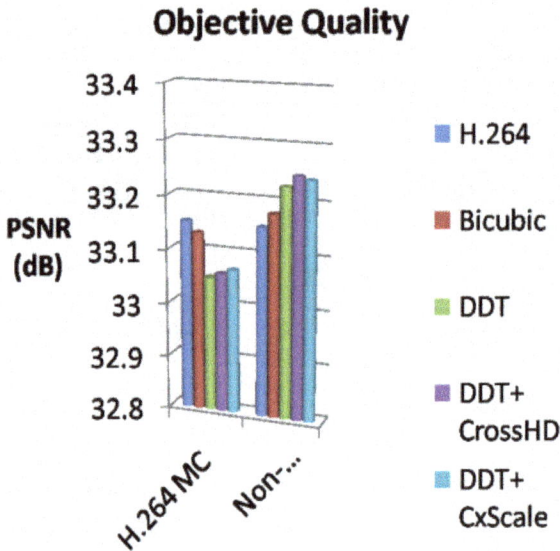

Figure 2. Comparison of objective quality for 5 distinct interpolation procedures. Objective quality is shown both for conventional H.264 and custom motion compensated prediction frames.

In Fig. 2 we show the results of the Objective quality both for conventional H.264 and custom motion compensated prediction frames. Custom motion compensation utilizes the interpolation filter used by the estimation procedure, whereas, conventional compensation uses the H.264 6-tap filter. Several videos of varying resolution were used (QCIF to 1080p). Moreover, Fig. 3 shows how the aforementioned techniques perform with respect to the execution time. Fig.2 shows that best results are achieved by the DDT (in computing HH and HV) with CrossHD (in computing HV). The fastest technique among all presented here, is the DDT with CxScale, which also results in the best PSNR when it is used with the H.264 standard compensation.

Figures 4 show interpolated images of the foreman cif sequence (352x288). We use four distinct interpolation methods at 4x in both directions to subjectively compare the quality of their results. In all four cases, the quarter pixels are calculated with a simple 2-tap bilinear (averaging) filter, which takes as input the two neighboring integer- or half-pixels (computed in a previous iteration by one of the four methods under evaluation).

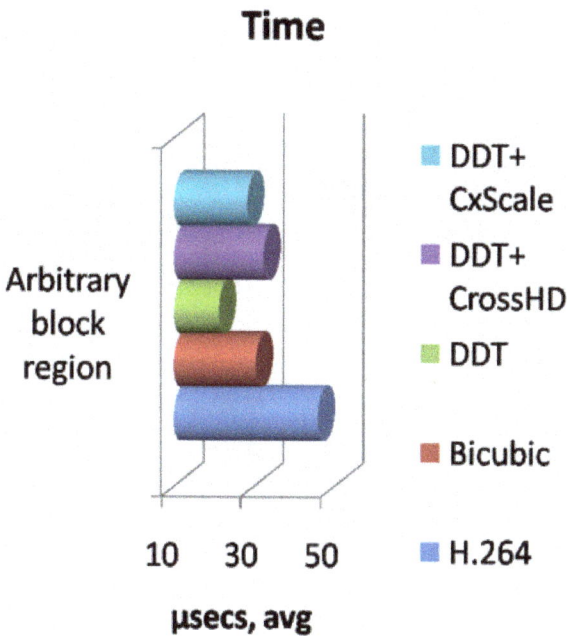

Figure 3. Comparison of execution time for 5 distinct interpolation procedures. Custom motion compensation utilizes the interpolation filter used by the estimation procedure, whereas, conventional compensation uses the H.264 6-tap filter. Several videos of varying resolution were used (QCIF to 1080p).

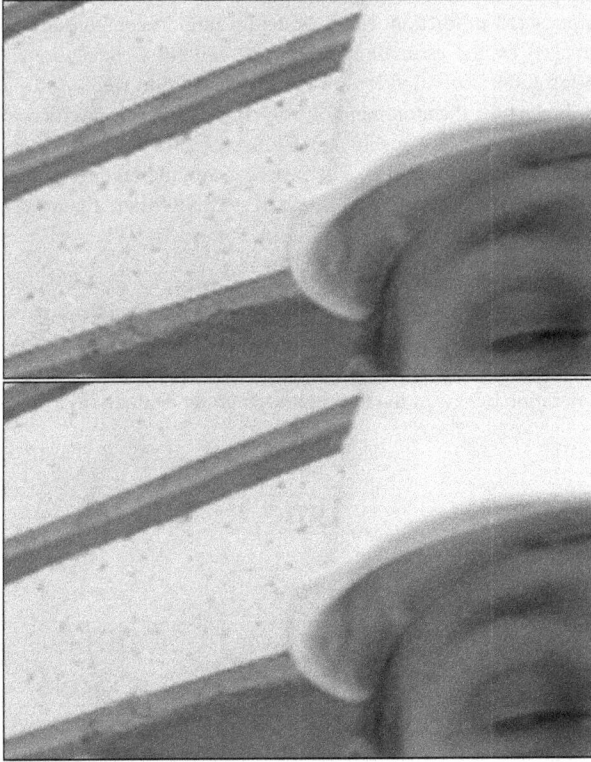

Figure 4.-5. Comparison of the H.264 filter (up) to the DDT ⊕ CxScale (down) on the "foreman" sequence. The example shows the two frames at their increased size (1408x1152) after interpolation from cif (352x288). DDT ⊕ CxScale (down) alleviates aliasing effects.

Figure 6 compares the 6-tap H.264 filter (up) to the combination of DDT and CxScale (down). Clearly, the latter produces much better images in terms of aliasing artifacts: the marquee indents on the wall look much sharper on the image below and the helmet is less jagged. Even though DDT ⊕ CxScale uses less taps, it achieves such aliasing reduction due to the employed edge detection mechanism. However, using a small number of taps and a large area as input to the proposed low-complexity comparison-based mechanism could obscure some finer details. Overall, DDT ⊕ CxScale improves the subjective quality of the enlarged image by using less execution time compared to the examined 6-tap filters. Figures 6 compare the combination of DDT ⊕ CrossHD (up) to the combination of DDT ⊕ [11] (down). Subjectively, the DDT ⊕ CrossHD method uses half the execution time of DDT ⊕ [11] to output images with very similar quality. Both methods reduce the aliasing artifacts compared to the examined 6-tap filters.

Figure 6.-7. Comparison of the DDT ⊕ CrossHD filter (up) to the DDT ⊕ [11] (down). Frames are shown at their increased size (1408x1152) after interpolation from "foreman" cif (352x288). DDT ⊕ CrossHD produces very similar subjective quality results to DDT ⊕ [11] in considerably less execution time.

4. Conclusion

Aiming at a significant complexity reduction under negligible video quality degradation, the paper proposed three novel interpolation techniques for use in the estimation process preceding the standard H.264/AVC motion compensation module of the encoder. Moreover, we evaluated their performance and compared their efficiency to three commonly used techniques. The results showed that the techniques using 4-tap Bicubic kernels constitute the most prominent substitute of the standard 6-tap filter. Further reduction of the estimation time was achieved via combinations of simple edge-detection based techniques. Future work includes parallelized implementations in VLSI/FPGA and cost-performance analysis.

Author details

Georgios Georgis, George Lentaris and Dionysios Reisis*

*Address all correspondence to: dreisis@phys.uoa.gr

Electronics Laboratory, Physics Deparment, National and Kapodistrian University of Athens (NKUA), Greece

References

[1] Yu-Wen, Huang, Bing-Yu, Hsieh, Shao-Yi, Chien, Shyh-Yih, Ma, & Liang-Gee, Chen. (2006). Analysis and Complexity Reduction of Multiple Reference Frames Motion Estimation in H.264/AVC. *IEEE Transactions on Circuits and Systems for Video Technology*, doi :10.1109/TCSVT.2006.872783, 16(4), 507-522.

[2] Tung-Chien, Chen, Yu-Wen, Huang, & Liang-Gee, Chen. (2004). Fully utilized and reusable architecture for fractional motion estimation of H.264/AVC. *IEEE Intl. Conf. on Acoustics, Speech, and Signal Processing (ICASSP)*, doi : 10.1109/ICASSP. 2004.1327034, 9-12.

[3] ITU, Telecommunication Standardization Sector. (2010). Advanced Video Coding for Generic Audiovisual Services. *ITU-T*, 167-169, Mar.

[4] Gupta, P. S. S. B. K., & Korada, R. (2004). Novel algorithm to reduce the complexity of quarter-pixel motion estimation. *Proc. of Visual Communications and Image Processing*, 5308, 31-36, Jan, doi: 10.1117/12.532336.

[5] Hyun, C. J., & Sunwoo, M. H. (2009). Low Power Complexity-Reduced ME and Interpolation Algorithms for H.264/AVC. *J. of Signal Processing Systems*, 56(2), 285-293, Sept, doi : 10.1007/s11265-008-0224-4.

[6] Changqi, Yang, Goto, S., & Ikenaga, T. (2006). High performance VLSI architecture of fractional motion estimation in H.264 for HDTV. *IEEE Intl. Symposium on Circuits and Systems (ISCAS)*, September, doi : 10.1109/ISCAS.2006.1693157.

[7] Chao-Yang , Kao, Cheng-Long, Wu, & Lin, Youn-Long. (2010). A high performance three-engine architecture for H.264/AVC fractional motion estimation. *IEEE Tran. On Very Large Scale Integration Systems*, April, doi : 10.1109/ICME.2008.4607389, 18(4), 662-666.

[8] Dodgson, N. A. (1997). Quadratic Interpolation for Image Resampling. *IEEE Trans. on Image Processing*, 6(9), 1322-1326, Sept, doi: 10.1109/83.623195.

[9] Keys, R.G. (1981). Cubic Convolution Interpolation for Digital Image Processing. *IEEE Transactions on Acoustics, Speech and Signal Processing*, Dec, doi : 10.1109/TASSP. 1981.1163711, 29(6), 1153-1160.

[10] Burger, W., & Burge, M. (2008). Digital Image Processing, an Algorithmic approach using Java. *1st ed. New York, USA: Springer.*

[11] Su, D., & Willis, P. (2004). Image Interpolation by Pixel-Level Data-Dependent Triangulation. *Computer Graph. For.*, doi : 10.1111/j.1467-8659.2004.00752.x, 23(2), 189-201.

[12] Chen, Tung-Chien, Huang, Yu-Wen, & Chen, Liang-Gee. (2004). Analysis and design of macroblock pipelining for H.264/AVC VLSI architecture. *IEEE Intl. Symp. on Circuits and Systems (ISCAS)*, 273-276, doi : 10.1109/ISCAS.2004.1329261.

[13] Hyun, C. J., Kim, S. D., & Sunwoo, M. H. (2006). Efficient memory reuse and sub-pixel interpolation algorithms for ME/MC of H.264/AVC. *IEEE Workshop on Signal Processing Systems Design and Implementation*, October, doi : 10.1109/SIPS.2006.352612, 377-38.

[14] Song, Y., Ma, Y., Liu, Z., Ikenaga, T., & Goto, S. (2008). Hardware-oriented direction-based fast fractional motion estimation algorithm in H.264/AVC. *IEEE International Conference on Multimedia and Expo*, 1009-1012, June, doi : 10.1109/ICME.2008.4607608.

[15] Vatis, Y, & Ostermann, J. (2009). Adaptive Interpolation Filter for H.264/AVC. *Circuits and Systems for Video Technology, IEEE Transactions on*, 19(2), 179-192, Feb., doi: 10.1109/TCSVT.2008.2009259.

[16] Hsueh-Ming, Hang, Peng, Wen-Hsiao, Chia-Hsin, Chan, & Chun-Chi, Chen. (2010). Towards the Next Video Standard: High Efficiency Video Coding. *Proceedings of the Second APSIPA Annual Summit and Conference*, 609-618, Biopolis, Singapore, 14-17 December.

[17] Dmytro, Rusanovskyy, Ugur, Kemal, Hallapuro, Antti, Lainema, Jani, & Gabbouj, Moncef. (2009). Video Coding With Low-Complexity Directional Adaptive Interpolation Filters. *IEEE Transactions on Circuits and Systems for Video Technology*, 19(8), August, doi : 10.1109/TCSVT.2009.2022708.

[18] Fuldseth, A., Bjontegaard, G., Rusanovskyy, D., Ugur, K., & Lainema, J. (2008). Low complexity directional interpolation filter. Berlin, Germany, ITU-T Q.6/SG16, VCEG-AI12, July.

[19] Zhang, Kai, Guo, Xun, An, Jicheng, Huang, Yu-Wen, Lei, S., & Gao, Wen. (2012). A Single-Pass-Based Localized Adaptive Interpolation Filter for Video Coding. *Circuits and Systems for Video Technology, IEEE Transactions on*, 22(1), 43-55, Jan., doi: 10.1109/TCSVT.2011.2157194.

[20] Cho, Jaehyun, Lee, Dong-Bok, Cheol, Shin Jeong, & Song, Byung Cheol. (2011). Block-adaptive interpolation filter for sub-pixel motion compensation. *19th European Signal Processing Conference (EUSIPCO)*, 2156-2160.

[21] Georgis, G, Lentaris, G., & Reisis, D. (2012). Study of Interpolation Filters for Motion Estimation with Application in H.264/AVC Encoders. *IEEE Intl. Conference on Circuits and Systems (ICECS)*, 9-12, Beirut, doi : 10.1109/ICECS.2011.6122201, 9-12.

Chapter 8

Algorithms for Efficient Computation of Convolution

Karas Pavel and Svoboda David

Additional information is available at the end of the chapter

1. Introduction

Convolution is an important mathematical tool in both fields of signal and image processing. It is employed in filtering [1, 2], denoising [3], edge detection [4, 5], correlation [6], compression [7, 8], deconvolution [9, 10], simulation [11, 12], and in many other applications. Although the concept of convolution is not new, the efficient computation of convolution is still an open topic. As the amount of processed data is constantly increasing, there is considerable request for fast manipulation with huge data. Moreover, there is demand for fast algorithms which can exploit computational power of modern parallel architectures.

The basic convolution algorithm evaluates inner product of a flipped kernel and a neighborhood of each individual sample of an input signal. Although the time complexity of the algorithms based on this approach is quadratic, i.e. $O(N^2)$ [13, 14], the practical implementation is very slow. This is true especially for higher-dimensional tasks, where each new dimension worsens the complexity by increasing the degree of polynomial, i.e. $O(N^{2k})$. Thanks to its simplicity, the naïve algorithms are popular to be implemented on parallel architectures [15–17], yet the use of implementations is generally limited to small kernel sizes. Under some circumstances, the convolution can be computed faster than as mentioned in the text above.

In the case the higher dimensional convolution kernel is *separable* [18, 19], it can be decomposed into several lower dimensional kernels. In this sense, a 2-D separable kernel can be split into two 1-D kernels, for example. Due to the associativity of convolution, the input signal can be convolved step by step, first with one 1-D kernel, then with the second 1-D kernel. The result equals to the convolution of the input signal with the original 2-D kernel. Gaussian, Difference of Gaussian, and Sobel are the representatives of separable kernels commonly used in signal and image processing. Respecting the time complexity, this approach keeps the higher dimensional convolution to be a polynomial of lower degree, i.e. $O(kN^{k+1})$. On the other hand, there is a nontrivial group of algorithms that use general kernels. For example, the deconvolution or the template matching algorithms based on correlation methods typically use kernels, which cannot be characterized by special properties like separability. In this case, other convolution methods have to be used.

There also exist algorithms that can perform convolution in time $O(N)$. In this concept, the repetitive application of convolution kernel is reduced due to the fact that neighbouring positions overlap. Hence, the convolution in each individual sample is obtained as a weighted sum of both input samples and previously computed output samples. The design of so called *recursive filters* [18] allows them to be implemented efficiently on streaming architectures such as FPGA. Mostly, the recursive filters are not designed from scratch. Rather the well-known 1-D filters (Gaussian, Difference of Gaussian, ...) are converted into their recursive form. The extenstion to higher dimension is straighforward due to their separability. Also this method has its drawbacks. The conversion of general convolution kernel into its recursive version is a nontrivial task. Moreover, the recursive filtering often suffers from inaccuracy and instability [2].

While the convolution in time domain performs an inner product in each sample, in the Fourier domain [20], it can be computed as a simple point-wise multiplication. Due to this convolution property and the fast Fourier transform the convolution can be performed in time $O(N \log N)$. This approach is known as a *fast convolution* [1]. The main advantage of this method stems in the fact that no restrictions are imposed on the kernel. On the other hand, the excessive memory requirements make this approach not very popular. Fortunately, there exists a workaround: If a direct computation of fast convolution of larger signals or images is not realizable using common computers one can reduce the whole problem to several subtasks. In practice, this leads to splitting the signal and kernel into smaller pieces. The signal and kernel decomposition can be perfomed in two ways:

- Data can be decomposed in Fourier domain using so-called decimation-in-frequency (DIF) algorithm [1, 21]. The division of a signal and a kernel into smaller parts also offers a straightforward way of parallelizing the whole task.

- Data can be split in time domain according to overlap-save and overlap-add scheme [22, 23], respectively. Combining these two schemes with fast convolution one can receive a quasi-optimal solution that can be efficiently computed on any computer. Again, the solution naturally leads to a possible parallelization.

The aim of this chapter is to review the algorithms and approaches for computation of convolution with regards to various properties such as signal and kernel size or kernel separability (when processing k-dimensional signals). Target architectures include superscalar and parallel processing units (namely CPU, DSP, and GPU), programmable architectures (e.g. FPGA), and distributed systems (such as grids). The structure of the chapter is designed to cover various applications with respect to the signal size, from small to large scales.

In the first part, the state-of-the-art algorithms will be revised, namely (i) naïve approach, (ii) convolution with separable kernel, (iii) recursive filtering, and (iv) convolution in the frequency domain. In the second part, will be described convolution decomposition in both the spatial and the frequency domain and its implementation on a parallel architecture.

1.1. Shortcuts and symbols

In the following list you will find the most commonly used symbols in this chapter. We recommend you to go through it first to avoid some misunderstanding during reading the text.

- $\mathcal{F}[.], \mathcal{F}^{-1}[.]$ ... Fourier transform and inverse Fourier transform of a signal, respectively
- W_k^i, W_k^{-i} ... k-th sample of i-th Fourier transform base function and inverse Fourier transform base function, respectively
- z^* ... complex conjugate of complex number z
- $*$ ... symbol for convolution
- e ... Euler number (e ≈ 2.718)
- j ... complex unit ($j^2 = -1$)
- f, g ... input signal and convolution kernel, respectively
- h ... convolved signal
- F, G ... Fourier transforms of input signal f and convolution kernel g, respectively
- N^f, N^g ... length of input signal and convolution kernel, respectively (number of samples)
- n, k ... index of a signal in the spatial and the frequency domain, respectively
- n', k' ... index of a signal of half length in the spatial and the frequency domain, respectively
- P ... number of processing units in use
- Φ ... computational complexity function
- $||s||$ ... number of samples of a discrete signal (sequence) s

2. Naïve approach

First of all, let us recall the basic definition of convolution:

$$h(t) = (f * g)(t) = \int_{-\infty}^{\infty} f(t - \tau)g(\tau)d\tau. \tag{1}$$

Respecting the fact that Eq. (1) is used mainly in the fields of research different from image and signal procesing we will focus on the alternative definition that the reader is likely to be more familiar with—the dicrete signals:

$$h(n) = (f * g)(n) = \sum_{i=-\infty}^{\infty} f(n - i)g(i). \tag{2}$$

The basic (or *naïve*) approach visits the individual time samples n in the input signal f. In each position, it computes inner product of current sample neighbourhood and flipped kernel g, where the size of the neighbourhood is practically equal to the size of the convolution kernel. The result of this inner product is a number which is simply stored into the position n in the output signal h. It is noteworthy that according to the definition (2), the size of output signal h is always equal or greater than the size of the input signal f. This fact is related to the boundary conditions. Let $f(n) = 0$ for all $n < 0 \lor n > N^f$ and also $g(n) = 0$ for all $n < 0 \lor n > N^g$. Then computing the expression (2) at the position $n = -1$ likely gives non-zero value, i.e. the output signal becomes larger. It can be derived that the size of output signal h is equal to $N^f + N^g - 1$.

2.0.0.1. Analysis of time complexity.

For the computation of $f * g$ we need to perform $N^f N^g$ multiplications. The computational complexity of this algorithm is polynomial [13], but we must keep in mind what happens when the N^f and N^g become larger and namely what happens when we extend the computation into higher dimensions. In the 3-D case, for example, the expression (2) is slightly changed:

$$h^{3d}(n_x, n_y, n_z) = \left(f^{3d} * g^{3d} \right) (n_x, n_y, n_z)$$
$$= \sum_{i=-\infty}^{\infty} \sum_{j=-\infty}^{\infty} \sum_{k=-\infty}^{\infty} f^{3d}(n_x - i, n_y - j, n_z - k) g^{3d}(i, j, k) \tag{3}$$

Here, f^{3d}, g^{3d} and h^{3d} have the similar meaning as in (2). If we assume $||f^{3d}|| = N_x^f \times N_y^f \times N_z^f$ and $||g^{3d}|| = N_x^g \times N_y^g \times N_z^g$, the complexity of our filtering will raise from $N^f N^g$ in the 1-D case to $N_x^f N_y^f N_z^f N_x^g N_y^g N_z^g$, which is unusable for larger signals or kernels. Hence, for higher dimensional tasks the use of this approach is becomes impractical, as each dimension increases the degree of this polynomial. Although the time complexity of this algorithm is polynomial the use of this solution is advantageous only if we handle with kernels with a small support. An example of such kernels are well-known filters from signal/image processing:

$$\begin{bmatrix} 1 & 2 & 1 \\ 0 & 0 & 0 \\ -1 & -2 & -1 \end{bmatrix} \begin{bmatrix} 1 & 2 & 1 \\ 2 & 4 & 2 \\ 1 & 2 & 1 \end{bmatrix}$$
$$\text{Sobel} \qquad \text{Gaussian}$$

For better insight, let us consider the convolution of two relatively small 3-D signals $1024 \times 1024 \times 100$ voxels and $128 \times 128 \times 100$ voxels—the example is shown in Fig. 1. When this convolution was performed in double precision on Intel Xeon QuadCore 2.83 GHz computer it lasted cca for 7 days if the computation was based on the basic approach.

2.0.0.2. Parallelization.

Due to its simplicity and no specific restrictions, the naïve convolution is still the most popular approach. Its computation is usually sped up by employing large computer clusters that significantly decrease the time complexity per one computer. This approach [15–17] assumes the availability of some computer cluster, however.

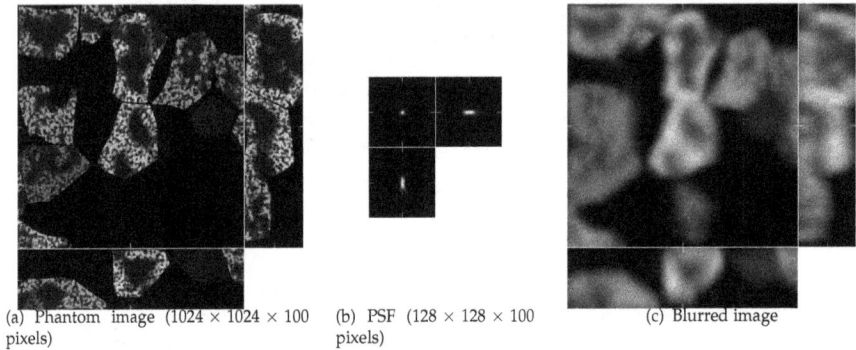

(a) Phantom image (1024 × 1024 × 100 pixels)   (b) PSF (128 × 128 × 100 pixels)   (c) Blurred image

Figure 1. Example of a 3-D convolution. The images show an artificial (phantom) image of a tissue, a PSF of an optical microscope, and blurred image, computed by the convolution of the two images. Each 3-D image is represented by three 2-D views (XY, YZ, and XZ).

2.1. Convolution on a custom hardware

Dedicated and configurable hardware, namely digital signal processors (DSP) or Field-programmable gate array (FPGA) units are very popular in the field of signal processing for their promising computational power at both low cost and low power consumption. Although the approach based on the Fourier transform is more popular in digital signal processing for its ability to process enormously long signals, the naïve convolution with a small convolution kernel on various architectures has been also well studied in the literature, especially in the context of the 2-D and multi-dimensional convolution.

Shoup [24] proposed techniques for automatic generation of convolution pipelines for small kernels such as of 3×3 pixels. Benedetti et al. [25] proposed a multi-FPGA solution by using an external memory to store a FIFO buffer and partitioning of data among several FPGA units, allowing to increase the size of the convolution kernel. Perri et al. [26] followed the previous work by designing a fully reconfigurable FPGA-based 2-D convolution processor. The core of this processor contains four 16-bit SIMD 3×3 convolvers, allowing real-time computation of convolution of a 8-bit or 16-bit image with a 3×3 or 5×5 convolution kernel. Recently, convolution on a custom specialized hardware, e.g. FPGA, ASIC, and DSP, is used to detect objects [27], edges [28], and other features in various real-time applications.

2.2. GPU-based convolution

From the beginning, graphics processing units (GPU) had been designed for visualisation purposes. Since the beginning of the 21st century, they started to play a role in general computations. This phenomenon is often referred to as general-purpose computing on graphics processing units (GPGPU) [29]. At first, there used to be no high-level programming languages specifically designed for general computation purposes. The programmers instead had to use shading languages such as Cg, High Level Shading Language (HLSL) or OpenGL Shading Language (GLSL) [29–31], to utilize texture units. Recently, two programming frameworks are widely used among the GPGPU community, namely CUDA [32] and OpenCL [33].

For their ability to efficiently process 2-D and 3-D images and videos, GPUs have been utilized in various image processing applications, including those based on the convolution. Several convolution algorithms including the naïve one are included in the CUDA Computing SDK [34]. The naïve convolution on the graphics hardware has been also described in [35] and included in the Nvidia Performance Primitives library [36]. Specific applications, namely Canny edge detection [37, 38] or real-time object detection[39] have been studied in the literature. It can be noted that the problem of computing a rank filter such as the median filter has a naïve solution similar to the one of the convolution. Examples can be found in the aforementioned CUDA SDK or in [40, 41].

Basically, the convolution is a *memory-bound* problem [42], i.e. the ratio between the arithmetic operations and memory accesses is low. The adjacent threads process the adjacent signal samples including the common neighbourhood. Hence, they should share the data via a faster memory space, e.g. *shared memory* [35]. To store input data, programmers can also use *texture memory* which is read-only but cached. Furthermore, the texture cache exhibits the 2-D locality which makes it naturally suitable especially for 2-D convolutions.

3. Separable convolution

3.1. Separable convolution

The naïve algorithm is of polynomial complexity. Furthermore, with each added dimension the polynomial degree raises linearly which leads to very expensive computation of convolution in higher dimensions. Fortunately, some kernels are so called *separable* [18, 19]. The convolution with these kernels can be simply decomposed into several lower dimensional (let us say "cheaper") convolutions. Gaussian and Sobel [4] are the representatives of such group of kernels.

Separable convolution kernel must fullfil the condition that its matrix has rank equal to one. In other words, all the rows must be linearly dependent. Why? Let us construct such a kernel. Given one row vector

$$\vec{u} = (u_1, u_2, u_3, \ldots, u_m)$$

and one column vector

$$\vec{v}^T = (v_1, v_2, v_3, \ldots, v_n)$$

let us convolve them together:

$$\vec{u} * \vec{v} = (u_1, u_2, u_3, \ldots, u_m) * \begin{pmatrix} v_1 \\ v_2 \\ v_3 \\ \vdots \\ v_n \end{pmatrix} = \begin{pmatrix} u_1v_1 & u_2v_1 & u_3v_1 & \ldots & u_mv_1 \\ u_1v_2 & u_2v_2 & u_3v_2 & \ldots & u_mv_2 \\ u_1v_3 & u_2v_3 & u_3v_3 & \ldots & u_mv_3 \\ \vdots & \vdots & \vdots & \ddots & \vdots \\ u_1v_n & u_2v_n & u_3v_n & \ldots & u_mv_n \end{pmatrix} = A \qquad (4)$$

It is clear that $rank(A) = 1$. Here, A is a matrix representing some separable convolution kernel while $\vec{u}$ and $\vec{v}$ are the previously referred lower dimensional (cheaper) convolution kernels.

3.1.0.3. Analysis of Time Complexity.

In the previous section, we derived the complexity of naïve approach. We also explained how the complexity worsens when we increase the dimensionality of the processed data. In case the convolution kernel is separable we can split the hard problem into a sequence of several simpler problems. Let us recall the 3-D naïve convolution from (3). Assume that g^{3d} is separable, i.e. $g^{3d} = g_x * g_y * g_z$. Then the expression is simplified in the following way:

$$h^{3d}(n_x, n_y, n_z) = \left(f^{3d} * g^{3d}\right)(n_x, n_y, n_z) \tag{5}$$

$$= \left(f^{3d} * \left(g_x * g_y * g_z\right)\right)(n_x, n_y, n_z) \quad /associativity/ \tag{6}$$

$$= \left(\left(\left(f^{3d} * g_x\right) * g_y\right) * g_z\right)(n_x, n_y, n_z) \tag{7}$$

$$= \left(\sum_{i=-\infty}^{\infty}\left(\sum_{j=-\infty}^{\infty}\left(\sum_{k=-\infty}^{\infty} f^{3d}(n_x-i, n_y-j, n_z-k)g_z(k)\right)g_y(j)\right)g_x(i)\right) \tag{8}$$

The complexity of such algorithm is then reduced from $N_x^f N_y^f N_z^f N_x^g N_y^g N_z^g$ to $N_x^f N_y^f N_z^f \left(N_x^g + N_y^g + N_z^g\right)$.

One should keep in mind that the kernel decomposition is usually the only one decomposition that can be performed in this task. It is based on the fact that many well-known kernels (Gaussian, Sobel) have some special properties. Nevertheless, the input signal is typically unpredictable and in higher dimensional cases it is unlikely one could separate it into individual lower-dimensional signals.

3.2. Separable convolution on various architectures

As separable filters are very popular in many applications, a number of implementations on various architectures can be found in the literature. Among the most favourite filters, the Gaussian filter is often used for pre-processing, for example in optical flow applications [43, 44]. Fialka et al. [45] compared the separable and the fast convolution on the graphics hardware and proved both the kernel size and separability to be the essential properties that have to be considered when choosing an appropriate implementation. They proved the separable convolution to be more efficient for kernel sizes up to tens of pixels in each dimension which is usually sufficient if the convolution is used for the pre-processing.

The implementation usually does not require particular optimizations as the separable convolution is intrinsically a sequence of 1-D basic convolutions. Programmers should nevertheless consider some tuning steps regarding the memory accesses, as mentioned in Section 2.2. For the case of a GPU implementation, this issue is discussed in [35]. The GPU implementation described in the document is also included in the CUDA SDK [34].

4. Recursive filtering

The convolution is a process where the inner product, whose size corresponds to kernel size, is computed again and again in each individual sample. One of the vectors (kernel), that enter this operation, is always the same. It is clear that we could compute the whole inner product only in one position while the neighbouring position can be computed as a slightly modified difference with respect to the first position. Analogously, the same is valid for all the following positions. The computation of the convolution using this difference-based approach is called *recursive filtering* [2, 18].

4.0.0.4. Example.

The well-known pure averaging filter in 1D is defined as follows:

$$h(n) = \sum_{i=0}^{n-1} f(n-i) \tag{9}$$

The performance of this filter worsen with the width of its support. Fortunately, there exists a recursive version of this filter with constant complexity regardless the size of its support. Such a filter is no more defined via standard convolution but using the recursive formula:

$$h(n) = h(n-1) + f(n) - f(n-n) \tag{10}$$

The transform of standard convolution into a recursive filtering is not a simple task. There are three main issues that should be solved:

1. replication – given slow (but correctly working) non-recursive filter, find its recursive version
2. stability – the recursive formula may cause the computation to diverge
3. accuracy – the recursion may cause the accumulation of small errors

The transform is a quite complex task and so-called Z-transform [22] is typically employed in this process. Each recursive filter may be designed as all other filters from scratch. In practice, the standard well-known filters are used as the bases and subsequently their recursive counterpart is found. There are two principal approaches how to do it:

• analytically – the filter is step by step constructed via the math formulas [46]
• numerically – the filter is derived using numerical methods [47, 48]

4.1. Recursive filters on various architectures

Streaming architectures.

The recursive filtering is a popular approach especially on streaming architectures such as FPGA. The data can be processed in a stream keeping the memory requirements on a minimum level. This allows moving the computation to relatively small and cheap embedded systems. The recursive filters are thus used in various real-time applications such as edge detection [49], video filtering [50], and optical flow [51].

Parallel architectures.

As for the parallel architectures, Robelley et al. [52] presented a mathematical formulation for computing time-invariant recursive filters on general SIMD DSP architectures. Authors also discuss the speed-up factor regarding to the level of parallelism and the filter order. Among the GPU implementations, we can mention the work of Trebien and Oliveira who implemented recursive filters in CUDA for the purpose of the realistic sound synthesis and processing [53]. In this case, recursive filters were computed in the frequency domain.

5. Fast convolution

In the previous sections, we have introduced the common approaches to compute the convolution in the time (spatial) domain. We mentioned that in some applications, one has to cope with signals of millions of samples where the computation of the convolution requires too much time. Hence, for long or multi-dimensional input signals, the popular approach is to compute the convolution in the frequency domain which is sometimes referred to as the *fast convolution*. As shown in [45], the fast convolution can be even more efficient than the separable version if the number of kernel samples is large enough. Although the concept of the fast Fourier transform [54] and the frequency-based convolution [55] is several decades old, with new architectures upcoming, one has to deal with new problems. For example, the efficient access to the memory was an important issue in 1970s [56] just as it is today [21, 23]. Another problem to be considered is the numerical precision [57].

In the following text, we will first recall the Fourier transform along with some of its important properties and the convolution theorem which provides us with a powerful tool for the convolution computation. Subsequently, we will describe the algorithm of the so-called fast Fourier transform, often simply denoted as FFT, and mention some notable implementations of the FFT. Finally, we will summarize the benefits and drawbacks of the fast convolution.

5.1. Fourier transform

The Fourier transform $F = \mathcal{F}[f]$ of a function f and the inverse Fourier transform $f = \mathcal{F}^{-1}[F]$ are defined as follows:

$$F(\omega) \equiv \int_{-\infty}^{+\infty} f(t) e^{-jt\omega} dt, \qquad f(t) \equiv \frac{1}{2\pi} \int_{-\infty}^{+\infty} F(\omega) e^{j\omega t} d\omega. \qquad (11)$$

The discrete finite equivalents of the aforementioned transforms are defined as follows:

$$F(k) \equiv \sum_{n=0}^{N-1} f(n) e^{-j(2\pi/N)nk}, \qquad f(n) \equiv \frac{1}{N} \sum_{k=0}^{N-1} F(k) e^{j(2\pi/N)kn} \qquad (12)$$

where $k, n = 0, 1, \ldots, N - 1$. The so-called normalization factors $\frac{1}{2\pi}$ and $\frac{1}{N}$, respectively, guarantee that the identity $f = \mathcal{F}^{-1}[\mathcal{F}[f]]$ is maintained. The exponential function $e^{-j(2\pi/N)}$ is called the base function. For the sake of simplicity, we will refer to it as W_K.

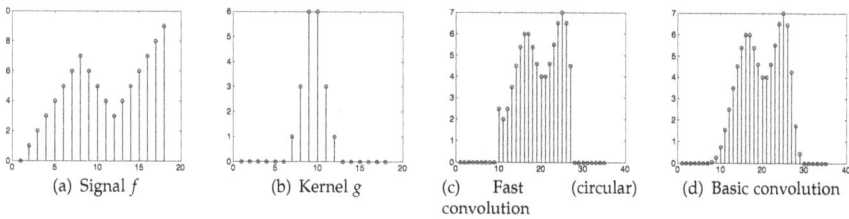

Figure 2. Example of the so-called windowing effect produced by signal f (a) and kernel g (b). The circular convolution causes border effects as seen in (c). The properly computed basic convolution is shown in (d).

If the sequence $f(n), n = 0, 1, \ldots, N - 1$, is real, the discrete Fourier transform $F(k)$ keeps some specific properties, in particular:

$$F(k) = F(N - k)^*. \tag{13}$$

This means that in the output signal F, only half of the samples are useful, the rest is redundant. As the real signals are typical for many practical applications, in most popular FT and FFT implementations, users are hence provided with special functions to handle real signals in order to save time and memory.

5.2. Convolution theorem

According to the convolution theorem, the Fourier transform convolution of two signals f and g is equal to the product of the Fourier transforms $\mathcal{F}[f]$ and $\mathcal{F}[g]$, respectively [58]:

$$\mathcal{F}[f * g] = \mathcal{F}[f]\mathcal{F}[g]. \tag{14}$$

In the following text, we will sometimes refer to the convolution computed by applying Eq. (14) as the "classical" fast convolution algorithm.

In the discrete case, the same holds for periodic signals (sequences) and is sometimes referred to as the circular or cyclic convolution [22]. However, in practical applications, one usually deals with non-periodic finite signals. This results into the so-called windowing problem [59], causing undesirable artefacts in the output signals—see Fig. 2. In practice, the problem is usually solved by either imposing the periodicity into the kernel, adding a so-called windowing function, or padding the kernel with zero values. One also has to consider the sizes of both the input signal and the convolution kernel which have to be equal. Generally, this is also solved by padding both the signal and the kernel with zero values. The size of both padded signals which enter the convolution is hence $N = N^f + N^g - 1$ where N^f and N^g is the number of signal and kernel samples, respectively. The equivalent property holds for the multi-dimensional case. The most time-demanding operation of the fast convolution approach is the Fourier transform which can be computed by the fast Fourier transform

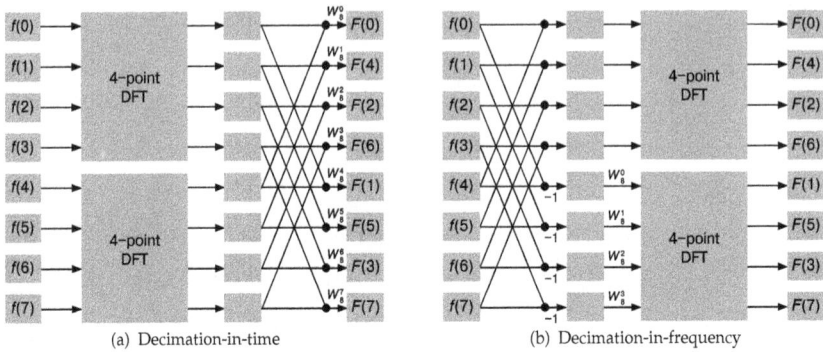

(a) Decimation-in-time   (b) Decimation-in-frequency

Figure 3. The basic two radix-2 FFT algorithms: decimation-in-time and decimation-in-frequency. Demonstration on an input signal of 8 samples.

algorithm. The time complexity of the fast convolution is hence equal to the complexity of the FFT, that is $O(N \log N)$. The detailed discussion on the complexity is provided in Section 6.

5.3. Fast Fourier transform

In 1965, Cooley and Tukey [60] proposed an algorithm for fast computation of the Fourier transform. The widely-known algorithm was then improved through years and optimized for various signal lengths but the basic idea remained the same. The problem is handled in a divide-and-conquer manner by splitting the input signal into N parts[1] and processing the individual parts recursively. Without loss of generality, we will recall the idea of the FFT for $N = 2$ which is the simplest situation. There are two fundamental approaches to split the signal. They are called *decimation in time (DIT)* and *decimation in frequency (DIF)* [58].

Decimation in time (DIT).

Assuming that N is even, the radix-2 decimation-in-time algorithm splits the input signal $f(n)$, $n = 0, 1, \ldots, N - 1$ into parts $f_e(n')$ and $f_o(n')$, $n' = 0, 1, \ldots, N/2 - 1$ of even and odd samples, respectively. By recursive usage of the approach, the discrete Fourier transforms F_e and F_o of the two parts are computed. Finally, the resulting Fourier transform F can be computed as follows:

$$F(k) = F_e(k) + W_N^k F_o(k) \tag{15}$$

where $k = 0, 1, \ldots, N - 1$. The signals F_e and F_o are of half length, however, they are periodic, hence

$$F_e(k' + N/2) = F_e(k'), \qquad F_o(k' + N/2) = F_o(k') \tag{16}$$

for any $k' = 0, 1, \ldots, N/2 - 1$. The algorithm is shown in Fig. 3(a).

---

[1] The individual variants of the algorithm for a particular N are called radix-N algorithms.

Decimation in frequency (DIF).

Having the signal f of an even length N, the sequences f_r and f_s of the half length are created as follows:

$$f_r(n') = f(n') + f(n' + N/2), \qquad f_s(n') = \left[f(n') - f(n' + N/2)\right] W_N^{-n'}. \tag{17}$$

Then, the Fourier transform F_r and F_s fulfill the following property: $F_r(k') = F(2k')$ and $F_s(k') = F(2k' + 1)$ for any $k' = 0, 1, \ldots, N/2 - 1$. Hence, the sequences f_r and f_s are then processed recursively, as shown in Fig. 3(b). It is easy to deduce the inverse equation from Eq. (17):

$$f(n') = \frac{1}{2}\left[f_r(n') + f_s(n')W_N^{n'}\right], \qquad f(n' + N/2) = \frac{1}{2}\left[f_r(n') - f_s(n')W_N^{n'}\right]. \tag{18}$$

5.4. The most popular FFT implementations

On CPU.

One of the most popular FFT implementations ever is so-called Fastest Fourier Transform in the West (FFTW) [61]. It is kept updated and available for download on the web page http://www.fftw.org/. According to the authors' comprehensive benchmark [62], it is still one of the fastest CPU implementations available. The top performance is achieved by using multiple CPU threads, the extended instruction sets of modern processors such as SSE/SSE2, optimized radix-N algorithms for N up to 7, optimized functions for purely real input data etc. Other popular CPU implementations can be found e.g. in the Intel libraries called Intel Integrated Performance Primitives (IPP) [63] and Intel Math Kernel Library (MKL) [64]. In terms of performance, they are comparable with the FFTW.

On other architectures.

For the graphics hardware, there exists several implementations in the literature [65–67]. Probably the most widely-used one is the CUFFT library by Nvidia. Although it is dedicated to the Nvidia graphics cards, it is popular due to its good performance and ease of use. It also contains optimized functions for real input data. The FFT has been also implemented on various architectures, including DSP [68] and FPGA [69].

5.5. Benefits and drawbacks of the fast convolution

To summarize this section, fast convolution is the most efficient approach if both signal and kernel contain thousands of samples or more, or if the kernel is slightly smaller but non-separable. Thanks to numerous implementations, it is accessible to a wide range of users on various architectures. The main drawbacks are the windowing problem, the relatively lower numerical precision, and considerable memory requirements due to the signal padding. In the following, we will examine the memory usage in detail and propose several approaches to optimize it on modern parallel architectures.

6. Decomposition in the time domain

In this section, we will focus on the decomposition of the fast convolution in the time domain. We will provide the analysis of time and space complexity. Regarding the former, we will focus on the number of additions and multiplications needed for the computation of studied algorithms.

Utilizing the convolution theorem and the fast Fourier transform the 1-D convolution of signal f and kernel g requires

$$(N^f + N^g) \left[\frac{9}{2} \log_2(N^f + N^g) + 1 \right] \tag{19}$$

steps [8]. Here, the term $(N^f + N^g)$ means that the processed signal f was zero padded[2] to prevent the overlap effect caused by circular convolution. The kernel was modified in the same way. Another advantage of using Fourier transform stems from its separability. Convolving two 3-D signals f^{3d} and g^{3d}, where $||f||^{3d} = N_x^f \times N_y^f \times N_z^f$ and $||g^{3d}|| = N_x^g \times N_y^g \times N_z^g$, we need only

$$(N_x^f + N_x^g)(N_y^f + N_y^g)(N_z^f + N_z^g) \left[\frac{9}{2} \log_2 \left((N_x^f + N_x^g)(N_y^f + N_y^g)(N_z^f + N_z^g) \right) + 1 \right] \tag{20}$$

steps in total.

Up to now, this method seems to be optimal. Before we proceed, let us look into the space complexity of this approach. If we do not take into account buffers for the input/output signals and serialize both Fourier transforms, we need space for two equally aligned Fourier signals and some negligible Fourier transform workspace. In total, it is

$$(N^f + N^g) \cdot C \tag{21}$$

bytes, where $(N^f + N^g)$ is a size of one padded signal and C is a constant dependent on the required algorithm precision (single, double or long double). If the double precision is required, for example, then $C = 2 \cdot \texttt{sizeof(double)}$, which corresponds to two Fourier signals used by real-valued FFT. In the 3-D case, when $||f^{3d}|| = N_x^f \times N_y^f \times N_z^f$ and $||g^{3d}|| = N_x^g \times N_y^g \times N_z^g$ the space needed by the aligned signal is proportionally higher: $(N_x^f + N_x^g)(N_y^f + N_y^g)(N_z^f + N_z^g) \cdot C$ bytes.

Keeping in mind that due to the lack of available memory, direct computation of fast convolution is not realizable using common computers we will try to split the whole task into several subtasks. This means that the input signal and kernel will be split into smaller pieces, so called *tiles* that need not be of the same size. Hence, we will try to reduce the memory requirements while keeping the efficiency of the whole convolution process as proposed in [23].

---

[2] The size of padded signal should be exactly $(N^f + N^g - 1)$. For the sake of simplicity, we reduced this term to $(N^f + N^g)$ as we suppose $N^f \gg 1$ and $N^g \gg 1$.

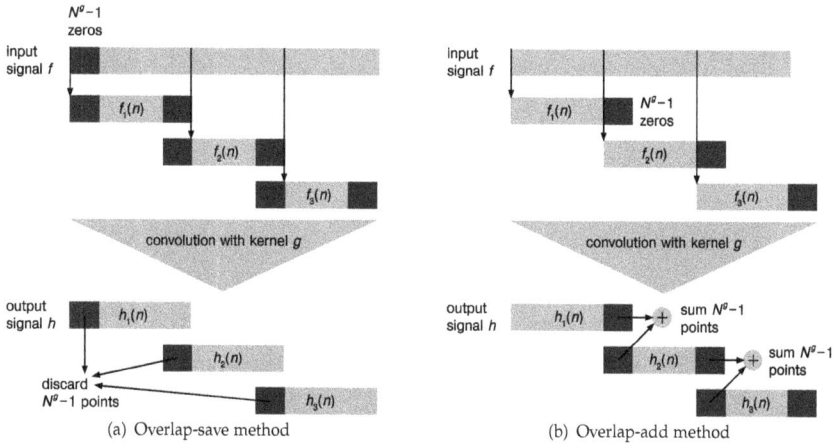

Figure 4. Using the overlap-save and overlap-add methods, the input data can be segmented into smaller blocks and convolved separately. Finally, the sub-parts are concatenated (a) or summed (b) together.

6.1. Signal tiling

Splitting the input signal f into smaller disjoint tiles $f_1, f_2, \ldots, f_m$, then performing m smaller convolutions $f_i * g, i = 1, 2, \ldots, m$ and finally concatenating the results together with discarding the overlaps is a well-known algorithm in digital signal processing. The implementation is commonly known as the *overlap-save method* [22].

6.1.0.5. Method.

Without loss of generality we will focus on the manipulation with just one *tile f_i*. The other tiles are processes in the same way. The tile f_i is uniquely determined by its size and shift with respect to the origin of f. Its size and shift also uniquely determine the area in the output signal h where the expected result of $f_i * g$ is going to be stored. In order to guarantee that the convolution $f_i * g$ computes correctly the appropriate part of output signal h, the tile f_i must be equipped with some overlap to its neighbours. The size of this overlap is equal to the size of the whole kernel g. Hence, the tile f_i is extended equally on both sides and we get f'_i. If the tile f_i is the boundary one, it is padded with zero values. As the fast convolution required both the signal and the kernel of the same size the kernel g must be also extended. It is just padded with zeros which produces g'. As soon as f'_i and g' are prepared, the convolution $f'_i * g'$ can be performed and the result is cropped to the size $\|f_i\|$. Then, all the convolutions $f'_i * g', i = 1, 2, \ldots, m$ are successively performed and the output signal h is obtained by *concatenating* the individual results together. A general form of the method is shown in Fig. 4(a).

6.1.0.6. Analysis of time complexity.

Let us inspect the memory requirements for this approach. As the filtered signal f is split into m pieces, the respective memory requirements are lowered to

$$\left(\frac{N^f}{m} + N^g\right) \cdot C \tag{22}$$

bytes. Concerning the time complexity, after splitting the signal f into m tiles, we need to perform

$$(N^f + mN^g)\left[\frac{9}{2}\log_2\left(\frac{N^f}{m} + N^g\right) + 1\right] \tag{23}$$

multiplications in total. If there is no division ($m = 1$) we get the time complexity of the fast approach. If the division is total ($m = N^f$) we get even worse complexity than the basic convolution has. The higher the level of splitting is required the worse the complexity is. Therefore, we can conclude that splitting only the input signal into tiles does not help.

6.2. Kernel tiling

From the previous text, we recognize that splitting only the input signal f might be inefficient. It may even happen that the kernel g is so large that splitting of only the signal f does not reduce the memory requirements sufficiently. As the convolution belongs to commutative operators one could recommend swapping the input signal and the kernel. This may help, namely when the signal f is small and the kernel g is very large. As soon as the signal and the kernel are swapped, we can simply apply the overlap-save method. However, this approach fails when both the signal and the kernel are too large. Let us decompose the kernel g as well.

6.2.0.7. Method.

Keeping in mind that the input signal f has already been decomposed into m tiles using overlap-save method, we can focus on the manipulation with just one tile $f_i, i = 1, 2, \ldots, m$. For the computation of convolution of the selected tile f_i and the large kernel g we will employ so called *overlap-add method* [22]. This method splits the kernel g into n disjoint (nonoverlapping) pieces $g_j, j = 1, 2, \ldots, n$. Then, it performs n cheaper convolutions $f_i * g_j$, and finally it adds the results together preserving the appropriate overruns.

Without loss of generality we will focus on the manipulation with just one *kernel tile* g_j. Prior to the computation, the selected *tile* g_j has to be aligned to the size $||f_i|| + ||g_j||$. It is done simply by padding g_j with zeros equally on both sides. In this way, we get the tile g'_j. The signal tile f_i is also aligned to the size $||f_i|| + ||g_j||$. However, f'_i is not padded with zeros. It is created from f_i by extending its support equally on both sides.

Each kernel tile g_j has its positive shift s_j with respect to the origin of g. This shift is very important for further computation and cannot be omitted. Before we perform the convolution $f'_i * g'_j$ we must shift the tile f'_i within f by s_j samples to the left. The reason originates from the idea of kernel decomposition and minus sign in Eq. (2) which causes the whole kernel to be flipped. As soon as the convolution $f'_i * g'_j$ is performed, its result is cropped to the size $||f_i||$ and *added* to the output signal h into the position defined by overlap-save method. Finally, all the convolutions $f'_i * g'_j, j = 1, 2, \ldots n$ are performed to get complete result for one given tile f_i. A general form of the method is shown in Fig. 4(b).

The complete computation of the convolution across all signal and kernel tiles is sketched in the Algorithm 1.

Algorithm 1. Divide-and-conquer approach applied to the convolution over large data.

$(f, g) \leftarrow$ (input signal, kernel)
$f \rightarrow f_1, f_2, \ldots, f_m$ {split $'f'$ into tiles according to overlap-save scheme}
$g \rightarrow g_1, g_2, \ldots, g_n$ {split $'g'$ into tiles according to overlap-add scheme}
$h \leftarrow 0$ {create the output signal $'h'$ and fill it with zeros}
for $i = 1$ to m **do**
  for $j = 1$ to n **do**
    $h_{ij} \leftarrow$ convolve(f_i, g_j)
    {use fast convolution}
    $h_{ij} \leftarrow$ discard_overruns(h_{ij})
    {discard h_{ij} overruns following overlap-save output rules}
    $h \leftarrow h +$ shift(h_{ij})
    {add h_{ij} to h following overlap-add output rules}
  end for
end for
Output $\leftarrow h$

6.2.0.8. Analysis of time complexity.

Let us suppose the signal f is split into m tiles and kernel g is decomposed into n tiles. The time complexity of the fast convolution $f_i * g_j$ is

$$\left(\frac{N^f}{m} + \frac{N^g}{n} \right) \left[\frac{9}{2} \log_2 \left(\frac{N^f}{m} + \frac{N^g}{n} \right) + 1 \right]. \tag{24}$$

We have m signal tiles and n kernel tiles. In order to perform the complete convolution $f * g$ we have to perform $m \times n$ convolutions (see the nested loops in Algorithm 1) of the individual signal and kernel tiles. In total, we have to complete

$$\left(nN^f + mN^g \right) \left[\frac{9}{2} \log_2 \left(\frac{N^f}{m} + \frac{N^g}{n} \right) + 1 \right] \tag{25}$$

steps. One can clearly see that without any division ($m = n = 1$) we get the complexity of fast convolution, i.e. the class $O((N^f + N^g) \log(N^f + N^g))$. For total division ($m = N^f$ and

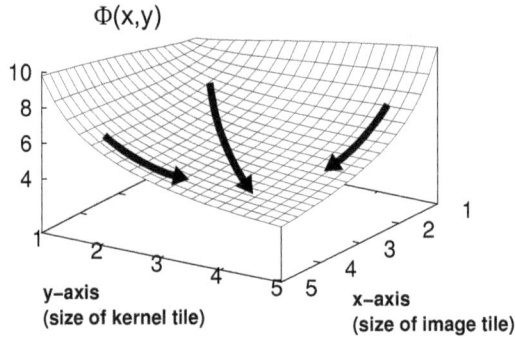

Figure 5. A graph of a function $\Phi(x, y)$ that represents the time complexity of tiled convolution. The x-axis and y-axis correspond to number of samples in signal and kernel tile, respectively. The evident minimum of function $\Phi(x, y)$ occurs in the location, where both variables (sizes of tiles) are maximized and equal at the same time.

$n = N^g$) we obtain basic convolution, i.e. the complexity class $O(N^f N^g)$. Concerning the space occupied by our convolution algorithm, we need

$$\left(\frac{N^f}{m} + \frac{N^g}{n} \right) \cdot C \tag{26}$$

bytes, where C is again the precision dependent constant and m, n are the levels of division of signal f and kernel g, respectively.

6.2.0.9. Algorithm optimality.

We currently designed an algorithm of splitting the signal f into m tiles and the kernel g into n tiles. Now we will answer the question regarding the optimal way of splitting the input signal and the kernel. As the relationship between m and n is hard to be expressed and N^f and N^g are constants let us define the following substitution: $x = \frac{N^f}{m}$ and $y = \frac{N^g}{n}$. Here x and y stand for the sizes of the signal and the kernel tiles, respectively. Applying this substitution to Eq. (25) and simplifying, we get the function

$$\Phi(x, y) = N^f N^g \left(\frac{1}{x} + \frac{1}{y} \right) \left[\frac{9}{2} \log_2(x + y) + 1 \right] \tag{27}$$

The plot of this function is depicted in Figure 5. The minimum of this function is reached if and only if $x = y$ and both variables x and y are maximized, i.e. the input signal and the kernel tiles should be of the same size (equal number of samples) and they should be as large as possible. In order to reach the optimal solution, the size of the tile should be the power of small primes [70]. In this sense, it is recommended to fulfill both criteria put on the tile size: the maximality (as stated above) and the capability of simple decomposition into small primes.

6.3. Extension to higher dimensions

All the previous statements are related only to a 1-D signal. Provided both signal and kernel are 3-dimensional and the tiling proces identical in all the axes, we can combine Eq. (20) and Eq. (25) in order to get:

$$\left(nN_x^f+mN_x^g\right)\left(nN_y^f+mN_y^g\right)\left(nN_z^f+mN_z^g\right)\left[\frac{9}{2}\log_2\left(\frac{N_x^f}{m}+\frac{N_x^g}{n}\right)\left(\frac{N_y^f}{m}+\frac{N_y^g}{n}\right)\left(\frac{N_z^f}{m}+\frac{N_z^g}{n}\right)+1\right] \quad (28)$$

This statement can be further generalized for higher dimensions or for irregular tiling process. The proof can be simply derived from the separability of multidimensional Fourier transform, which guarantees that the time complexity of the higher dimensional Fourier transform depends on the amount of processed samples only. There is no difference in the time complexity if the higher-dimensional signal is elongated or in the shape of cube.

6.4. Parallelization

6.4.0.10. On multicore CPU.

As the majority of recent computers are equipped with multi-core CPUs the following text will be devoted to the idea of parallelization of our approach using this architecture. Each such computer is equipped with two or more cores, however both cores share one memory. This means that execution of two or more huge convolutions concurrently may simply fail due to lack of available memory. The possible workaround is to perform one more division, i.e. signal and kernel tiles will be further split into even smaller pieces. Let p be a number that defines how many sub-pieces the signal and the kernel tiles should be split into. Let P be a number of available processors. If we execute the individual convolutions in parallel we get the overall number of multiplications

$$\frac{npN^f+mpN^g}{P}\left[\frac{9}{2}\log\left(\frac{N^f}{mp}+\frac{N^g}{np}\right)+1\right] \quad (29)$$

and the space requirements

$$\left(\frac{N^f}{mp}+\frac{N^g}{np}\right)\cdot C\cdot P \quad (30)$$

Let us study the relationship p versus P:

- $p < P$ ... The space complexity becomes worse than in the original non-parallelized version (26). Hence, there is no advantage of using this approach.

- $p > P$ ... There are no additional memory requirements. However, the signal and kernel are split into too small pieces. We have to handle large number of overlaps of tiles which will cause the time complexity (29) to become worse than in the non-parallelized case (25).

- $p = P$ ... The space complexity is the same as in the original approach. The time complexity is slightly better but practically it brings no advantage due to lots of memory accesses. The efficiency of this approach would be brought to evidence only if $P \gg 1$. As the standard multi-core processors are typically equipped with only 2, 4 or 8 cores, neither this approach was found to be very useful.

6.4.0.11. On computer clusters.

Regarding computer clusters the problem with one shared memory is solved as each computer has its private memory. Therefore, the total number of multiplications (see Eq. (25)) is modified by factor $\frac{B}{P}$, where P is the number of available computers and B is a constant representing the overheads and the cost of data transmission among the individual computers. The computation becomes effective only if $P > B$. The memory requirements for each node remain the same as in the non-parallelized case as each computer takes care of its own private memory space.

7. Decomposition in the frequency domain

Just as the concept of the decomposition in the spatial (time) domain, the decomposition in the frequency domain can be used for the fast convolution algorithm, in order to (i) decrease the required amount of memory available per processing unit, (ii) employ multiple processing units without need of extensive data transfers between the processors. In the following text, we introduce the concept of the decomposition [21] along with optimization steps suitable for purely real data [71]. Subsequently, we present the results on achieved on a current graphics hardware. Finally, we conclude the applications and architectures where the approach can be used.

7.1. Decomposition using the DIF algorithm

In Section 5.3, the decimation-in-frequency algorithm was recalled. The DIF can be used not only to compute FFT itself but also to decompose the fast convolution. This algorithm can be divided into several phases, namely (i) so-called *decomposition* into parts using Eq. (17), (ii) the Fourier transforms of the parts, (iii) the convolution by pointwise multiplication itself, (iv) the inverse Fourier transforms, and (v) so-called *composition* using Eq. (18). In the following paragraph, we provide the mathematical background for the individual phases. The scheme description of the algorithm is shown in Fig. 6(a).

By employing Eq. (17), both the input signal f and g can be divided into sub-parts f_r, f_s and g_r, g_s, respectively. As the Fourier transforms F_r and F_s satisfy $F_r(k') = F(2k')$ and $F_s(k') = F(2k' + 1)$ and the equivalent property is held for G_r and G_s, by applying FFT on F_r F_s, G_r, and G_s, individually, we obtain two separate parts of both the signal and the kernel. Subsequently, by computing the point-wise multiplication $H_r = F_r G_r$ and $H_s = F_s G_s$, respectively, we obtain two separate parts of the Fourier transform of the convolution $h = f * g$. Finally, the result h is obtained by applying Eq. (18) to the inverse Fourier transforms h_r and h_s.

In the first and the last phase, it is inevitable to store the whole input signals in the memory. Here, the memory requirements are equal to those in the classical fast convolution algorithm. However, in the phases (ii)–(iv) which are by far the most computationally extensive, the

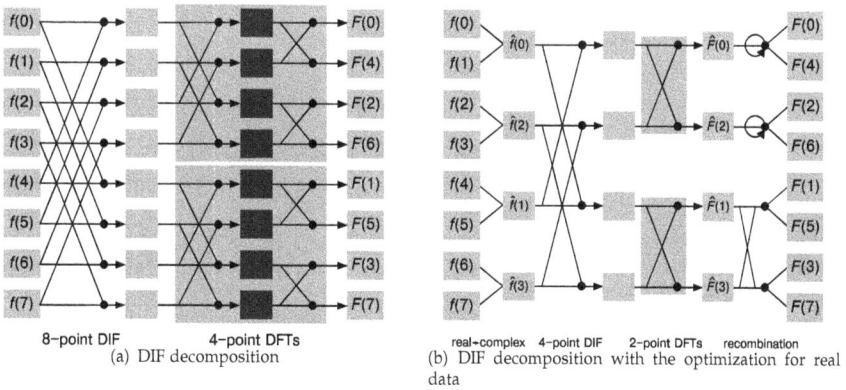

(a) DIF decomposition

(b) DIF decomposition with the optimization for real data

Figure 6. A scheme description of the convolution algorithm with the decomposition in the frequency domain [71]. An input signal is decomposed into 2 parts by the decimation in frequency (DIF) algorithms. The parts are subsequently processed independently using the discrete Fourier transform (DFT).

memory requirements are inversely proportional to the number of parts d the signals are divided into. The algorithm is hence suitable for architectures with the star topology where the central node is relatively slow but has large memory, and the end nodes are fast but have small memory. The powerful desktop PC with one or several GPU cards is a typical example of such architecture.

It can be noted that the decimation-in-time (DIT) algorithm can also be used for the purpose of decomposing the convolution problem. However, its properties make it sub-efficient for practical use. Firstly, its time complexity is comparable with the one of DIF. Secondly and most important, it requires significantly more data transfers between the central and end nodes. In Section 7.5, the complexity of the individual algorithms is analysed in detail.

7.2. Optimization for purely real signals

In most practical applications, users work with purely real input signals. As described in Section 5.1, the Fourier transform is complex but satisfies specific properties when applied on such data. Therefore, it is reasonable to optimize the fast convolution algorithm in order to reduce both the time and the memory complexity. In the following paragraphs, we will describe three fundamental approaches to optimize the fast convolution of real signals.

Real-to-complex FFT.

As described in Section 5.4, most popular FFT implementations offer specialized functions for the FFT of purely real input data. With the classical fast convolution, users are advised to use specific functions of their preferred FFT library. With the DIF decomposition, it is nevertheless no more possible to use such functions as the decomposed signals are no more real.

Combination of signal and kernel.

It is possible to combine the two real input signals $f(n)$ and $g(n)$, $n = 0, 1, \ldots, N - 1$, into one complex signal $f(n) + jg(n)$ of the same length. However, this operation requires an additional buffer of length at least N. This poses significantly higher demands on the memory available at the central node.

"Complexification" of input signals.

Provided that the length N of a real input signal f is even, we can introduce a complex signal $\hat{f}(n') \equiv f(2n') + jf(2n' + 1)$ for any $n' = 0, 1, \ldots, N/2 - 1$. As the most common way of storing the complex signals is to store real and complex components, alternately, a real signal can be turned into a complex one by simply over-casting the data type, avoiding any computations or data transfers. The relationship between the Fourier transforms F and $\hat{F}$ is given by following:

$$F(k') = \frac{1}{2}\left(\alpha_+(k') - jW_N^{k'}\alpha_-(k')\right), \qquad F(k' + N/2) = \frac{1}{2}\left(\alpha_+(k') + jW_N^{k'}\alpha_-(k')\right), \qquad (31)$$

where

$$\alpha_\pm(k') \equiv \hat{F}(k') \pm \hat{F}^*(N/2 - k'). \qquad (32)$$

As the third approach yields the best performance, it is used in the final version of the algorithm. The computation of Eq. (31), (32) will be further referred to as the *recombination* phase. The scheme description of the algorithm is shown in Fig. 6(b).

7.3. Getting further

The algorithm can be used not only in 1D but generally for any n-dimensional input signals. To achieve maximum data transfer efficiency, it is advisable to perform the decomposition in the first (y in 2D or z in 3D) axis so that the individual sub-parts form the undivided memory blocks, as explained in [21].

Furthermore, the input data can be decomposed into generally d parts using an appropriate radix-d algorithm in both the decomposition and the composition phase. It should be noted, however, that due to the recombination phase, the algorithm requires twice more memory space per end node for $d > 2$. This is due to fact that some of the parts need to be recombined with others—refer to Fig. 6(b). To be more precise, the memory requirements are $2(N^f + N^g)/d$ for $d = 2$ and $4(N^f + N^g)/d$ for $d > 2$.

7.4. GPU and multi-GPU implementation

As Nvidia provides users with the CUFFT library [32] for the efficient computation of the fast Fourier transform, the GPU implementation of the aforementioned algorithm is quite straightforward. The scheme description of the implementation is shown in Fig. 7. It should be noted that the significant part of the computation time is spent for the data transfers between the computing nodes (CPU and GPU, in this case). The algorithm is designed to keep the number of data transfers as low as possible. Nevertheless, it is highly

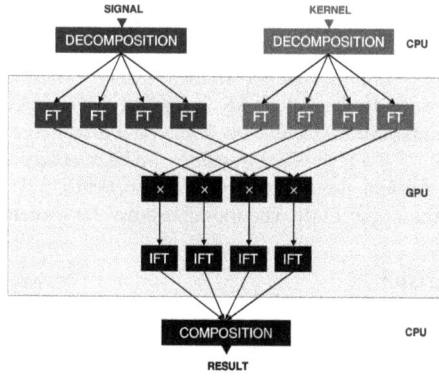

Figure 7. A scheme description of the proposed algorithm for the convolution with the decomposition in the frequency domain, implemented on GPU [21]. The example shows the decomposition into 4 parts.

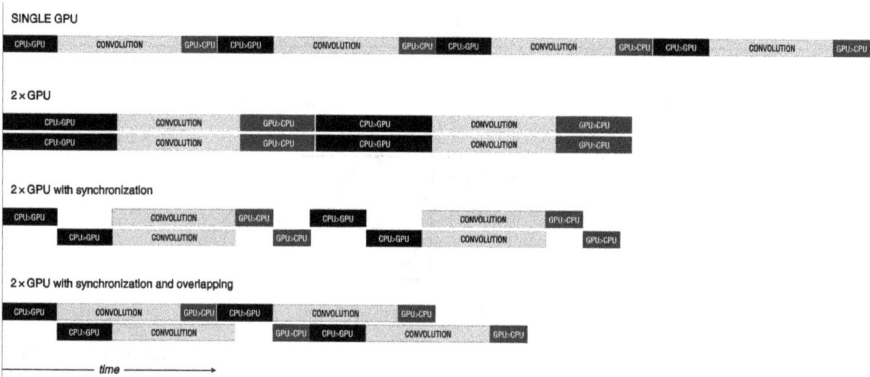

Figure 8. A model timeline of the algorithm workflow [21]. The dark boxes denote data transfers between CPU and GPU while the light boxes represent convolution computations. The first row shows the single-GPU implementation. The second row depicts parallel usage of two GPUs. The data transfers are performed concurrently but through a common bus, therefore they last twice longer. For the third row, the data transfers are synchronized so that only one transfer is made at a time. In the last row, the data transfers are overlapped with the convolution execution.

recommendable to overlap the data transfers with some computation phases in order to keep the implementation as efficient as possible.

To prove the importance of the overlapping, we provide a detailed analysis of the algorithm workflow. The overall computation time T required by the algorithm can be expressed as follows:

$$T = \max(t_p + t_d, t_a) + t_{h \to d} + \frac{t_{conv}}{P} + t_{d \to h} + t_c, \tag{33}$$

where t_p is the time required for the initial signal padding, t_d for decomposition, t_a for allocating memory and setting up FFT plans on GPU, $t_{h \to d}$ for data transfers from CPU to GPU (host to device), t_{conv} for the convolution including the FFT, recombination phase, point-wise multiplication, and the inverse FFT, $t_{d \to h}$ for data transfers from GPU to CPU (device to host) and finally t_c for composition. The number of end nodes (GPU cards) is denoted by P. It is evident that in accordance with the famous Amdahl's law [72], the speed-up achieved by multiple end nodes is limited to the only parallel phase of the algorithm which is the convolution itself. Now if the data are decomposed into d parts and sent to P end units and if $d > P > 1$, the data transfers can be overlapped with the convolution phase. This means that the real computation time is shorter than T as in Eq. (33). Eq. (33) can be hence viewed as the upper limit. The model example is shown in Fig. 8.

7.5. Algorithm comparison

In the previous text, we mentioned three approaches for the decomposition of the fast convolution: Tiling (decomposition in the time domain), the DIF-based, and the DIT-based algorithm. For fair comparison of the three, we compute the number of arithmetic operations, the number of data transfers, and the memory requirements per end node, with respect to the input signal length and the d parameter, i.e. the number of parts the data are divided into. As for the tiling method, the computation is based on Eq. (27) while setting $d = m = n$ (the optimum case). The results are shown in Table 1.

Method	# of operations	# of data transfers	Memory required per node
DIF	$(N^f + N^g) \left\lceil \frac{9}{2} \log_2(N^f + N^g) + 1 \right\rceil$	$3(N^f + N^g)$	$4(N^f + N^g)/d$
DIT	$(N^f + N^g) \left\lceil \frac{9}{2} \log_2(N^f + N^g) + 2 \right\rceil$	$(d+1)(N^f + N^g)$	$4(N^f + N^g)/d$
Tiling	$d(N^f + N^g) \left\lceil \frac{9}{2} \log_2(\frac{N^f + N^g}{d}) + 1 \right\rceil$	$(d+1)(N^f + N^g)$	$(N^f + N^g)/d$

Table 1. Methods for decomposition of the fast convolution and their requirements

To conclude the results, it can be noted that the tiling method is the best one in terms of memory demands. It requires 4× less memory per end node than the DIF-based and the DIT-based algorithms. On the other hand, both the number of the operations and the number of data transfers are dependent on the d parameter which is not the case of the DIF-based method. By dividing the data into more sub-parts, the memory requirements of the DIF-based algorithm decrease while the number of operations and memory transactions remain constant. Hence, the DIF-based algorithm can be generally more efficient than the tiling.

7.6. Applications and architectures

Both the tiling and the DIF-based algorithm can be used to allow the computation of the fast convolution in the applications where the convolving signals are multi-dimensional and/or contain too many samples to be handled efficiently on a single computer. We already mentioned the application of the optical microscopy data where the convolution is used to simulate the image degradation introduced by an optical system. Using the decomposition methods, the computation can be distributed over (a) a computer grid, (b) multiple CPU and

GPU units where CPU is usually provided with more memory, hence it is used as a central node for the (de)composition of the data.

8. Conclusions

In this text, we introduce the convolution as an important tool in both signal and image processing. In the first part, we mention some of the most popular applications it is employed in and recall its mathematical definition. Subsequently, we present a number of common algorithms for an efficient computation of the convolution on various architectures. The simplest approach—so-called naïve convolution—is to perform the convolution straightly using the definition. Although it is less efficient than other algorithms, it is the most general one and is popular in some specific applications where small convolution kernels are used, such as edge or object detection. If the convolution kernel is multi-dimensional and can be expressed as a convolution of several 1-D kernels, then the naïve convolution is usually replaced by its alternative, so-called separable convolution. The lowest time complexity can be achieved by using the recursive filtering. Here, the result of the convolution at each position can be obtained by applying a few arithmetical operations to the previous result. Besides the efficiency, the advantage is that these filters are suitable for streaming architectures such as FPGA. On the other hand, this method is generally not suitable for all convolution kernels as the recursive filters are often numerically unstable and inaccurate. The last algorithm present in the chapter is the fast convolution. According to the so-called convolution theorem, the convolution can be computed in the frequency domain by a simple point-wise multiplication of the Fourier transforms of the input signals. This approach is the most suitable for long signals and kernels as it yields generally the best time complexity. However, it has non-trivial memory demands caused by the fact that the input data need to be padded.

Therefore, in the second part of the chapter, we describe two approaches to reduce the memory requirements of the fast convolution. The first one, so-called tiling is performed in the spatial (time) domain. It is the most efficient with respect to the memory requirements. However, with a higher number of sub-parts the input data are divided into, both the number of arithmetical operations and the number of potential data transfers increase. Hence, in some applications or on some architectures (such as the desktop PC with one ore multiple graphics cards) where the overhead of data transfers is critical, one can use a different approach, based on the decomposition-in-frequency (DIF) algorithm which is widely known from the concept of the fast Fourier transform. We also mention the third method based on the decomposition-in-time (DIT) algorithm. However, the DIT-based algorithm is sub-efficient from every point of view so there is no reason for it to be used instead of the DIF-based one. In the end of the chapter, we also provide a detailed analysis of (i) the number of arithmetical operations, (ii) the number of data transfers, (iii) the memory requirements for each of the three methods.

As the convolution is one of the most extensively-studied operations in the signal processing, the list of the algorithms and implementations mentioned in this chapter is not and cannot be complete. Nevertheless, we tried to include those that we consider to be the most popular and widely-used. We also believe that the decomposition tricks which are described in the second part of the chapter and are the subject of the authors' original research can help readers to improve their own applications, regardless of target architecture.

Acknowledgments

This work has been supported by the Grant Agency of the Czech Republic (Grant No. P302/12/G157).

Author details

Pavel Karas* and David Svoboda

* Address all correspondence to: xkaras1@fi.muni.cz

Centre for Biomedical Image Analysis, Faculty of Informatics, Masaryk University, Brno, Czech Republic

References

[1] J. Jan. *Digital Signal Filtering, Analysis and Restoration (Telecommunications Series)*. INSPEC, Inc., 2000.

[2] S. W. Smith. *Digital Signal Processing*. Newnes, 2003.

[3] A. Foi. Noise estimation and removal in MR imaging: The variance stabilization approach. In *IEEE International Symposium on Biomedical Imaging: from Nano to Macro*, pages 1809—1814, 2011.

[4] J. R. Parker. *Algorithms for Image Processing and Computer Vision*. Wiley Publishing, 2nd edition, 2010.

[5] J. Canny. A computational approach to edge detection. *IEEE T-PAMI*, 8:769–698, 1986.

[6] D. H. Ballard. Generalizing the Hough transform to detect arbitrary shapes. *Pattern Recognition*, 13(2):111–122, 1981.

[7] D. Salomon. *Data Compression: The Complete Reference*. Springer-Verlag, 2007.

[8] R. C. Gonzalez and R. E. Woods. *Digital Image Processing*. Prentice Hall, 2002. ISBN: 0-201-18075-8.

[9] K. R. Castleman. *Digital Image Processing*. Prentice Hall, 1996.

[10] P. J. Verveer. Computational and optical methods for improving resolution and signal quality in fluorescence microscopy. 1998. PhD Thesis.

[11] A. Lehmussola, J. Selinummi, P. Ruusuvuori, A. Niemistö, and O. Yli-Harja. Simulating fluorescent microscope images of cell populations. In *Proceedings of the 27th Annual International Conference of the IEEE Engineering in Medicine and Biology Society (EMBC'05)*, pages 3153–3156, 2005.

[12] D. Svoboda, M. Kozubek, and S. Stejskal. Generation of Digital Phantoms of Cell Nuclei and Simulation of Image Formation in 3D Image Cytometry. *Cytometry part A*, 75A(6):494–509, JUN 2009.

[13] W. K. Pratt. *Digital Image Processing*. Wiley, 3rd edition edition, 2001.

[14] T. Bräunl. *Parallel Image Processing*. Springer, 2001.

[15] H.-M. Yip, I. Ahmad, and T.-C. Pong. An Efficient Parallel Algorithm for Computing the Gaussian Convolution of Multi-dimensional Image Data. *The Journal of Supercomputing*, 14(3):233–255, 1999. ISSN: 0920-8542.

[16] O. Schwarzkopf. Computing Convolutions on Mesh-Like Structures. In *Proceedings of the Seventh International Parallel Processing Symposium*, pages 695–699, 1993.

[17] S. Kadam. Parallelization of Low-Level Computer Vision Algorithms on Clusters. In *AMS '08: Proceedings of the 2008 Second Asia International Conference on Modelling & Simulation (AMS)*, pages 113–118, Washington, DC, USA, 2008. IEEE Computer Society. ISBN: 978-0-7695-3136-6.

[18] B. Jähne. *Digital Image Processing*. Springer, 5th edition edition, 2002.

[19] Robert Hummel and David Loew. Computing Large-Kernel Convolutions of Images. Technical report, New York University, Courant Institute of Mathematical Sciences, 1986.

[20] R. N. Bracewell. *Fourier Analysis and Imaging*. Springer, 2006.

[21] P. Karas and D. Svoboda. Convolution of large 3D images on GPU and its decomposition. *EURASIP Journal on Advances in Signal Processing*, 2011(1):120, 2011.

[22] A.V. Oppenheim, R.W. Schafer, J.R. Buck, et al. *Discrete-time signal processing*, volume 2. Prentice hall Upper Saddle River^ eN. JNJ, 1989.

[23] D. Svoboda. Efficient computation of convolution of huge images. *Image Analysis and Processing–ICIAP 2011*, pages 453–462, 2011.

[24] R. G. Shoup. Parameterized convolution filtering in an FPGA. In *Selected papers from the Oxford 1993 international workshop on field programmable logic and applications on More FPGAs*, pages 274–280, Oxford, UK, UK, 1994. Abingdon EE&CS Books.

[25] A. Benedetti, A. Prati, and N. Scarabottolo. Image convolution on FPGAs: the implementation of a multi-FPGA FIFO structure. In *Euromicro Conference, 1998. Proceedings. 24th*, volume 1, pages 123–130 vol.1, Aug 1998.

[26] S. Perri, M. Lanuzza, P. Corsonello, and G. Cocorullo. A high-performance fully reconfigurable FPGA-based 2D convolution processor. *Microprocessors and Microsystems*, 29(8—9):381–391, 2005. Special Issue on FPGAs: Case Studies in Computer Vision and Image Processing.

[27] A. Herout, P. Zemcik, M. Hradis, R. Juranek, J. Havel, R. Josth, and L. Polok. *Low-Level Image Features for Real-Time Object Detection*. InTech, 2010.

[28] H. Shan and N. A. Hazanchuk. Adaptive Edge Detection for Real-Time Video Processing using FPGAs. Application notes, Altera Corporation, 2005.

[29] J.D. Owens, D. Luebke, N. Govindaraju, M. Harris, J. Krüger, A.E. Lefohn, and T.J. Purcell. A Survey of General-Purpose Computation on Graphics Hardware. pages 21–51, August 2005.

[30] D. Castaño-Díez, D. Moser, A. Schoenegger, S. Pruggnaller, and A. S. Frangakis. Performance evaluation of image processing algorithms on the GPU. *Journal of Structural Biology*, 164(1):153–160, 2008.

[31] S. Ryoo, C.I. Rodrigues, S.S. Baghsorkhi, S.S. Stone, D.B. Kirk, and Wen-mei W. Hwu. Optimization principles and application performance evaluation of a multithreaded GPU using CUDA. In *PPoPP '08: Proceedings of the 13th ACM SIGPLAN Symposium on Principles and practice of parallel programming*, pages 73–82, New York, NY, USA, 2008. ACM.

[32] NVIDIA Developer Zone. http://developer.nvidia.com/category/zone/cuda-zone, Apr 2012.

[33] Khronos Group. OpenCL. http://www.khronos.org/opencl/, 2011.

[34] CUDA Downloads. http://developer.nvidia.com/cuda-downloads, Apr 2012.

[35] V. Podlozhnyuk. Image Convolution with CUDA. http://developer.download.nvidia.com/assets/cuda/files/convolutionSeparable.pdf, Jun 2007.

[36] NVIDIA Performance Primitives. http://developer.nvidia.com/npp, Feb 2012.

[37] Y. Luo and R. Duraiswami. Canny edge detection on NVIDIA CUDA. In *Computer Vision and Pattern Recognition Workshops, 2008. CVPRW '08. IEEE Computer Society Conference on*, pages 1–8, Jun 2008.

[38] K. Ogawa, Y. Ito, and K. Nakano. Efficient Canny Edge Detection Using a GPU. In *Networking and Computing (ICNC), 2010 First International Conference on*, pages 279–280, Nov 2010.

[39] A. Herout, R. Jošth, R. Juránek, J. Havel, M. Hradiš, and P. Zemčík. Real-time object detection on CUDA. *Journal of Real-Time Image Processing*, 6:159–170, 2011. 10.1007/s11554-010-0179-0.

[40] Ke Zhang, Jiangbo Lu, G. Lafruit, R. Lauwereins, and L. Van Gool. Real-time accurate stereo with bitwise fast voting on CUDA. In *IEEE 12th International Conference on Computer Vision Workshops (ICCV Workshops)*, pages 794 –800, Oct 2009.

[41] Wei Chen, M. Beister, Y. Kyriakou, and M. Kachelries. High performance median filtering using commodity graphics hardware. In *Nuclear Science Symposium Conference Record (NSS/MIC), 2009 IEEE*, pages 4142–4147, Nov 2009.

[42] S. Che, M. Boyer, J. Meng, D. Tarjan, J.W. Sheaffer, and K. Skadron. A performance study of general-purpose applications on graphics processors using CUDA. *Journal of parallel and distributed computing*, 68(10):1370–1380, 2008.

[43] Zhaoyi Wei, Dah-Jye Lee, B. E. Nelson, J. K. Archibald, and B. B. Edwards. FPGA-Based Embedded Motion Estimation Sensor. 2008.

[44] XinXin Wang and B.E. Shi. GPU implemention of fast Gabor filters. In *Circuits and Systems (ISCAS), Proceedings of 2010 IEEE International Symposium on*, pages 373–376, Jun 2010.

[45] O. Fialka and M. Čadík. FFT and Convolution Performance in Image Filtering on GPU. In *Information Visualization, 2006. IV 2006. Tenth International Conference on*, pages 609–614, 2006.

[46] J. S. Jin and Y. Gao. Recursive implementation of LoG filtering. *Real-Time Imaging*, 3(1):59–65, February 1997.

[47] R. Deriche. Using Canny's criteria to derive a recursively implemented optimal edge detector. *The International Journal of Computer Vision*, 1(2):167–187, May 1987.

[48] I. T. Young and L. J. van Vliet. Recursive implementation of the Gaussian filter. *Signal Processing*, 44(2):139–151, 1995.

[49] F.G. Lorca, L. Kessal, and D. Demigny. Efficient ASIC and FPGA implementations of IIR filters for real time edge detection. In *Image Processing, 1997. Proceedings., International Conference on*, volume 2, pages 406–409 vol.2, Oct 1997.

[50] R.D. Turney, A.M. Reza, and J.G.R. Delva. FPGA implementation of adaptive temporal Kalman filter for real time video filtering. In *Acoustics, Speech, and Signal Processing, 1999. Proceedings., 1999 IEEE International Conference on*, volume 4, pages 2231–2234 vol.4, Mar 1999.

[51] J. Diaz, E. Ros, F. Pelayo, E.M. Ortigosa, and S. Mota. FPGA-based real-time optical-flow system. *Circuits and Systems for Video Technology, IEEE Transactions on*, 16(2):274–279, Feb 2006.

[52] J. Robelly, G. Cichon, H. Seidel, and G. Fettweis. Implementation of recursive digital filters into vector SIMD DSP architectures. In *Acoustics, Speech, and Signal Processing, 2004. Proceedings. (ICASSP '04). IEEE International Conference on*, volume 5, pages V – 165–8 vol.5, may 2004.

[53] F. Trebien and M. Oliveira. Realistic real-time sound re-synthesis and processing for interactive virtual worlds. *The Visual Computer*, 25:469–477, 2009. 10.1007/s00371-009-0341-5.

[54] E.O. Brigham and R.E. Morrow. The fast Fourier transform. *Spectrum, IEEE*, 4(12):63–70, 1967.

[55] H.J. Nussbaumer. Fast Fourier transform and convolution algorithms. *Berlin and New York, Springer-Verlag(Springer Series in Information Sciences.*, 2, 1982.

[56] Donald Fraser. Array Permutation by Index-Digit Permutation. *J. ACM*, 23(2):298–309, April 1976.

[57] G.U. Ramos. Roundoff error analysis of the fast Fourier transform. *Math. Comp*, 25:757–768, 1971.

[58] R. N. Bracewell. *The Fourier Transform and Its Applications*. McGraw-Hill, 3rd edition, 2000.

[59] F.J. Harris. On the use of windows for harmonic analysis with the discrete Fourier transform. *Proceedings of the IEEE*, 66(1):51–83, 1978.

[60] J.W. Cooley and J.W. Tukey. An algorithm for the machine calculation of complex Fourier series. *Math. Comput*, 19(90):297–301, 1965.

[61] M. Frigo and S.G. Johnson. The Fastest Fourier Transform in the West. 1997.

[62] M. Frigo and S.G. Johnson. benchFFT. http://www.fftw.org/benchfft/, 2012.

[63] Intel Integrated Performance Primitives. http://software.intel.com/en-us/articles/intel-ipp/, 2012.

[64] Intel Integrated Performance Primitives. http://software.intel.com/en-us/articles/intel-mkl/, 2012.

[65] A. Nukada, Y. Ogata, T. Endo, and S. Matsuoka. Bandwidth intensive 3-D FFT kernel for GPUs using CUDA. In *SC '08: Proceedings of the 2008 ACM/IEEE conference on Supercomputing*, pages 1–11, Piscataway, NJ, USA, 2008. IEEE Press.

[66] N.K. Govindaraju, B. Lloyd, Y. Dotsenko, B. Smith, and J. Manferdelli. High performance discrete Fourier transforms on graphics processors. In *SC '08: Proceedings of the 2008 ACM/IEEE conference on Supercomputing*, pages 1–12, Piscataway, NJ, USA, 2008. IEEE Press.

[67] R. Tsuchiyama, T. Nakamura, T. Iizuka, A. Asahara, and S. Miki. The OpenCL Programming Book. *Group*, 2009.

[68] Z. Li, H. Sorensen, and C. Burrus. FFT and convolution algorithms on DSP microprocessors. In *Acoustics, Speech, and Signal Processing, IEEE International Conference on ICASSP'86.*, volume 11, pages 289–292. IEEE, 1986.

[69] I.S. Uzun, A. Amira, and A. Bouridane. FPGA implementations of fast Fourier transforms for real-time signal and image processing. In *Vision, Image and Signal Processing, IEE Proceedings-*, volume 152, pages 283–296. IET, 2005.

[70] M. Heideman, D. Johnson, and C. Burrus. Gauss and the history of the fast Fourier transform. *ASSP Magazine, IEEE*, 1(4):14–21, Oct 1984. ISSN: 0740-7467.

[71] P. Karas, D. Svoboda, and P. Zemčík. GPU Optimization of Convolution for Large 3-D Real Images. In *Advanced Concepts for Intelligent Vision Systems (ACIVS), 2012*. Springer, 2012. Accepted.

[72] G.M. Amdahl. Validity of the single processor approach to achieving large scale computing capabilities. In *Proceedings of the April 18-20, 1967, spring joint computer conference*, pages 483–485. ACM, 1967.

Advanced Architectures and Implementations

A Digital Signal Processing Architecture for Soft-Output MIMO Lattice Reduction Aided Detection

Alan T. Murray and Steven R. Weller

Additional information is available at the end of the chapter

1. Introduction

Many wireless communication standards now include the use of multiple transmit and receive antennas as a means of achieving increased throughput or spectral efficiency, including LTE, WiMAX and WiFi (IEEE 802.11n). The task of a detector for a multi-input multi-output (MIMO) communications channel is to separate the spatially mixed and noise-corrupted data streams, and to produce reliable estimates of the transmitted bits. The brute-force maximum-likelihood (ML) detector provides optimal error-rate performance, but is computationally infeasible when either dense symbol constellations or large numbers of antennas are used. Hardware implementation of ML receivers is therefore very challenging, leading to linear detectors based on well-known approaches such as zero forcing (ZF) or minimum mean-square error (MMSE) detection, or nonlinear methods such as successive interference cancellation (SIC), which offer manageable receiver complexity at the expense of highly suboptimal error-rate performance.

One powerful class of receivers which have been developed over the past decade is based on the highly developed mathematical theory of point lattices, which are periodic arrangements of discrete points. The basic idea is to consider the distortion introduced by the noise-free part of a MIMO channel as a representation of a lattice, then to perform suboptimal detection on an "improved" representation of the channel matrix based derived from a "reduced" lattice. The suitably reduced lattice facilitates the search for the lattice point closest to the received vector, shifting most of the computational complexity to a pre-processing step before linear detection. Such lattice reduction aided detection (LRAD) based approaches to MIMO receiver design have significantly closed the gap between feasible yet high-performance MIMO detection, and optimal (but impractical) ML detection.

To date, most LRAD-based MIMO detectors produce hard outputs, in which an estimate of the most likely vector of transmitted symbols is generated. For high-performance wireless communication systems, however, it is commonplace that the information transmitted over the air is coded, thereby containing not only raw data, but also the redundant information needed to perform forward error correction (FEC) at the receiver. State-of-the-art FEC codes such as turbo codes and low-density parity-check (LDPC) codes [1], which require estimates of the *probability* that a given transmitted bit was a 1 or a 0, therefore call for *soft output* detectors. The extension of hard-output LRAD detectors to the soft-output case is therefore of high practical relevance, but also recognized as a difficult problem [2, p16]. In this chapter, we present what is believed to be the first digital signal processing (DSP) implementation of a soft-output lattice reduction aided MIMO detector, based on an approach to MIMO detection known as subspace LRAD (SLRAD) proposed by Windpassinger [3, 4].

The chapter is organized as follows. In Section 2 we present the wireless MIMO system model, with an emphasis on how transmitted symbols are drawn from point sets consistent with the lattice theoretic approach to follow. In Section 3 we formally define lattices, and present the most celebrated algorithm for lattice reduction, known as the Lenstra-Lenstra-Lovász (LLL) algorithm. We then show how hard-output lattice-based detection can be used in conjunction with commonly used linear MIMO detectors in Section 4. In Section 5 we outline Windpassinger's subspace-based approach to LRAD in which a list of candidate symbols is produced, thereby facilitating soft-output LRAD. Finally in Section 6 we present a detailed description of our hardware implementation of a soft-output lattice reduction aided MIMO detector.

2. System model

We consider a MIMO wireless communication system with n_T transmit and n_R receive antennas. The complex baseband model for this MIMO system is

$$\mathbf{y} = \mathbf{Hx} + \mathbf{n}, \tag{1}$$

where $\mathbf{y} \in \mathbb{C}^{n_R}$ is the received vector, $\mathbf{H} \in \mathbb{C}^{n_R \times n_T}$ is the channel matrix, $\mathbf{n} \in \mathbb{C}^{n_R}$ is the channel noise, and $\mathbf{x} \in \mathbb{C}^{n_T}$ is the vector of transmitted symbols, as shown in Fig. 1.

We assume that the noise $\mathbf{n} \triangleq [n_1, n_2, \ldots, n_{n_R}]^T$ contains independent and identically distributed (i.i.d.) elements $n_m \sim \mathcal{CN}(0, \sigma^2)$, $m = 1, \ldots, n_R$. The channel matrix $\mathbf{H}$ has i.i.d. entries $h_{m,n} \sim \mathcal{CN}(0, 1)$, for $m = 1, \ldots, n_R$ and $n = 1, \ldots, n_T$, where it is assumed that there are at least as many receive antennas as transmit antennas: $n_R \geq n_T$.

An uncorrelated Rayleigh fading propagation environment is therefore assumed in this chapter, though it should be noted that lattice reduction aided detection receivers similar to those presented later in this chapter have been proposed for environments in which there is either temporal [5] or frequency-selective [6] fading.

The task of the MIMO receiver is to recover $\mathbf{x}$ from $\mathbf{y}$, based on knowledge of both the channel realization $\mathbf{H}$ and the channel noise variance σ^2.

The vector of transmitted symbols is denoted $\mathbf{x} \triangleq [x_1, x_2, \ldots, x_{n_T}]^T$. In this chapter we restrict attention to transmit symbols drawn from finite sets of points, known as *constellations*,

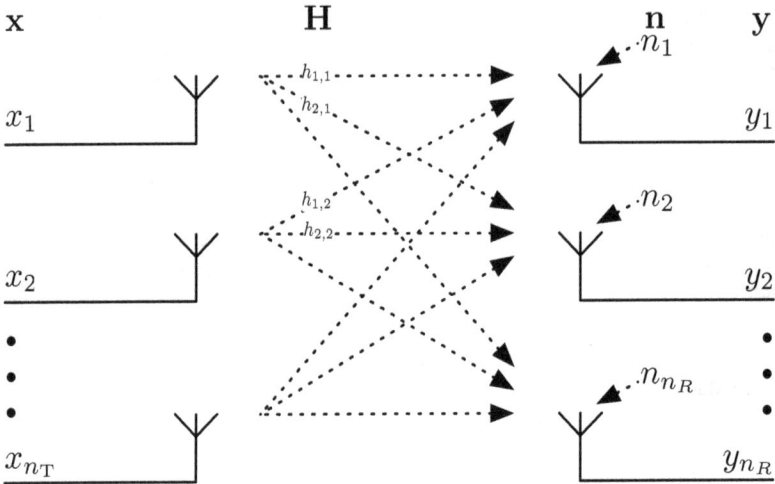

Figure 1. MIMO Wireless Channel

drawn from a square grid, and in particular the quadrature phase-shift keying (QPSK) and, 16-quadrature amplitude modulation (16-QAM) and 64-QAM constellations depicted in Fig. 2. We do not consider non-rectangular constellations, such as 8-PSK, due to an inherent incompatibility with the lattice-theoretic framework exploited by lattice reduction aided detection, and also the limited applicability of non-rectangular constellations in emerging wireless communication standards.

The symbol transmitted from the nth antenna, denoted x_n, is drawn from a constellation $\mathcal{A}_n$:

$$x_n \in \sqrt{E_{sn}}\mathcal{A}_n, \tag{2}$$

where the scalar E_{sn} is the average transmitted symbol power. We define the vector $\mathbf{E_s} \triangleq [E_{s1}, E_{s2}, \ldots, E_{sn_T}]$ so that

$$\mathbb{E}\left[\mathbf{x}\mathbf{x}^H\right] = \text{diag}\left(\mathbf{E_s}\right). \tag{3}$$

The selection of $\mathbf{E_s}$ depends on the particular objective of transmit power scaling and indeed varies in practical implementations. In this chapter, to enable fair comparison between systems employing differing modulation formats, we constrain average unity power per information bit ($E_b = 1$).

The constellations considered in this chapter are formed from a subset of scaled and shifted Gaussian integers $\mathbb{Z}[i] \triangleq \{a + ib \mid a, b \in \mathbb{Z}\}$ [7, p. 230]:

$$\mathbb{X} \triangleq \left\{a + ib + \frac{1+i}{2} \mid a, b \in \mathbb{Z}\right\}. \tag{4}$$

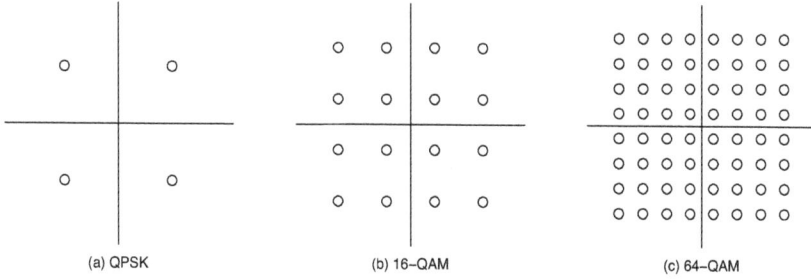

Figure 2. The three constellations used in this chapter

In this chapter, we restrict attention to the three subsets $\mathcal{X}_n \subset \mathbb{X}$ shown in Fig. 3, where the introduction of the offset term in (4) maintains symmetry of each constellation with respect to the axes. We refer to constellations formed in this manner as *Gaussian integer constellations*.

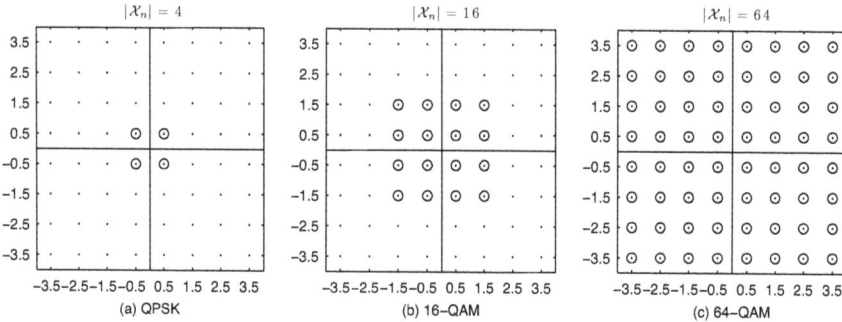

Figure 3. Three Gaussian integer constellations

The constellation $\mathcal{A}_n$ with elements $\alpha_n \in \mathcal{A}_n$ employed at the nth transmit antenna is:

$$\mathcal{A}_n = \frac{\mathcal{X}_n}{\sqrt{c_n}}, \tag{5}$$

where c_n is the average energy of $\mathcal{X}_n$. Dividing each element of $\mathcal{X}_n$ by $\sqrt{c_n}$ ensures that $\mathcal{A}_n$ has unity average energy and is referred to as normalized constellations. For square QAM constellations such as those in Fig. 2, $c_n = (|\mathcal{X}_n| - 1)/6$.

It is important to note that we deliberately allow each transmit antenna to be independently mapped to a constellation set. In summary, the transmitted symbols x_n are formed by scaling of the elements $\bar{x}_n \in \mathcal{X}_n$:

$$x_n = \sqrt{\frac{E_{sn}}{c_n}}\bar{x}_n. \tag{6}$$

The effect of a given channel realization $\mathbf{H}$ is to rotate and stretch (or contract) the axes of the otherwise square decision regions of the optimal, maximum-likelihood (ML) receiver. The error probability of a detector is determined by the distance of constellation points (mapped by $\mathbf{H}$) from the associated decision boundaries. The essential idea of LR-aided detectors is to obtain a "more orthogonal" representation for the channel realization $\mathbf{H}$, before detection using a low-complexity (sub-optimal) receiver. In the following section we make these ideas precise, drawing on the well-established mathematical literature on point lattices to formalize what is meant by the notion of a "more orthogonal" representation, and how it can be achieved and quantified.

3. Lattice reduction

3.1. Lattices

A *complex lattice* consists of all linear combinations of the set of linearly independent basis column vectors $\mathbf{b}_k, 1 \leq k \leq M$ of the basis matrix $\mathbf{B} \in \mathbb{C}^{N \times M}, M \leq N$. A complex lattice formed from basis matrix $\mathbf{B}$ is therefore the set of points

$$\mathcal{L}(\mathbf{B}) \triangleq \left\{ \sum_{k=1}^{M} s_k \mathbf{b}_k \mid s_k \in \mathbb{Z}[i] \right\},$$

where $\mathbb{Z}[i] \triangleq \{a + ib \mid a, b \in \mathbb{Z}\}$ is the ring of Gaussian integers [7].

The number of possible bases for a given lattice $\mathcal{L}$ is infinite, since any basis $\tilde{\mathbf{B}} = \mathbf{BT}$ forms the same lattice $\mathcal{L}(\tilde{\mathbf{B}}) = \mathcal{L}(\mathbf{B})$ when the transformation matrix $\mathbf{T}$ is unimodular, i.e. $\det(\mathbf{T}) = \pm 1$ and $\mathbf{T} \in \mathbb{Z}[i]^{M \times M}$. Finding a basis in which the basis vectors are (roughly speaking) reasonably short and almost orthogonal is known as lattice *basis reduction*, which we now describe formally.

3.2. Lenstra-Lenstra-Lovász (LLL) algorithm

The Lenstra-Lenstra-Lovász (LLL) algorithm was originally published as a lattice reduction algorithm operating on real-valued matrices [8]. Many works use the real decomposition of the complex-valued MIMO transmission model [3, 9]. Lattice reduction methods can operate on both real and complex integer lattices and in particular the LLL algorithm has been extended for complex lattice reduction [10]. The complex LLL (CLLL) algorithm can be summarized as follows. We make the following definitions:

- $\mathcal{H}_i$ is the squared Euclidean norm of the orthogonal vectors produced by the Gram-Schmidt orthogonalization (GSO) of $\mathbf{H}$

- μ_{ij} is the ratio of the length of the orthogonal projection of the ith basis onto the jth orthogonal vector and the length of the jth orthogonal vector
- $\mathbf{H}^i{}_L$ and $\mathbf{T}^i$ represent the values of the reduced basis and transform after the ith step of the LLL algorithm
- Initially, $\mathbf{H}_L{}^0 = \mathbf{H}$ and $\mathbf{T}^0 = \mathbf{I}_{n_T}$
- k is the index of the current column of $\mathbf{H}$ being processed such that $2 \leq k \leq n_T$

The LLL algorithm consists of three basic steps:

1. $\mathcal{H}$ and μ are computed using a modified GSO procedure [11]
2. Size reduction aims to make basis vectors shorter and more orthogonal by asserting the condition that $|\Re(\mu_{k,j})| \leq 0.5$ and $|\Im(\mu_{k,j})| \leq 0.5$ for all $j < k$
3. Basis vectors $\mathbf{h}_{k-1}$ and $\mathbf{h}_k$ are swapped if a so-called *swapping condition* is satisfied such that size reduction can be repeated to make basis vectors shorter

Size reduction and basis vector swapping iterates until the swapping condition is no longer satisfied by any pair of $\mathbf{h}_{k-1}$ and $\mathbf{h}_k$. The resultant basis is then said to be *reduced*. The swapping condition for LLL reduction, also called the *Lovász condition*, is:

$$\mathcal{H}_k < (\delta - |\mu_{k,k-1}|^2)\mathcal{H}_{k-1}, \tag{7}$$

where δ satisfying $\frac{1}{4} < \delta < 1$ is a factor selected to achieve an acceptable quality-complexity trade off [8].

After each swapping step, $\mathcal{H}_{k-1}$, $\mathcal{H}_k$ and some of the $\mu_{i,j}$ values needed to be updated. Techniques can be employed to minimize the number and frequency of recalculations of $\mathcal{H}$ and μ elements [11]. The LLL algorithm is detailed in Algorithm 1.

Example 3.1. *Suppose $\delta = \frac{3}{4}$ and*

$$\mathbf{H} = \begin{bmatrix} 0.75 & -0.5 \\ 0.5 & -0.5 \end{bmatrix}.$$

Then

$$\mathbf{H}^0{}_L = \begin{bmatrix} 0.75 & -0.5 \\ 0.5 & -0.5 \end{bmatrix} \text{ and } \mathbf{T}^0 = \begin{bmatrix} 1 & 0 \\ 0 & 1 \end{bmatrix},$$

and from the modified GSO

$$\mu = \begin{bmatrix} 1.0000 & 0.0000 \\ -0.7692 & 1.0000 \end{bmatrix} \text{ and } \mathcal{H} = \begin{bmatrix} 0.8125 \\ 0.0192 \end{bmatrix}.$$

Starting with columns 1 and 2, as $|\mu_{2,1}| > 0.5$, size reduction is performed on these columns adding the first column to the second and yielding the following partially reduced matrix and corresponding transform:

Algorithm 1 $[\mathbf{H}, \mathbf{T}] \Leftarrow \mathrm{LLL}(\mathbf{H}, \delta)$

Input: $\mathbf{H} \in \mathbb{C}^{n \times m}$ and $\delta \in \mathbb{R}^n$

$\quad \mathbf{T} \Leftarrow \mathbf{I}_n, k \Leftarrow 2$

$\quad$ **for** $j = 1$ to n **do**

$\quad\quad \mathcal{H}_j = \left\langle \mathbf{H}_j, \mathbf{H}_j \right\rangle$

$\quad$ **end for**

$\quad$ **for** $j = 1$ to n **do**

$\quad\quad$ **for** $i = j + 1$ to n **do**

$\quad\quad\quad \mu_{i,j} \Leftarrow \frac{1}{\mathcal{H}_j} \left(\left\langle \mathbf{h}_i, \mathbf{h}_j \right\rangle - \sum_{k=1}^{j-1} \mu_{j,k}^H \mu_{i,k} \mathcal{H}_k \right)$

$\quad\quad\quad \mathcal{H}_i \Leftarrow \mathcal{H}_i - |\mu_{i,j}|^2 \mathcal{H}_j$

$\quad\quad$ **end for**

$\quad$ **end for**

$\quad$ **while** $k \leq n$ **do**

$\quad\quad [\mathbf{H}, \mathbf{T}, \mu] \Leftarrow \mathrm{Reduce}\,(\mathbf{H}, \mathbf{T}, \mu, k, k-1)$ $\qquad\qquad$ // Size Reduction

$\quad\quad$ **if** $\mathcal{H}_k < (\delta - |\mu_{k,k-1}|^2)\mathcal{H}_{k-1}$ **then** $\qquad\qquad$ // Lovász condition check

$\quad\quad\quad$ Swap columns k and $k - 1$ of $\mathbf{H}$ and $\mathbf{T}$

$\quad\quad\quad$ Update $\mathcal{H}$ and μ where $\dot{\mathcal{H}}$ and $\dot{\mu}$ denote the new values

$\quad\quad\quad \dot{\mathcal{H}}_{k-1} = \mathcal{H}_k + |\mu_{k,k-1}|^2 \mathcal{H}_{k-1}$

$\quad\quad\quad \dot{\mu}_{k,k-1} = \mu_{k,k-1}^H \left(\frac{\mathcal{H}_{k-1}}{\dot{\mathcal{H}}_{k-1}} \right)$

$\quad\quad\quad \dot{\mathcal{H}}_k = \left(\frac{\mathcal{H}_{k-1}}{\dot{\mathcal{H}}_{k-1}} \right) \mathcal{H}_k$

$\quad\quad\quad \dot{\mu}_{i,k-1} = \mu_{i,k-1} \dot{\mu}_{k,k-1} + \mu_{i,k} \left(\frac{\mathcal{H}_k}{\dot{\mathcal{H}}_{k-1}} \right)$ $\qquad$ // $i = k+1$ to n

$\quad\quad\quad \dot{\mu}_{i,k} = \mu_{i,k-1} - \mu_{i,k} \mu_{k,k-1}$ $\qquad\qquad\qquad$ // $i = k+1$ to n

$\quad\quad\quad \dot{\mu}_{k-1,j} = \mu_{k,j}$ $\qquad\qquad\qquad\qquad\qquad$ // $j = 1$ to $k-2$

$\quad\quad\quad \dot{\mu}_{k,j} = \mu_{k-1,j}$ $\qquad\qquad\qquad\qquad\qquad$ // $j = 1$ to $k-2$

$\quad\quad\quad k = \max(2, k-1)$

$\quad\quad$ **else**

$\quad\quad\quad$ **for** $j = k - 2$ **downto** 1 **do**

$\quad\quad\quad\quad [\mathbf{H}, \mathbf{T}, \mu] \Leftarrow \mathrm{Reduce}\,(\mathbf{H}, \mathbf{T}, \mu, k, j)$ $\qquad$ // Size Reduction

$\quad\quad\quad$ **end for**

$\quad\quad\quad k = k + 1$

$\quad\quad$ **end if**

$\quad$ **end while**

$$\mathbf{H}^1{}_L = \begin{bmatrix} 0.75 & 0.25 \\ 0.5 & 0 \end{bmatrix} \text{ and } \mathbf{T}^1 = \begin{bmatrix} 1 & 1 \\ 0 & 1 \end{bmatrix},$$

$$\mu = \begin{bmatrix} 1.0000 & 0.0000 \\ 0.2308 & 1.0000 \end{bmatrix} \text{ and } \mathcal{H} = \begin{bmatrix} 0.8125 \\ 0.0192 \end{bmatrix}.$$

Algorithm 2 $[\mathbf{H}, \mathbf{T}, \mu] \Leftarrow \text{Reduce}\,(\mathbf{H}, \mathbf{T}, \mu, k, j)$

---

if $|\Re\left(\mu_{k,j}\right)| > \frac{1}{2}$ **or** $|\Im\left(\mu_{k,j}\right)| > \frac{1}{2}$ **then**

$\quad c \Leftarrow \lfloor \mu_{k,j} \rceil$

$\quad \mathbf{H}_k \Leftarrow \mathbf{H}_k - c\mathbf{H}_j$

$\quad \mathbf{T}_k \Leftarrow \mathbf{T}_k - c\mathbf{T}_j$

$\quad$ **for** $l = 1$ **to** j **do**

$\quad\quad \mu_{k,l} \Leftarrow \mu_{k,l} - c\mu_{j,l}$

$\quad$ **end for**

end if

Next the Lovász condition is checked and, since $\mathcal{H}_2 < (\delta - |\mu_{2,1}|^2)\mathcal{H}_1$, the two columns are swapped, yielding:

$$\mathbf{H}^2{}_L = \begin{bmatrix} 0.25 & 0.75 \\ 0 & 0.5 \end{bmatrix} \text{ and } \mathbf{T}^2 = \begin{bmatrix} 1 & 1 \\ 1 & 0 \end{bmatrix},$$

$$\mu = \begin{bmatrix} 1.0000 & 0.0000 \\ 3.0000 & 1.0000 \end{bmatrix} \text{ and } \mathcal{H} = \begin{bmatrix} 0.0625 \\ 0.2500 \end{bmatrix}.$$

Size reduction is then performed on the columns once more; this time by subtracting three times the first column from the second we have:

$$\mathbf{H}_L = \begin{bmatrix} 0.25 & 0 \\ 0 & 0.5 \end{bmatrix} \text{ and } \mathbf{T} = \begin{bmatrix} 1 & -2 \\ 1 & -3 \end{bmatrix},$$

$$\mu = \begin{bmatrix} 1.0000 & 0.0000 \\ 0.0000 & 1.0000 \end{bmatrix} \text{ and } \mathcal{H} = \begin{bmatrix} 0.0625 \\ 0.2500 \end{bmatrix}.$$

The Lovász condition (7) is now satisfied, and the algorithm terminates. ∎

3.3. Orthogonality defect

The orthogonality of a matrix $\mathbf{H}$ can be quantified using the *orthogonality defect*, defined as [4, §4.6.2]:

$$\delta(\mathbf{H}) = \frac{\prod_{k=1}^{n_T} \|\mathbf{h}_k\|}{\left|\sqrt{\det(\mathbf{H}^H\mathbf{H})}\right|}, \tag{8}$$

where $\mathbf{h}_k$ is the kth column of $\mathbf{H}$, $\delta(\mathbf{H}) \geq 1$ for all $\mathbf{H}$ and $\delta(\mathbf{H}) = 1$ if and only if the columns of $\mathbf{H}$ are orthogonal. When the number of columns and rows of $\mathbf{H}$ are equal, the denominator can be simplified to $|\det(\mathbf{H})|$. From (8), matrices with correlated columns or larger column norms will result in higher orthogonality defects. This also causes their inverse or generalized inverse to have larger row norms, leading to noise enhancement. As will be shown in Section 4, matrices with a lower orthogonality defect therefore induce less noise enhancement in ZF- or MMSE-based detectors as the probability of error, for example as calculated in (15), can be reduced.

To illustrate the impact of lattice reduction on orthogonality defect, we generated 10^6 randomly chosen $\mathbf{H} \in \mathbb{C}^{4\times4}$, and computed the lattice reduced equivalent $\mathbf{H}_L$. The orthogonality defect was calculated using (8) both before and after lattice reduction. The results are presented in the form of cumulative distributions in Fig. 4, where the effect of lattice reduction on orthogonality defect is clearly apparent. Lattice basis reduction has also been shown to improve matrix conditioning [12]. It is this improvement that reduces noise enhancement in linear detection methods and reduces the error rate of LRAD-based systems.

Numerous researchers have investigated and compared the application of various lattice reduction algorithms for MIMO detection. In addition to the LLL algorithm, these include Korkine–Zolotarev (KZ) [13], and Seysen's [14] lattice reduction algorithms; see [2] and the references therein for applications to MIMO detection. In this chapter we restrict attention to the LLL algorithm, since numerous simulation studies suggest that lattice-reduction-aided detection is well suited to low-complexity MIMO receivers when large constellations are used [15, 16].

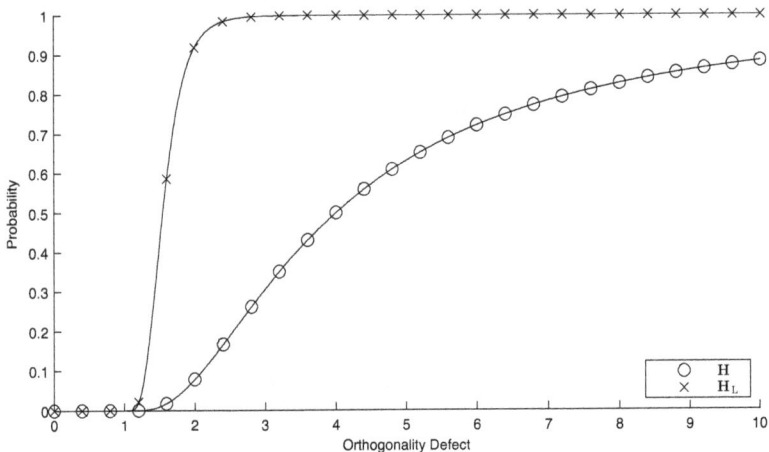

Figure 4. Cumulative distributions of the orthogonality defect for non-reduced and reduced basis channel matrices

4. Hard detection using lattice reduction

Detectors which output an estimate of the most likely vector of transmitted symbols are said to be *hard output* detectors. Hard estimates are denoted $\hat{\mathbf{b}}$ for bit vector estimates and $\hat{\mathbf{x}}$ for symbol vector estimates. Detectors which generate not just a vector of bit estimates but also an estimate of the *probability* that a given transmitted bit was a 1 or a 0 are said to be *soft output* detectors. Soft output detectors provide a significant benefit when combined with channel coding schemes which make use of soft information, such as turbo codes or low-density parity-check (LDPC) codes, but typically increase receiver complexity by a significant degree.

4.1. Maximum-Likelihood detection

The maximum-likelihood (ML) detector selects from the set of possible transmitted symbol vectors $\mathbf{x} \in \mathcal{A}^{n_T}$ the vector $\hat{\mathbf{x}}_{\mathrm{ML}}$ which minimizes the Euclidean distance to the receive vector:

$$\hat{\mathbf{x}}_{\mathrm{ML}} = \arg\min_{\mathbf{x} \in \mathcal{A}^{n_T}} \|\mathbf{y} - \mathbf{H}\mathbf{x}\|^2. \tag{9}$$

This is achieved by exhaustively examining all possible transmit vectors; see Algorithm 3. Whilst the ML detection algorithm is conceptually simple, its complexity is exponential in the size of the constellation and number of transmit antennas, and is therefore practical for real-time hardware implementation only in the simplest of settings. As the optimal detector, the performance of the the ML detector serves as a benchmark for the detection schemes of the following sections.

Algorithm 3 ML Algorithm - $[\hat{\mathbf{x}}, \mathbf{b}] \Leftarrow \mathrm{MLdetect}\left(\mathbf{H}, \sigma^2, \mathbf{y}\right)$

$e_{min} \Leftarrow \inf$
for $b = 0$ to $2^{n_{\mathrm{BPT}}} - 1$ **do** $// \ 2^{n_{\mathrm{BPT}}}$ iterations
  $\hat{\mathbf{x}} \Leftarrow \mathrm{mod}\left(\mathbf{b}\right)$
  $e \Leftarrow \|\mathbf{y} - \mathbf{H}\hat{\mathbf{x}}\|^2$ $// \ 4n_R n_T + 2n_R \ \mathrm{M}, \ 4n_R n_T + 4n_R \ \mathrm{A}$
  if $e < e_{min}$ **then**
    $e_{min} = e$
    $\hat{\mathbf{x}}_{ml} = \hat{\mathbf{x}}$
    $\mathbf{b}_{ml} = \mathbf{b}$
  end if
end for

4.2. Zero Forcing estimation

The most straightforward linear detection scheme is *zero forcing* (ZF), also known as least squares estimation, which works to reverse the effect of the MIMO channel matrix on the transmitted symbols. By finding the least squares solution to (1), it is referred to as zero forcing as the interference caused by $\mathbf{H}$ is forced to zero by multiplication of the received vector $\mathbf{y}$ by $\mathbf{W}_{\mathrm{ZF}}$, the inverse (or generalized inverse) of the channel matrix:

$$\tilde{\mathbf{x}}_{\mathrm{ZF}} = \mathbf{W}_{\mathrm{ZF}}\mathbf{y}. \tag{10}$$

We use the notation $\tilde{\mathbf{x}}$ to represent an unconstrained estimate of the vector of transmitted symbols. The likelihood that $\tilde{\mathbf{x}}$ actually maps to a constellation point is negligibly small and so the nearest valid constellations point must be found. ZF finds the estimate of the vector of transmitted symbols $\hat{\mathbf{x}}_{\mathrm{ZF}}$ as follows:

$$\hat{\mathbf{x}}_{\mathrm{ZF}} = \arg\min_{\mathbf{x} \in \mathcal{A}^{n_T}} \|\mathbf{W}_{\mathrm{ZF}}\mathbf{y} - \mathbf{x}\|^2, \tag{11}$$

where $\hat{x}_{ZF}$ is found by independently rounding each element of $\tilde{x}$ to the nearest constellation point. The vector $\hat{x}_{ZF}$ can then be demodulated to find b_{ZF}, an estimate of the vector of transmitted bits, as shown in Algorithm 4.

There are numerous methods to find the least squares solution to (1), including those that directly calculate the matrix W_{ZF}. In this chapter, we utilize the well known Moore-Penrose pseudoinverse:

$$W_{ZF} = \left(H^H H\right)^{-1} H^H. \tag{12}$$

Algorithm 4 ZF Algorithm - $[\hat{x}, b] \Leftarrow$ ZFdetect $\left(H, \sigma^2, y\right)$

$\quad W_{ZF} = \left(H^H H\right)^{-1} H^H$
$\quad \tilde{x} \Leftarrow W_{ZF} y$ $// \ 4n_R n_T \ M, \ 4n_R n_T \ A$
$\quad \delta \Leftarrow [1+i]/2$
$\quad \hat{x}_{ZF} \Leftarrow \text{round} \left(\tilde{x}, \delta\right)$
$\quad b_{ZF} \Leftarrow \text{demod} \left(\hat{x}_{ZF}\right)$

4.3. Noise enhancement

Whilst ZF completely reverses the effects of the MIMO channel matrix, if the columns of H are correlated, ZF will amplify or enhance the noise. By identifying that $W_{ZF} H = I$ and then multiplying (1) by W_{ZF} we can calculate the effective additive noise component of the estimated vector of transmitted symbols:

$$\tilde{x}_{ZF} = x + W_{ZF} n. \tag{13}$$

It is intuitive that the noise existing in the unconstrained transmit symbol estimate $\tilde{x}_{ZF}$ is $W_{ZF} n$. When the rows of W_{ZF} have a large Euclidean distance, multiplication of the received vector leads to the additive noise component in y being amplified. We can now show how a poorly conditioned or correlated channel matrix will result in significant noise enhancement in ZF by examining the probability of error:

$$e = \tilde{x} - x$$
$$= W_{ZF} n \tag{14}$$
$$p_e = \text{diag} \left(\epsilon_n (ee^H)\right)$$
$$= \sigma^2 \text{diag} \left(\left(H^H H\right)^{-1}\right) \tag{15}$$

Existing work [17] has looked at the statistical properties of the channel matrix, and in particular the effect of this noise enhancement, leading to a tight analytical bound of the performance of ZF detectors in Rayleigh fading channels.

4.4. Minimum Mean-Square Error (MMSE) estimation

MMSE estimation acts to balance the reduction of the interference caused by $\mathbf{H}$ and the noise enhancement due to correlation of the columns in $\mathbf{H}$. Rather than completely remove the effect of the MIMO channel, MMSE estimation works to find a coefficient which minimizes the criterion:

$$\mathbf{W}_{\text{MMSE}} = \arg\min_{\mathbf{W}} \|\mathbf{W}\mathbf{y} - \mathbf{x}\|^2. \tag{16}$$

The solution to (16) is the well-known MMSE estimator, also known as the Wiener filter:

$$\mathbf{W}_{\text{MMSE}} = \left(\mathbf{H}^H\mathbf{H} + \sigma^2\mathbf{I}_{n_T}\right)^{-1}\mathbf{H}^H \tag{17}$$

$$= \begin{bmatrix} \mathbf{H} \\ \sigma\mathbf{I} \end{bmatrix}^\dagger \tag{18}$$

The shorthand notation of (18) was first proposed in [18] and is referred to as the *extended channel matrix*, which in this chapter is denoted

$$\overline{\mathbf{H}} = \begin{bmatrix} \mathbf{H} \\ \sigma\mathbf{I} \end{bmatrix}. \tag{19}$$

Similarly to ZF detection, MMSE detection finds the estimate of the vector of transmitted symbols $\widehat{\mathbf{x}}_{\text{MMSE}}$ as follows:

$$\widehat{\mathbf{x}}_{\text{MMSE}} = \arg\min_{\mathbf{x} \in \mathcal{A}^{n_T}} \|\mathbf{W}_{\text{MMSE}}\mathbf{y} - \mathbf{x}\|^2, \tag{20}$$

where $\widehat{\mathbf{x}}_{\text{MMSE}}$ is found by independently rounding each element of $\tilde{\mathbf{x}}$ to the nearest constellation point. It is well-known that as the noise term approaches zero (at high signal-to-noise ratios), the MMSE estimator becomes equivalent to a ZF estimator.

Compared to ZF detection, MMSE results on average in less noise enhancement, as $\overline{\mathbf{H}}$ is better conditioned. This can be seen intuitively as a result of adding a diagonal matrix relating to the noise variance as in (17) or alternatively due to the stacked structure of (18) resulting in a decrease in correlation. Unlike ZF, however, MMSE does not perfectly reverse or remove the interference of $\mathbf{H}$, leading to interference between the otherwise independent transmit antennas. As with ZF, analytical performance bounds for MMSE detectors have been developed [17, 19] for various channel models.

Utilizing the shorthand notation of the extended channel matrix of (18), ZF detection can be readily extended to perform MMSE detection, as shown in Algorithm 5. Note that due to the extra rows of $\overline{\mathbf{H}}$ as compared to $\mathbf{H}$, the computational complexity of calculating $\mathbf{W}_{\text{MMSE}}$ is roughly double that of $\mathbf{W}_{\text{ZF}}$.

Algorithm 5 MMSE Algorithm - $[\hat{\mathbf{x}}, \mathbf{b}] \Leftarrow$ MMSEdetect $(\mathbf{H}, \sigma^2, \mathbf{y})$

$$\overline{\mathbf{H}} \Leftarrow \begin{bmatrix} \mathbf{H} \\ \sigma\mathbf{I} \end{bmatrix}$$

$[\hat{\mathbf{x}}_{\text{MMSE}}, \mathbf{b}_{\text{MMSE}}] \Leftarrow$ ZFdetect $(\overline{\mathbf{H}}, \sigma^2, \mathbf{y})$

4.5. Detection using Lattice Reduction

Lattice basis reduction [20, §2.6.1] reduces the orthogonality defect, thereby reducing noise enhancement. This is achieved by finding a closer to orthogonal set of basis vectors. This reduced lattice basis is found by optimizing the generating matrix, which in the present application is a MIMO channel matrix realization. This closer-to-orthogonal set is found using elementary operations on basis vectors. Complex integer linear combinations of the column vectors of $\mathbf{H}$ are taken to form the reduced matrix $\mathbf{H}_L$ which spans the same set of points $\mathbf{H}\mathbb{X}^{n_T} \equiv \mathbf{H}_L\mathbb{X}^{n_T}$ and so

$$\mathbf{H}_L = \mathbf{HT} \text{ or } \mathbf{H} = \mathbf{H}_L\mathbf{T}^{-1}, \tag{21}$$

where $\mathbf{T}$ is a unimodular matrix with complex integer entries and $\det(\mathbf{T}) = \pm1$, therefore $\mathbf{T}^{-1}$ also contains only complex integer entries.

As in [3], by finding an equivalent and closer to orthogonal set of the basis vectors, $\mathbf{H}_L$, noise enhancement is reduced when quantization is performed. Importantly, as $\mathbf{T}^{-1}$ and $\bar{\mathbf{x}}$ both contain only integer spaced entries, so does $\mathbf{T}^{-1}\bar{\mathbf{x}}$ and so symbol detection or quantization is merely rounding to the grid $\mathbb{X}$.

Once the lattice reduced channel matrix is found, we then calculate the pseudoinverse as would be done in ZF or MMSE detection. LRAD therefore operates using the following steps, which are adapted from [3] and detailed in [21]:

1. Find the reduced lattice basis
2. Use the pseudoinverse of the reduced basis to form estimates
3. Quantize estimates to $\mathbb{X}$
4. Transform and bound points to constellation points

As shown in Algorithm 6, received vectors $\mathbf{y}$ are multiplied with the pseudoinverse of the reduced basis $\mathbf{H}_L$ to find a soft estimate of the vector of transmitted symbols in the reduced domain. These symbols are then quantized to an integer grid. (Depending on the transform generated, this integer grid may be offset by a half in both real and imaginary dimensions.) These hard estimates are then transformed, using the transform matrix $\mathbf{T}$ generated by the LR algorithm, to find an estimate of the vector of transmitted symbols. However, as these symbols may fall outside the range of constellation points invalid constellation points are clipped back to the nearest constellation point.

Algorithm 6 LRAD Algorithm - $[\hat{\mathbf{x}}, \mathbf{b}] \Leftarrow$ LRADdetect $(\mathbf{H}, \sigma^2, \mathbf{y})$

$[\mathbf{H}_L, \mathbf{T}] \Leftarrow$ LR$(\mathbf{H})$ // LR is a lattice reduction algorithm such as Algorithm 1

$\delta \Leftarrow \mathbf{T}[1+i]/2$

$\tilde{\mathbf{x}} \Leftarrow \bar{\mathbf{H}}_L^\dagger \mathbf{y}$

$\hat{\mathbf{x}}_{LRZF} \Leftarrow \mathbf{T}\left(\text{round}\left(\tilde{\mathbf{x}}, \delta\right)\right)$ // $4n_T^2$ M, $4n_T^2$ A

$\mathbf{b}_{ZF} \Leftarrow$ demod $\left(\hat{\mathbf{x}}_{LRZF}\right)$

---

5. Subspace-based LRAD

5.1. Hard-output SLRAD

For hard estimation, quantization of the ZF or MMSE estimate in the transmit constellation domain is replaced by the same quantization in the lattice reduced domain. The equivalent for soft estimation calls for the calculation of the error induced by quantization in the lattice reduced domain. Unfortunately, just as it is hard to ensure quantization to valid symbols in the lattice reduced domain, it is equally hard to iterate over all possible valid symbols in the lattice reduced domain in order to estimate each bit probability.

Whilst Zhang et al. [22] present a detailed comparison of various soft output based detectors and proposes several powerful methods for generating soft output information, there are some key shortcomings, and the performance of the detectors in [22] are only evaluated using QPSK constellations. This is problematic in that a range of wireless communication standards are moving to denser constellations, such as 16-QAM and 64-QAM. This motivates the investigation of lattice reduction based detectors capable of producing *candidate lists*.

The subspace lattice reduction aided detection (SLRAD) approach of Windpassinger [3] forms a subspace of the channel matrix $\mathbf{H}$ by removing a single column from the channel matrix. This column removal allows the corresponding transmit antenna's symbol estimate to be constrained in order to calculate an estimate for what the other transmit antennae sent. For each transmit antenna a number of symbols is systematically proposed and for each proposal the set of most likely symbols transmitted on the other antennae is calculated, as shown in Algorithm 7.

The SLRAD algorithm therefore creates a list of candidate symbols, the Euclidean distance of each of these candidates from the origin being used to determine the most likely vector of transmitted symbols for a hard-output detector.

Whilst performance of SLRAD is close to that of ML (see Fig. 5), the complexity is proportional only to the sum of the size of the constellations employed on each transmit antenna. Therefore only a modest number of candidate symbols needs to be investigated, even for dense constellations. For example, a system with 4 transmit antennas each utilizing 64-QAM results in only $4 \times 64 = 256$ candidates.

5.2. Soft-output SLRAD

As a candidate-based detector, the hard-output SLRAD detector can be extended to generate soft output information. The probability of all the candidates where a bit is one is divided by the probability of all candidates where the bit is zero. An attractive property of subspace

---

Algorithm 7 SLRAD Algorithm - $[\hat{\mathbf{x}}, \mathbf{b}] \Leftarrow$ SLRADdetect $(\mathbf{H}, \sigma^2, \mathbf{y})$

$e_{min} \Leftarrow \inf$
for $k = 1$ to n_T **do**
   $\mathbf{H}_s \Leftarrow \mathbf{H}_{[1...(k-1)(k+1)...n_T]}$
   for all s in $\mathcal{A}_k$ **do**
      $\mathbf{y}_s \Leftarrow \mathbf{y} - \mathbf{h}_k s$ // $4n_R$ M, $4n_R$ A
      $\hat{\mathbf{x}}_s \Leftarrow$ LRADdetect $(\mathbf{H}_s, \mathbf{y}_s, \sigma^2)$
      $\hat{\mathbf{x}}_{[1...(k-1)(k+1)...n_T]} = \hat{\mathbf{x}}_s$
      $\hat{\mathbf{s}}_k = s$
      $e \Leftarrow \|\mathbf{y} - \mathbf{H}\hat{\mathbf{x}}\|^2$
      if $e < e_{min}$ **then**
         $e_{min} = e$
         $\hat{\mathbf{x}}_{SLR} = \hat{\mathbf{x}}$
      end if
   end for
end for
$\mathbf{b}_{SLR} \Leftarrow$ demod $(\hat{\mathbf{x}}_{SLR})$

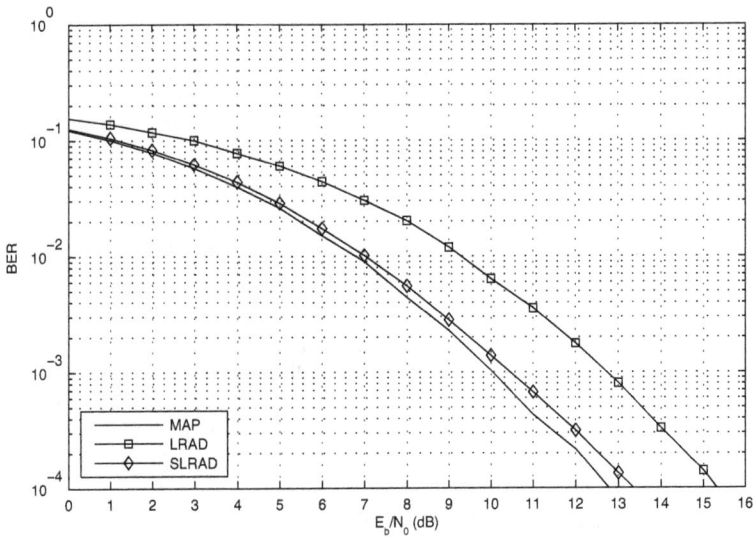

Figure 5. Bit error rate (BER) Performance of ML, LRAD and SLRAD for 4×4 MIMO with 16-QAM

detectors is that every bit is guaranteed to have at least one candidate where the bit is a one and likewise a candidate where it is zero. Without this property, it is not possible to accurately form an estimate for the ratio of the bit's value probabilities.

Algorithm 8 Soft Output SLRAD Algorithm - $[\mathbf{L}_e] \Leftarrow$ SLRADdetect-soft $(\mathbf{H}, \sigma^2, \mathbf{y})$

$\mathbf{n}_{bit} \Leftarrow 0$
$\mathbf{d}_{bit} \Leftarrow 0$
for $k = 1$ to n_T **do**
  $\mathbf{H}_s \Leftarrow \mathbf{H}_{[1\ldots(k-1)(k+1)\ldots n_T]}$
  for all s in $\mathcal{A}_k$ **do**
    $\mathbf{y}_s \Leftarrow \mathbf{y} - \mathbf{h}_k s$
    $\widehat{\mathbf{x}}_s \Leftarrow$ LRADdetect $(\mathbf{H}_s, \mathbf{y}_s, \sigma^2)$
    $\widehat{\mathbf{x}}_{[1\ldots(k-1)(k+1)\ldots n_T]} = \widehat{\mathbf{x}}_s$
    $\widehat{\mathbf{s}}_k = s$
    $\mathbf{b} \Leftarrow$ demod $(\widehat{\mathbf{x}})$
    $e \Leftarrow \exp\left(\frac{-\|\mathbf{y} - \mathbf{H}\widehat{\mathbf{x}}\|^2}{\sigma^2}\right)$
    for all bits in current bit vector **do**
      if the current bit is a '1' **then**
        $\mathbf{n}_{bit} = \mathbf{n}_{bit} + e$
      else
        $\mathbf{d}_{bit} = \mathbf{d}_{bit} + e$
      end if
    end for
  end for
end for
$\mathbf{L}_e \Leftarrow \log[\mathbf{n}] - \log[\mathbf{d}]$

The soft-output SLRAD algorithm is shown in Algorithm 8. This algorithm leads in a natural fashion to the top-level data flow diagram in Fig. 6. The candidate chain block in Fig. 7 performs the following key steps (once for each submatrix of **H** formed by deleting one column from **H**):

1. subspace candidate estimate generation;
2. lattice reduced domain quantization;
3. reversal of the lattice basis transform;
4. bounding to ensure valid constellation symbols; and
5. demodulation and Euclidean distance calculation.

6. Hardware implementation

6.1. Existing work

The first published VLSI implementation of a lattice reduction aided detector [23] is based on Brun's algorithm for finding integer relations [24]. Brun's algorithm offers lower complexity at a performance cost when compared to the commonly utilized complex LLL algorithm.

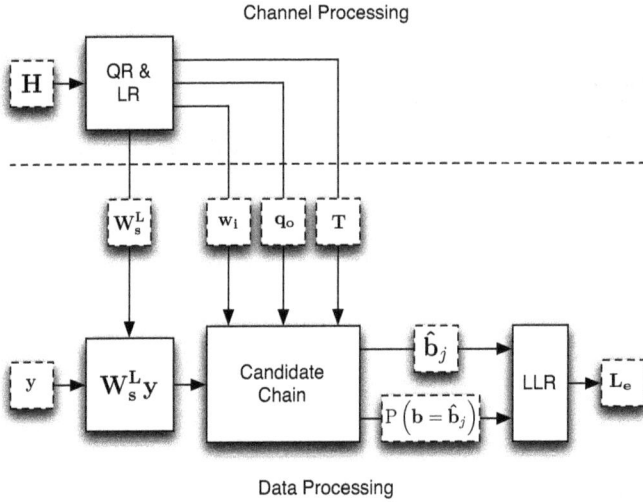

Figure 6. Top Level Data Flow Diagram

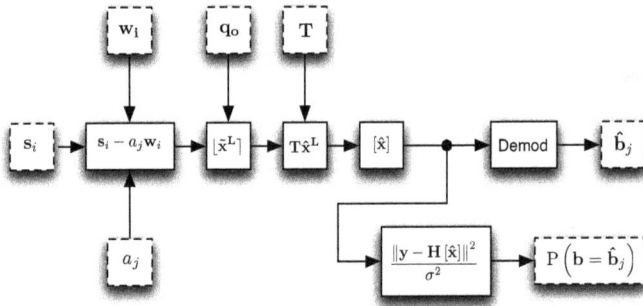

Figure 7. Candidate Chain Data Flow Diagram

Brun's algorithm is criticised in [25] as it achieves inferior performance and no analytical result has been reported to prove the level of diversity that can be achieved. This work applies a uniform scaling factor to the elements of the same matrix or vector to ensure that the magnitudes of the largest real and imaginary parts are as close as possible to, but smaller than one. This pragmatic approach offers a good compromise between true floating-point arithmetic, with its computational overhead, and a simple fixed point arithmetic with significantly reduced dynamic range. However, it appears that no active scaling is performed in the algorithm to prevent numeric overflow. Instead, it is claimed without substantiation that a bound exists which is used to calculate the required number of integer bits.

The work in [26] implements a sorted QR decomposition using Householder CORDIC units to reduce the number of LLL iterations needed. The complex LLL algorithm is used but, as with most LLL implementations, requires the use of divisions, using the Newton–Raphson algorithm, throughout the LLL iterations.

The work of [27] builds on [26] and discusses novel search based extensions to LRAD introduced in [28] which generate a candidate list and therefore soft outputs. However, the hardware implementation does not discuss this and therefore it is presumed that the hardware implementation is hard output. Due to the time-multiplexed complex multiplier pipeline, this approach is forced to rely on the use of priority inversion to prevent deadlocks due to data dependencies. Analysis is not performed on the precision required and in particular magnitude bounding is not performed which results in a large number of integer bits being required.

In [29], the authors build on their prior work [26, 27] by offering several improvements. This revision implements Sorted QRD to reduce the number of LLL swapping steps. Once again, the hardware implementation is presumed to only offer hard outputs as no mention is made of the candidate generation required to form soft outputs nor the hardware required to calculate LLRs. Unlike the prior works, an upper bound of 4 integer bits is identified for the elements of the R matrix which offers a significant reduction in the precision required.

Several works [30, 31] make use of systolic arrays in their implementation. This requires careful scheduling to maximize component utilization. The former work makes use of the Complex LLL algorithm whereas the later extends the LLL algorithm through the use of the Siegel condition to avoid the requirement for division operations.

The field-programmable gate array (FPGA) implementation of [32] implements the Clarkson's algorithm variant of LLL [33]. However this implementation only considers slower off-the-shelf FPGA components, including the use of square root and division operations that have not been optimized. The FPGA and application-specific integrated circuit (ASIC) implementation [34] claims to achieve a "fivefold improvement in terms of throughput at the cost of only slightly more FPGA resources" over [26] and [32]. This work uses CORDIC units along with a modification of the LLL algorithm by replacing the size-reduction criterion with the reverse Siegel condition. The hard output performance of this implementation is also enhanced by the use of soft interference cancellation (SIC), which requires the use of the sorted QR decomposition.

6.2. Architecture for Subspace Lattice Reduction Aided Detection

Our proposed architecture implements a soft-output lattice reduction-aided detector based on the subspace LRAD (SLRAD) approach of Windpassinger [3, 4]. The top-level schematic layout is shown in Fig. 8. A key feature of the detector is the separation of channel and data processing sections, shown above and below the dashed line in Fig. 8, respectively. Channel processing is computationally expensive, and includes the decomposition and lattice reduction of the MIMO channel matrix $\mathbf{H}$. The separation of channel and data processing therefore enables the receiver to exploit the typically slow variation in channel gains relative to the symbol rate, whereby the output of the computationally expensive channel processing step is used is used in processing the data spanning multiple data frames.

The channel processing section in Fig. 8 is fed with elements h_{in} of the estimated MIMO channel matrix $\mathbf{H}$ generated by an external MIMO channel estimator (not shown), while channel multiply and accumulator (CMAC) units perform rotations under control of the Givens control unit. Data processing involves the subspace-based detection of incoming received values, in addition to the calculation of soft outputs in the form of log-likelihood ratio (LLR) values. The data processing section is fed elements of the received vector $\mathbf{y}$, scaled by automatic gain control (AGC) to ensure that analog-to-digital converters (ADCs) are not saturated, and therefore that fixed-point inputs are within a defined range. The data multiply and accumulate (DMAC) and detection (DET) blocks in Fig. 8 are described in Section 6.4. The outputs of the data processing section are LLR values for the bits corresponding to each vector of transmitted symbols $\mathbf{x}$.

Unlike [26, 27], this work implements the Scaled and Decoupled QR (SDQR) Decomposition [35]. The use of the SDQR provides a definitive bound on the required integer precision and allows the number of fractional bits to be varied with a constant and small number of integer bits.

6.3. Channel processing

6.3.1. Givens Control Unit

The calculation of the SDQR rotation values is performed by a Givens Control unit. This unit is a single cycle processor which generates the Givens rotation $\mathbf{G}$ which zeros the element $\mathbf{P}_{j,i}$ by rotating the jth row with the ith row of $\mathbf{P}$ and $\mathbf{\Phi}$. The Givens Control unit is capable of a throughput of one rotation variable per cycle by calculating a Givens rotation every four cycles. Two rotation variables are emitted in the third cycle (values $\mathbf{G}_{1,1}$ and $\mathbf{G}_{1,2}$) and fourth cycle (values $\mathbf{G}_{2,1}$ and $\mathbf{G}_{2,2}$) of each Givens rotation calculation. The Givens Control Unit also maintains the decoupled k values and also dynamically scales $\mathbf{G}$ to maintain scaling of not only k but indirectly $\mathbf{P}$ and the rotated $\mathbf{y}$. This processor implements the reciprocal function required for division through the use of Newton–Raphson iterations.

6.3.2. Channel MAC (CMAC) Unit

The application of rotation operations are performed by processor units referred to as Channel Multiply and Accumulator (CMAC) units. Each CMAC unit includes sufficient register space to store a full column of the MIMO channel $\mathbf{H}$ as well as necessary intermediate values. All input, output and stored register values are complex numbers specified using

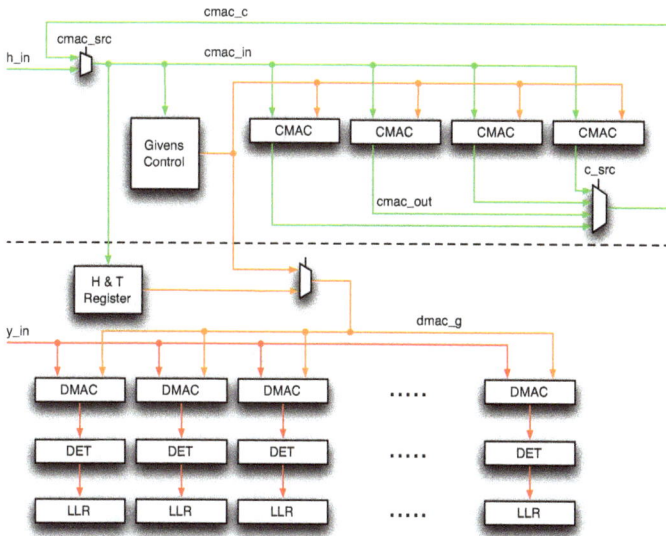

Figure 8. Top-level schematic layout. Channel and data processing sections are shown above and below the dashed line, respectively

custom extensions to the VHDL fixed point math package. Arithmetic implemented includes a complex multiplier and complex addition unit with the output of the multiplier being one of the operands of the adder, as shown in Fig. 9.

Whilst this architecture greatly simplifies the challenge of processor unit scheduling, the units are still unavoidably under-utilized. The CMAC units become unused once the their corresponding column of H is fully zeroed. As a result, the CMAC unit corresponding to the ith column of H is in use for i/n_T of the SDQR execution period.

The CMAC units provide outputs which feed a multiplexer, as shown in Fig. 8. Required in order to perform back substitution, this allows the transfer of register values between CMAC units by feeding the output of a unit to the input of another.

6.4. Data processing

6.4.1. Data MAC (DMAC) Unit

Processor units referred to as Data Multiply and Accumulator (DMAC) units, are implemented to apply Givens rotation operations on the received vector y. Each DMAC unit includes sufficient register space to store a full vector of received values y as well as necessary intermediate values.

Multiple DMAC units are implemented so that the necessary rotations required to apply a Givens rotation to a full row of H can be performed in parallel. This avoids the need to stall not only the Givens Control unit but also stalling of DMAC units that would otherwise need to occur whilst each row element of H is rotated.

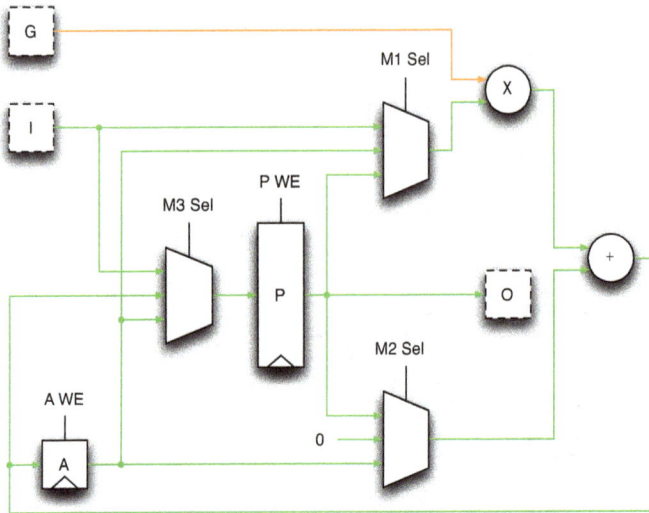

Figure 9. Custom Multiply and Accumulate Schematic Layout

Multiple DMAC units are implemented to achieve the necessary data throughput rate such that a single rotation operation can be applied to multiple received vectors in parallel. This builds on the presumption that the MIMO channel is approximately constant for multiple symbol periods. Given a sufficiently static MIMO channel, any number of DMAC units can be implemented. This allows a linear scaling of data throughput by simply adding more DMAC units, a key design feature of the the proposed architecture.

6.4.2. H&T Register File

As well as being loaded into CMAC units, when a new **H** is loaded into the processor, it is cached in the H&T register file. This is done to provide a copy of **H** for use when calculating the Euclidean distance of candidate estimates. The H&T register file is also used to store **T**, the lattice basis required to translate candidate estimates from the reduced basis prior to demodulation.

6.4.3. Candidate Detection (DET)

Each DMAC unit feeds a symbol detection chain which performs candidate generation and finally bitwise log-likelihood accumulation. This implements the data flow detailed in Fig. 7.

6.4.4. Log-likelihood ratio (LLR) Accumulator

Once a list of vectors of transmit symbol candidates has been generated, the probability of each of these vectors needs to be generated. Many approaches exist that avoid the need to implement the required log operations inherit in the calculation of log-likelihood ratio (LLR)

values. We implement the shifting method Log-MAP algorithm presented in [36], which utilizes the following piecewise linear approximation:

$$f(x) = \begin{cases} 0.70 - x/2 & 0.00 \leq x < 0.51 \\ 0.57 - x/4 & 0.51 \leq x < 1.44 \\ 0.39 - x/8 & 1.44 \leq x < 2.88 \\ 0.03 & 2.88 \leq x < 4.00 \\ 0.00 & 4.00 \leq x \end{cases} \tag{22}$$

The schematic for the LLR block is shown in Fig. 10.

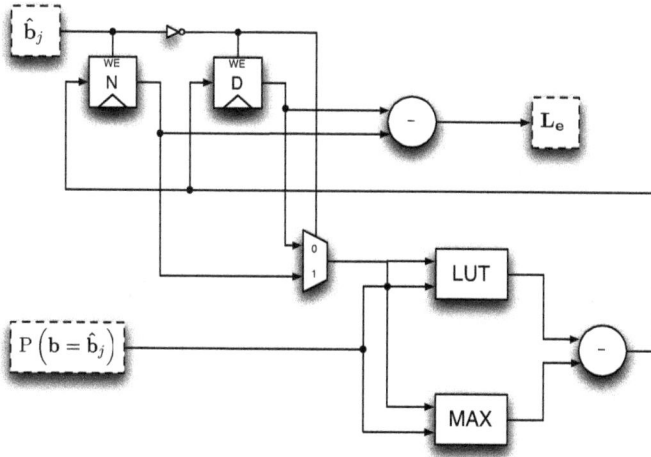

Figure 10. Log-likelihood ratio (LLR) Marginalization Schematic Layout

6.5. Processor instruction set

The overall architecture is a microcode-based system with detailed low level micro-operations that combine to implement higher level complex machine instructions. Each component including the Givens Control Unit, CMACs, DMACs and Detection Chains have their own micro-operations. The benefit is the provision of a flexible architecture capable of implementing the SLRAD algorithm, but which is also able to switch to simpler LRAD or even ZF algorithms based on the prevailing channel conditions.

6.5.1. Control Unit Micro-operations

The bulk of the channel processing involves the execution of the four operations that generate the Givens rotation $\mathbf{G}$. The first two, C1 and C2, calculate the new values for $\mathbf{P}_{j,i}$ and k_j;

the third C3 updates k_i and calculates $\mathbf{G}_{2,1}$ and $\mathbf{G}_{2,2}$; and the fourth C4 updates $\mathbf{P}_{i,i}$ and calculate $\mathbf{G}_{1,1}$ and $\mathbf{G}_{1,2}$. The control unit also includes an operation CR which performs the reciprocation the value k_j as needed. This supports back substitution as well as part of the LLL algorithm and is implemented using the Newton–Raphson algorithm. Finally, the control unit also performs other operations to marshal data between channel processing and data processing.

6.5.2. CMAC and DMAC Micro-operations

For the CMAC and DMAC units, the micro-operations, and their corresponding complex machine instructions are detailed in Table 1. In this table, the first column lists the micro-operation code, the third column lists the value provided on the input I of the CMAC and DMAC units and the final two columns detail the implemented function.

Description	I	CMAC	DMAC
		Data Load	
LC Load elements of $\mathbf{H}$	$\mathbf{H}_{m,n}$	$\mathbf{P}_{m,n} = I$	-
LD Load elements of $\mathbf{y}$	$\mathbf{y}_x$	-	$\Phi_x = I$
	Givens Rotation for SDQR and Basis Vector Swap Update		
G1 Multiply by $\mathbf{G}_{2,1}$	-	$A = G \times \mathbf{P}_i + 0$	$A = G \times \Phi_i + 0$
G2 Multiply by $\mathbf{G}_{2,2}$ and add	-	$A = G \times \mathbf{P}_j + A$	$A = G \times \Phi_j + A$
G3 Multiply by $\mathbf{G}_{1,2}$	-	$A = G \times \mathbf{P}_j + 0; \mathbf{P}_j = A$	$A = G \times \Phi_j + 0; \Phi_j = A$
G4 Multiply by $\mathbf{G}_{2,2}$ and add	-	$\mathbf{P}_i = G \times \mathbf{P}_i + A$	$\Phi_i = G \times \Phi_i + A$
		Back Substitution	
B1 Multiply Φ row by $-\mathbf{P}_{j,i}$	$-\mathbf{P}_{j,i}$	-	$A = G \times \Phi_i + 0$
B2 Accumulate with j^{th} row	-	-	$\Phi_j = 1 \times \Phi_j + A$
		Lattice Size Reduction	
R1 $G = \text{round}\left(\mathbf{P}_{l,j}\right)$	$\mathbf{P}_{l,j}$	$O = \mathbf{P}_{l,j}$	$A = G \times \Phi_j + 0$
R2 Get reduced row of Φ	-	-	$\Phi_l = 1 \times \Phi_l + A$
R3 $G = -\text{round}\left(\mathbf{P}_{l,j}\right)$	$\mathbf{P}_{x,l}$	$\mathbf{P}_{x,j} = G \times I + \mathbf{P}_{x,j}$	-

Table 1. CMAC and DMAC Instruction Set

The CMAC and DMAC units implement the micro-operations LC and LD to perform data load operations that load the channel matrix and received vector; G1 to G4 which implement the Givens rotations for not only the SDQR but also the zeroing step of LLL algorithm; B1 and B2 that performs back substitution operations; and R1 to R3 which perform the LLL column swap step.

The DMAC units require more or less the same operations as their CMAC counterparts, however, for the case of back substitution and lattice size reduction the implementation differs. For back substitution, the CMAC units must pass off-diagonal elements $\mathbf{P}_{j,i}$ to the DMAC units. For lattice size reduction, the CMAC units add an integer multiple of one column of $\mathbf{P}$ to another by iteratively executing R3 which passes elements between CMAC units where the target unit performs the multiplication and addition. On the other hand, DMAC units are able to perform the equivalent reduction operation in the two micro-operations R1 and R2 as lattice size reduction is performed in a row-wise fashion.

6.6. Comparisons with previously published work

The results in this section represent the first known digital signal processing architecture for a soft-output lattice reduction aided MIMO detector. For this reason we are unable to provide a direct comparison of our architecture with previously published work. Nevertheless, it is still possible to compare our implementation with three state-of-the-art VLSI implementations of hard-output LRAD-based MIMO detectors [32], [26], [34].

For $n_T = n_R = 4$, the combination of the CMAC micro-operations leads to the system latency outlined in Table 2. This table assumes a MIMO system represented by an extended channel matrix, requiring the zeroing of 16 elements of $\overline{\overline{H}}$. The majority of these elements require 4 cycles with the exception being the final element of each column requiring a 5th cycle due the the extra cycle required to compute the Newton-Raphson based reciprocal. An overhead of 12 cycles exists to load data into the processor.

For the LLL algorithm, column swap operations require 5 cycles to perform the single Givens rotation. Size reduction requires at most 3 cycles per pass over the full matrix. As with prior works, a simple strategy is used to fix the number of iterations of the LLL algorithm which caps the number of swaps and size reduction passes to 3. This yields 24 cycles per subspace or 96 cycles for the four subspaces.

Component	Latency
QR decomposition	80 cycles
Subspace Generation	30 cycles
Subspace Back-substitution	16 cycles
Subspace Lattice Reduction	96 cycles
Total for SLRAD	222 cycles

Table 2. Latency of Channel Processor

To provide context for the results in Table 2, we compare in Table 3 the latency of the proposed architecture with the latencies of three hard-output LRAD-based MIMO detectors for a 4-input, 4-output MIMO system employing QPSK modulation.

	[32]	[26]	[34]	this work
average cycles per matrix	420	130	14	222
soft outputs?	No	No	No	Yes

Table 3. Latency comparison between the proposed architecture and three state-of-the-art implementations

While the latency of the proposed architecture compares favourably with Barbero et al.'s solution [32], the significant performance penalty for generating soft outputs is apparent in comparison with the results of Gestner et al. [26] and (esp.) Bruderer et al. [34]. We caution that the results in Table 3 need to be interpreted carefully, however, since it is well known that hard-output MIMO detectors such as [32], [26] and [34] do not facilitate high-performance iterative receivers involving joint detection and decoding when error-control codes such as turbo codes and LDPC codes are employed [37], [22]. The proposed approach therefore trades off increased latency for improved BER performance and the ability to readily deal with dense constellations, e.g. 64-QAM.

7. Conclusion

In this chapter we have presented the first known digital signal processing implementation of a soft-output MIMO wireless communications receiver based on lattice reduction aided detection (LRAD). Further research is needed to provide the ASIC and FPGA synthesis results needed to facilitate a comprehensive comparison with prior works providing only hard outputs.

Author details

Alan T. Murray and Steven R. Weller

School of Electrical Engineering and Computer Science, University of Newcastle, Callaghan, NSW 2308, Australia

References

[1] S.J. Johnson. *Iterative Error Correction: Turbo, Low-Density Parity-Check and Repeat-Accumulate Codes.* Cambridge University Press, 2009.

[2] D. Wübben, D. Seethaler, J. Jaldén, and G. Matz. Lattice reduction. *IEEE Signal Process. Mag.*, 28(3):70–91, May 2011. DOI: 10.1109/MSP.2010.938758.

[3] C. Windpassinger, L.H.J. Lampe, and R. Fischer. From lattice-reduction-aided detection towards maximum-likelihood detection in MIMO systems. In *Int. Conf. on Wireless and Optical Commun. (WOC'03)*, July 2003.

[4] C. Windpassinger. *Detection and Precoding for Multiple Input Multiple Output Channels.* PhD thesis, Universität Erlangen-Nürnberg, 2004.

[5] A.T. Murray and S.R. Weller. Performance and complexity of adaptive lattice reduction in fading channels. In *Proc. Australian Comms. Workshop (AusCTW'09)*, pages 17–22, Sydney, Australia, February 2009. DOI: 10.1109/AUSCTW.2009.4805593.

[6] W. Liu, K.Choi, and H.Liu. Computationally efficient lattice reduction for MIMO-OFDM systems. In *Proc. 6th IEEE Int. Conf. on Wireless and Mobile Computing, Networking and Communications (WiMob'10)*, pages 264–267, October 2010. DOI: 10.1109/WIMOB.2010.5645056.

[7] G.H. Hardy, E.W. Wright, and J.H. Silverman. *An Introduction to the Theory of Numbers.* Oxford University Press, 6th edition, 2008.

[8] A.K. Lenstra, H.W. Lenstra, and L. Lovász. Factoring polynomials with rational coefficients. *Math. Ann.*, 261(4):515–534, 1982.

[9] P. Silvola, K. Hooli, and M. Juntti. Suboptimal soft-output MAP detector with lattice reduction. *IEEE Sig. Proc. Letters*, 13(6):321–324, June 2006. DOI: 10.1109/LSP.2006.871726.

[10] Y.H. Gan and W.H. Mow. Complex lattice reduction algorithms for low-complexity MIMO detection. In *Proc. IEEE Global Telecommunications Conf. (GLOBECOM '05)*, pages 2953–2957, St. Louis, MO, 28 November–2 December 2005. DOI: 10.1109/GLOCOM.2005.1578299.

[11] W.H. Mow. Universal lattice decoding: Principles and recent advances. *Wirel. Commun. Mob. Com.*, 3(5):553–569, August 2003. DOI: 10.1002/wcm.140.

[12] F.T. Luk and S. Qiao. Conditioning properties of the LLL algorithm. In M.S. Schmalz, G.X. Ritter, J. Barrera, J.T. Astola, and F.T. Luk, editors, *Mathematics for Signal and Information Processing*, August 2009.

[13] A. Korkine and G. Zolotarev. Sur les formes quadratiques. *Math. Ann.*, 6:366–389, 1873.

[14] M. Seysen. Simultaneous reduction of a lattice basis and its reciprocal basis. *Combinatorica*, 13(3):363–376, September 1993. DOI: 10.1007/BF01202355.

[15] C. Windpassinger, L.H.J. Lampe, R. Fischer, and T. Hehn. A performance study of MIMO detectors. *IEEE Trans. Wireless Commun.*, 5(8):2004–2008, August 2006. DOI: 10.1109/TWC.2006.1687712.

[16] X. Ma and W. Zhang. Performance analysis for MIMO systems with lattice-reduction aided linear equalization. *IEEE Trans. Commun.*, 56(2):309–318, February 2008. DOI: 10.1109/TCOMM.2008.060372.

[17] X. Li and Z. Nie. Performance losses in V-BLAST due to correlation. *IEEE Antennas Wireless Propag. Lett.*, 3(1):291–294, January 2004. DOI: 10.1109/LAWP.2004.838813.

[18] B. Hassibi. An efficient square-root algorithm for BLAST. In *Proc. IEEE Int. Conf. on Acoustics, Speech, and Signal Processing (ICASSP '00)*, volume 2, pages 737–740, 2000. 10.1109/ICASSP.2000.859065.

[19] M.R. McKay, I.B. Collings, and A.M. Tulino. Achievable sum rate of MIMO MMSE receivers: A general analytic framework. *IEEE Trans. Inform. Theory*, 56(1):396–410, January 2010. DOI: 10.1109/TIT.2009.2034893.

[20] H. Cohen. *A Course in Computational Algebraic Number Theory*. Springer, December 1993.

[21] E. Agrell, T. Eriksson, A. Vardy, and K. Zeger. Closest point search in lattices. *IEEE Trans. Inform. Theory*, 48(8):2201–2214, August 2002. DOI: 10.1109/TIT.2002.800499.

[22] W. Zhang X. Ma. Low-complexity soft-output decoding with lattice-reduction-aided detectors. *IEEE Trans. Commun.*, 58(9):2621–2629, September 2010. DOI: 10.1109/TCOMM.2010.080310.070641.

[23] A. Burg, D. Seethaler, and G. Matz. VLSI implementation of a lattice-reduction algorithm for multi-antenna broadcast precoding. In *Proc. IEEE Int. Symp. on Circuits and Systems (ISCAS'07)*, pages 673–676, New Orleans, LA, 27–30 May 2007. DOI: 10.1109/ISCAS.2007.377898.

[24] D. Seethaler and G. Matz. Efficient vector perturbation in multi-antenna multi-user systems based on approximate integer relations. In *Proc. European Signal Proc. Conf. (EUSIPCO'06)*, pages 4–8, Florence, Italy, 4–8 September 2006.

[25] W. Zhang, X. Ma, B. Gestner, and D.V. Anderson. Designing low-complexity equalizers for wireless systems. *IEEE Comms. Mag.*, 47(1):56–62, January 2009. DOI: 10.1109/MCOM.2009.4752677.

[26] B. Gestner, W. Zhang, X. Mai, and D.V. Anderson. VLSI implementation of a lattice reduction algorithm for low-complexity equalization. In *IEEE Int. Conf. on Circuits and Systems for Communications (ICCSC'08)*, pages 643–647, Shanghai, China, 26–28 May 2008. DOI: 10.1109/ICCSC.2008.142.

[27] W. Zhang. *Wireless Receiver Designs: From Information Theory to VLSI Implementation*. PhD thesis, Georgia Institute of Technology, December 2009.

[28] W. Zhang and X. Ma. Approaching optimal performance by lattice-reduction aided soft detectors. In *Proc. 41st Annual Conf. on Information Sciences and Systems (CISS '07)*, pages 818–822, 2007. DOI: 10.1109/CISS.2007.4298422.

[29] B. Gestner, W. Zhang, X. Ma, and D.V. Anderson. VLSI implementation of an effective lattice reduction algorithm with fixed-point considerations. In *Proc. IEEE Int. Conf. on Acoustics, Speech and Signal Processing (ICASSP 2009)*, pages 577–580, 2009. DOI: 10.1109/ICASSP.2009.4959649.

[30] J. Soler-Garrido, H. Vetter, M. Sandell, D. Milford, and A. Lillie. Implementation of a reduced-lattice MIMO detector for OFDM systems. In *Design, Automation & Test in Europe Conference & Exhibition (DATE '09)*, pages 1626–1631, 2009.

[31] N.-C. Wang, E. Biglieri, and K. Yao. Systolic arrays for lattice-reduction-aided MIMO detection. *J. Commun. Netw.*, 13(5):481–493, October 2011. DOI: 10.1109/JCN.2011.6112305.

[32] L.G. Barbero, D.L. Milliner, T. Ratnarajah, J.R. Barry, and C. Cowan. Rapid prototyping of Clarkson's lattice reduction for MIMO detection. In *Proc. IEEE Int. Conf. on Communications (ICC'09)*, pages 1–5, Dresden, Germany, 14–18 June 2009. DOI: 10.1109/ICC.2009.5199388.

[33] I.V.L. Clarkson. *Approximation of Linear Forms by Lattice Points with Applications to Signal Processing*. PhD thesis, Australian National University, January 1997.

[34] L. Bruderer, C. Studer, M. Wenk, D. Seethaler, and A. Burg. VLSI implementation of a low-complexity LLL lattice reduction algorithm for MIMO detection. In *Proc. IEEE Int. Symp. on Circuits and Systems (ISCAS'10)*, pages 3745–3748, Paris, France, 30 May–2 June 2010. DOI: 10.1109/ISCAS.2010.5537742.

[35] L.M. Davis. Scaled and decoupled Cholesky and QR decompositions with application to spherical MIMO detection. In *Proc. IEEE Wireless Communications and Networking (WCNC'2003)*, volume 1, pages 326–331, New Orleans, LA, 16–20 March 2003. DOI: 10.1109/WCNC.2003.1200369.

[36] L. Zhong, M. Gang, T. Yi-Zheng, and C. Yan-Min. A simplification of the log-MAP algorithm for turbo decoding. In *Proc. IEEE Asia-Pacific Conference on Circuits and Systems*, volume 2, pages 1057–1060, 2004. DOI: 10.1109/APCCAS.2004.1413065.

[37] D.L. Milliner and J.R. Barry. A lattice-reduction-aided soft detector for multiple-input multiple-output channels. In *Proc. IEEE Global Telecommunications Conference (GLOBECOM '06)*, 2006. DOI: 10.1109/GLOCOM.2006.84.

Chapter 10

Self-Organizing Architectures for Digital Signal Processing

Daniele Peri and Salvatore Gaglio

Additional information is available at the end of the chapter

1. Introduction

Technological bounds in digital circuits integration in the last decades have been fostering the development of massively parallel architectures for tasks that had not been touched before by traditional parallel paradigms. Even in personal computers, as well as in consumer and mobile devices, it is common to find powerful processing units composed of processing elements in the range of the hundreds to the thousands.

The request for mobile devices, that are self-powered, almost permanently switched on and connected through wireless networks, as well as environmental friendliness constraints, obviously urges to reduce energy consumption of processing units.

On the other hand, applications continuously keep pushing forward computing power needs. A number of such applications are actually performed on application specific or on almost general-purpose parallel multi-core unit, as in the case of 3D graphics, sound processing, and the like, in the multimedia arena.

The current industrial trend aims to increase computing power and energetic efficiency by adding cores to both main processors and specialized units. A number of experimental architectures have been proposed that try to achieve the same goal by exploiting different designs. Coarse and fine grained architectures, and more in general, reconfigurable architectures have been proposed to make the hardware adapt to the required tasks instead of using specialized software running on general purpose processing elements. This has especially been the case in computer vision, and intelligent systems in general.

More interestingly, in these fields, the quest for massively parallel and energy efficient hardware implementations, coupled with biological models of reference, may pour interest in reviewing well and lesser studied approaches that are centered on self-organizing processing structures. Indeed, current research on pattern recognition shows significant interest in highly structured models built on large numbers of processing nodes trained by demanding algorithms that can, at least partially, be implemented in a parallel fashion.

In this chapter we provide a review of self-organization as it may appears at the various abstract levels of computational architectures, and including applications to real-world problems.

We start outlining the properties related to complexity and self-organization in natural and artificial systems. Then we de-scribe the computational models that are better suited to study self-organizing systems. We then discuss self-organization at the hardware level. Finally, we look at networked systems paying particular attention to distributed sensing networks.

2. Self-organization and self-organizing systems

Human speculation on the visible order of things, either living or not, is so ancient to be considered one of the fundamental questions of mankind. Science has always been exploring the complex structure of Nature, adding pieces to pieces to its infinite puzzle.

Meanwhile, technologies evolve benefiting from new findings, sometimes trying either successfully or ingenuously to duplicate Nature's work. Improvements in technologies then reflect on further science advancements, closing the loop.

Order, self-organization, adaptation, evolution, emergence and several other terms reminds us that as artificial systems advance, gaining complexity, it is expected for them to be compared to natural ones in both structure and function.

At the time of the vacuum tube digital computer introduction in the 1940s, McCulloch and Pitts had already proposed their neuron model, while cyberneticists were starting to recognize their interdisciplinary studies on natural and artificial systems as a brand new field.

With their simple and primitive circuits, made of few thousands discrete components, digital computers were certainly "complex" with respect to the available technology, but orders of magnitude simpler and unstructured than their biological computational counterparts made of billions of neurons arranged by some "Self-Organization" process.

Anyway, it did not took much to Von Neumann to start exploring the theory of Cellular Automata and self-reproducing machines, attempting to bridge natural and artificial computational models.

Rosenblatt's perceptron was another attempt to propose a biologically inspired computational framework. As a confirmation of the difficulties in reverse-engineering Nature, it took a few decades for Artificial Neural Networks built with perceptrons to become viable means to tackle useful computational tasks.

Of all the connotations given to perceptron networks, "adaptive" has been certainly one of the most adopted, however many of the terms cited before have found some kind of use with ANNs [1].

With respect to the "self-organization" term, its use has been used so widespread in computer and information processing literature that the effort to determine its introduction is quite pointless.

One of the most well known uses of the term can be traced back to Kohonen's "Self-Organizing Maps" (SOMs) [2]. Differently from perceptron based Artificial Neural Networks, trained with supervised algorithms, Kohonen proposed an unsupervised method whose geometrical representation is that of a continuous rearrangement of points of the feature space around auto-determined cluster centers represented as cells in a two-dimensional array. In the topological representation of the evolving map during learning, centers can be visualized as they move forcing the two-dimensional map to stretch in the effort to cover the feature space. Even if SOMs are not a derivation of any biological model, some parallelism with the visual neocortex both in terms of function and structure has been drawn [3].

Given the impact of SOMs in machine learning and the excitement produced by a simple and effective unsupervised learning algorithm, it is not surprising that a large number of papers followed in Kohonen's, and that research on SOMs is still carried on actively. Fritzke proposed structures that grow from a small number of cells [4] modulating the network structure accordingly to the unknown probability distribution of the input.

Other research on SOMs, similarly to the evolution of multi-layered supervised neural networks, introduced some hierarchical organization, as Choi and Park did with their "Self-Creating and Organizing Neural Network" [5], or Rauber et al. with their "Growing Hierarchical SOM" [6], or followed the path of hardware implementation of SOMs either in analog [7], or digital form [8]. A surveillance application was proposed by Chacon-Murguia and Gonzalez-Duarte that mixes SOMs with neuro-fuzzy networks to detect objects in dynamic background for surveillance purposes [9].

At some point, Dingle and Jones proposed the Chaotic Self Organizing Feature Map [10] based on recurrent functions leading to chaotic behavior. Continuing this research, more recently Silva et al. proposed a self-organizing recursive architecture for continuous learning [11]. The importance of chaotic dynamics in self-organization will re-emerge in the discussion about computational frameworks.

The relevance of the previously cited work notwithstanding, "self-organization" –in the biological sense– capabilities, should rather be attributed to systems capable of self-assembling from simple elementary units, finding their coordination by direct interactions governed by simple mathematical rules – intrinsic of Nature, it could be stated. Such systems should rather find their biological model in the "prebiotic soup", in which chemical interactions between atoms and then compounds, lead to the organization of cells, tissues and complex organisms.

Random Boolean Networks (RBNs) were originally introduced by Kauffmann to model gene regulation mechanisms [12]. In Kauffman's Biology-centered view, evolution is resulting from the Darwinian selection mechanisms coupled with self-organization, adding the concept of "anti-chaos" to the already large and unsettled realm of complex systems.

Cells in biological systems share the same genetic information, nevertheless they differentiate on the basis of specific activation patterns of that very same information. From a computer engineering point of view, this is an upside-down perspective, as changing the software has been the way to make machines adapt to problems.

Indeed, the dichotomy of hardware and software has been the key of early digital computers evolution, permitting to get rid of the hand-wired logic of primordial calculators. Incidentally, if we put apart for a moment most of the involved technological considerations, Von Neumann's pioneering work on self-reproducing automata was a fifty years forward leap to meet the biological research at some common point.

More or less in the same years as Kauffman, Wolfram meticulously described the complex behavior of one-dimensional cellular automata (Figure 1) pointing out the emergence of self-organization [13, 14].

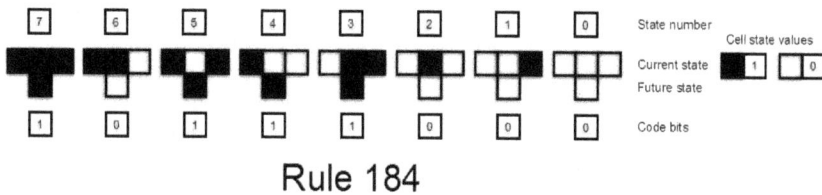

Rule 184

Figure 1. Simple one-dimensional binary cellular automata with two-cell neighborhood are named after the code introduced by Wolfram. The eight possible states for each cell and its neighborhood are arranged from right to left accordingly to the decimal interpretation of the three bit current values. The eight bits describing the future states are then interpreted as a decimal number. The code may be generalized to elementary cellular automata with any number of states, dimensionality, and neighborhood size.

Coincidentally, at that time the influence of Mandelbrot's work on fractals was at its peak, as well as the interest for simple formulae able to produce results so similar to those of natural processes [15]. That was certainly a rather inspiring time for those who are subject to the fascination of complexity arising from simplicity but, indeed, this feeling has been pervading the research in information systems for decades.

It was also the time of the advent of networking and – a few more years would have taken the Web to be brought to life – Internet. The latter has the mark of a "self-organizing" system well in its roots, and even in its name, in some way. Then, in a short time lapse, wireless networks broadened the communication horizon once again providing us with mobile systems.

The realm of computers thus has reached a point where interconnected systems at macro scale coexist with the micro scale of the circuits they are built upon, while the nano-dimensionality is being intensively explored. Compared to the many scales adopted to observe biological systems at their molecular, cellular, tissutal and macroscopic levels, this is still a very coarse and rigid stratification, nevertheless provides an interesting parallelism and some points to look at in the distance.

Either in the biological or in the computer realm, more levels bring more complexity. Systems are of different kinds and it needs some "handshaking", as in the networking jargon, to allow communication. Systems need to share resources, and then some sort of arbitration is needed. A large part of the engineering of information systems has thus become the design of communication and arbitration protocols to make system "self-organize".

Heylighen and Gershenson invoked self-organization in computers as a necessity to cope with an increasing complexity in information systems that creates a "bottleneck" limiting further progress [16]. They discussed inter-networking, and the rapid changes in hardware, software and protocols to deal with it, as only exacerbating the difficulties for human developers to keep everything under their own control. They then described a few qualitative principles to introduce self-organization in highly engineered and inter-networked information systems, with some references to current applications such as the hyperlinks-based Web, and, with some projections to the future, even software development paradigms.

Kohonen's networks, the medium access control and routing protocols of the many computer network types, and Kauffman's RBNs, all of them express self-organization of some degree. The heterogeneity of the three examples is evident, though. Some effort has been taken to formalize this hardly sought property of systems. Gershenson and Heylighen, moving from classical considerations based on thermodynamics, and then considering statistical entropy, provided an insight on what conditions should describe the emergence of self-organization in observed systems.

They concluded that the term "self-organization" may rather describe a way to look at systems than a class of systems [17].

3. Computational models

As anticipated, Random Boolean Networks (RBNs) trace in their biological model of inspiration their self-organizing abilities. RBNs consist in a network of N nodes with Boolean state, each having K of Boolean input connections. Both parameters N and K are fixed. Because of these characterizing parameters RBNs have also been called NK networks. Each node state is updated at discrete time steps accordingly to a Boolean function of the inputs. A variable number of Boolean outputs, propagating the node state, may departs from each node towards other nodes' inputs, arbitrarily. Indeed, both connections and the Boolean state update function are chosen randomly during initialization and never changed (Figure 2).

Kauffman discussed RBNs as finite discrete dynamical systems in terms of the sequences of states the networks run through. Given that 2^N states can be assumed by RBN, and that for each state there is only one possible successor, the network will run through finite-length cyclic sequences called *state cycles*, that are the *dynamical attractors* of the system.

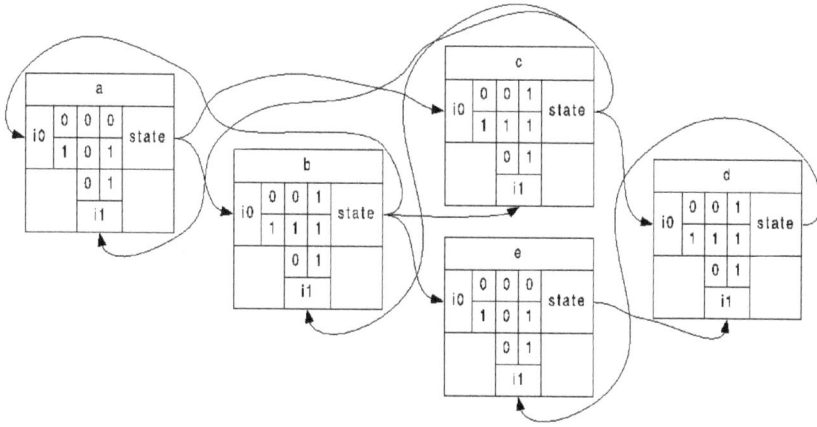

Figure 2. A Random Boolean Network with two input nodes. After each discrete time step, each node state is updated accordingly to the Boolean function of the inputs i0 and i1. Node state is fed to the output at the following step.

The behavior of a RBN can be represented by the state transition diagram, a directed graph having a connected component for each state cycle. Not all the states in each of these subgraphs are part of the respective cycle, as states having no antecedents – the so-called *garden-of-Eden* states – may be present; they instead compose the state cycle's *basin of attraction* (Figure 3).

Properties of the *state cycles*, such as cycle length, asymptotic patterns, and *basins of attraction* were used to classify the interesting complex behaviors of these simple models for different values of K [12]. Some basic findings, still providing some insights into the self-organization abilities of RBNs are reported in Table 1. When the network is completely interconnected ($K = N$), and the sensitivity to initial conditions is at its maximum, *state cycle* lengths become large as N increase, yet their number keeps being comparatively small.

When K is equal or greater than 5, RBNs keeps showing chaotic behavior. A few concepts need to be introduced to analyze these results. The *internal homogeneity* P of a Boolean function of K inputs is defined as the ratio $M / 2^K$, with M beeing the maximum between the number of 1's and 0's in the output column of the function's truth table. The *bias* B is then defined as $1 / \sqrt{P}$.

In contrast with the first two *chaotic* cases, when $K = 2$, RBNs show the emergence of "spontanous order" as both the cycle length and number of attractors scales with the square root of N. Moreover, these networks show other important properties that result in higher stability over perturbations of the activity of the nodes. Indeed, more recently, a linear dependance was found sampling larger networks [18]. For $K = 1$ the RBNs show a similar growth of the cycle length and an exponential rise of the number of attractors.

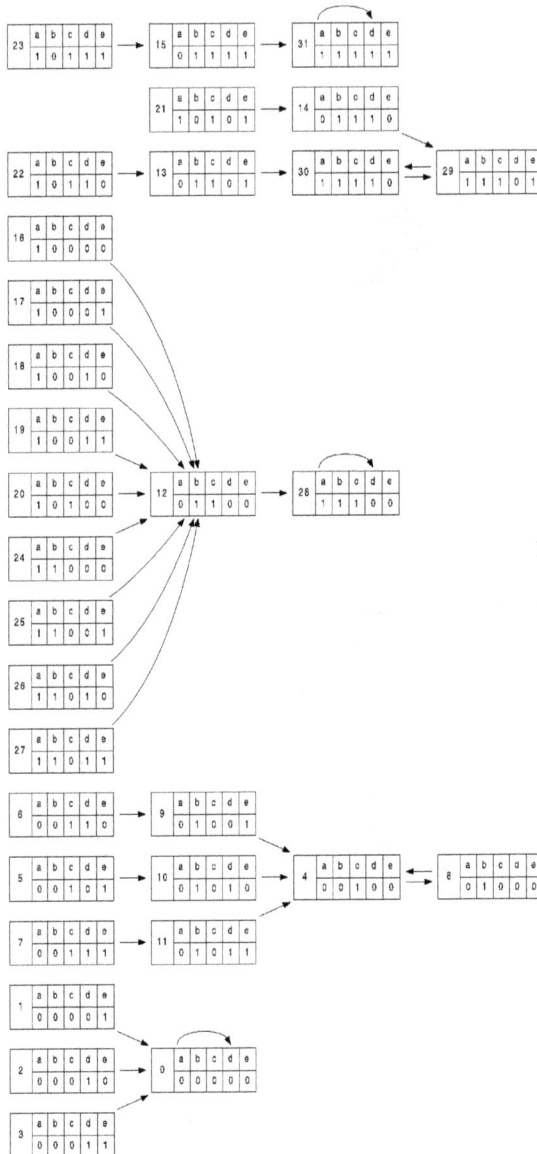

Figure 3. The state transition diagram of the RBN showed in Figure 2. States are numbered from 0 to 31 according to the binary representation of the five network nodes' values. The graph is partitioned into four unconnected components, one for each *state cycle*. The network will finally be attracted into one of the four *state cycles*: (31), (30, 29), (28), (4, 5), (0). On the left side, the 17 states that are unreachable from any other state are showed.

	State cycle length	Number of state cycle attractors
$K = N$	$\dfrac{2^{N/2}}{2}$	$\dfrac{N}{e}$
$K \geq 5$	$\dfrac{2^{\frac{BN}{2}}}{2}$ $(B > 1)$	$\sim N\left[\dfrac{\log\left(\frac{1}{1/2 \pm (P(K)-1/2)}\right)}{2}\right]$
$K = 2$	$\sqrt{N}$	$\sqrt{N}$
$K = 1$	$\sqrt{\frac{\pi}{2}}\sqrt{N}$	Exponential in N

Table 1. Properties of RBNs for different values of K. The state cycle length is the median length of the state cycles. B is the network *bias* while $P(K)$ is the mean *internal homogeneity* of all the Boolean functions of K inputs.

The boundaries among the ordered, critical and chaotic dynamical phases, yet not quite analytically assessed, still inspire new studies. An updated introduction to RBNs, including references to Asynchronous RBNs (ARBNs), Deterministic Asynchronous RBNs (DARBNs) and other variants of RBNs, can be found in form of a tutorial in [18]. Gershenson described several methods to guide the self-organization of RBNs [19]. The need of a "guiding" process seems somewhat contradictory with the premise in the title. Indeed, he investigated the mechanisms through which natural selection may intervene in the self-organization of biological structures, suggesting engineers may use the same parameters characterized in computational frameworks, such as RBNs and Cellular Automata.

Getting back to the gene regulation mechanisms the RBNs were designed to model, "self-organization" succeeded in "reducing the complexity" provided by the tens of thousands genes in the human genome to the mere hundreds types of human cells. RBNs are finite state space, deterministic, dynamical, discrete time systems whose self-organizing property derives from having attractors, i.e. states that can be revisited by the network. RBN can be in either a ordered or a chaotic dynamical phase, transitions are possible and the transition from one to the other is "characterized by its criticality".

A static, stable phase preserves information but is prevented from computing or adapting. A chaotic phase provides the requested variability for computing and adapting, but is incapable of preserving information. As the critical "interface" between the two phases provides the advantages of both phases, guiding the RBN towards self-organizations means finding the necessary conditions to make RBNs evolve towards the critical regime. Gershenson then considered several factors that can induce such evolution and gives a few hints on how criticality could help improve adaptability, ability to evolve, and robustness of RBNs.

4. Cellular automata

From definition it is evident that cellular automata are a special case of RBNs in which each node receives inputs only from neighbors. Their simpler topology and consequent implementation has given an appeal to these minimal models manifesting self-organization that goes beyond "recreational" applications such as Conway's "Game of Life".

Some theoretical extensions include Probabilistic Cellular Automata (PCA), and Fault-tolerant Cellular Automata (FCA), studied by Gács [20] in the context of the problem of reliable computation with unreliable components. In particular, the error probability of each component is not required to decrease as the number of components increases, and the faults affect the local state but not the transition function.

Even if they are purely theoretical, such models may be useful in designing massively parallel "self-organizing" architectures, as due to the distributed nature of information in cellular automata, "self-stabilization" is required beside traditional error-correction techniques.

Cellular automata have had many applications to real-world problems. No surprisingly, several biological models have been simulated with cellular automata. Shumate et al. described a simulation of Ductal Carcinoma in Situ [21]. Chaudary et al. proposed a simulation model for Tumorigenesis [22]. Shimokawa and Muraki investigated a simple model of nerve excitement propagation [23]. Sakamoto et al. proposed a method for surgery simulation based on voxel automata [24].

Cellular automata have also been used to model complex dynamics such as those of urban traffic [25–27]. Recent applications of cellular automata to image processing include super pixel segmentation [28], image compression [29], and computer graphics [30].

Cellular automata also continue to be used in more theoretical studies on algorithms [31, 32].

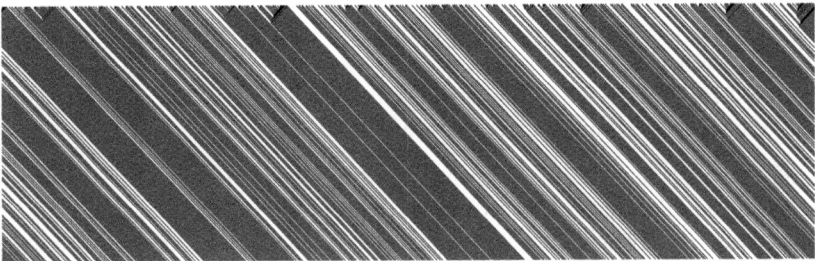

Figure 4. Rule 184 (see Figure 1) is one of most used cellular automaton in traffic simulation. The distribution of vehicles and spaces in a road lane is modeled as black and white cells, respectively, in each image row. The topmost cell row depicts the initial distribution (with a black/white ratio of 3/8) let evolve over 300 iterations. After few iterations still presenting random behavior, visible in form of triangular structures, the regularization ability of the rule is manifest as vehicles move to the right at constant speed.

5. Hardware

The balance between the hardware and software components in signal processing applications has always been a trade-off between the flexibility of the microprocessor-based solutions and the performance of ASIC implementations.

Taking the aforementioned SOMs and ANNs into account, literature abounds in hardware implementations that are motivated by scarce performance of the analogous sequential methods. In the 1980s, aiming at parallel real-time signal processing with the then available analog very-large-scale integration (VLSI) technology, Chua introduced his Cellular Neural Networks (CNN) [33], describing some application to image processing tasks [34]. Subsequently, Yang et al. showed a VLSI implementation of CNNs [35].

A couple of decades later, Ruckert et al. discussed massively parallel implementations of artificial neural networks at ultra-large-scale integration (ULSI) [36], later showing a massively parallel architecture for SOMs [37].

Hopfield had started its seminal work on recurrent neural networks posing himself the question whether "the ability of large collection of neurons to perform computational tasks may in part be a spontaneous collective consequence of having a large number of interacting simple neurons". He concluded that this is actually the case, and that implementation of such models could lead to integrated circuits that are more fault-tolerant than normal circuits [38].

Weightless Neural Networks (WNNs), being based on random access memories, provide another ANN paradigm inherently tied to circuit-level implementation whose origins trace back to Alexander's "Self-adaptive universal logic circuits" [39].

Though all of these are examples of systems having self-adapting qualities, self-organization at the hardware level – the "microscopic" layer in the our biological analogy– had simply not been possible until the advent of reconfigurable circuits, such as field programmable gate arrays (FPGAs), and coarse-grain reconfigurable arrays, added a new degree of configurability, and related complexity, to computer systems. A survey on reconfigurable hardware with emphasis on real-time configuration is provided by Shoa and Shirani [40].

5.1. Coarse-grained and fine-grained architectures

Hartenstein, reviewing most of the "coarse-grained reconfigurable architectures" of a decade (circa 2000) [41], suggested that with the explosion of design costs and reduction of production life cycles, performance becomes relatively less important in the design of computing devices. Instead, extension of product longevity, "reduction of support turnaround, in-system debugging, profiling, verification, tuning, field-maintenance, and field-upgrade" time by employing reconfigurable arrays is much more important. Hartenstein, dismissing "von Neumann" architectures as obsolete, in the light of the dominance of host/accelerators designs, proposed a new coarse-grained soft machine paradigm, in which a so called "co-compilation" provides instructions for the host and data-path configuration information at the same time.

Other approaches to the mapping of high-level coarse-grained mapping from high-level synthesis are present in literature [42]. Recently, implementations as System On a Chip design of highly reconfigurable arrays have been described with applications to face detection [43], Internet protocols processing [44], FIR filters and ICA [45]. Several SOMs, CNNs and derivative neural network models have been designed for reconfigurable hardware [46, 47].

Fine-grained systems bring configurability close to the gate or transistor level, permitting analog, digital, and hybrid implementations. In contrast to coarse-grained systems, data path width is reduced to the bare minimum, with the advantages of increased flexibility and lower costs, but the general purpose routing is generally less energy efficient.

Nanotechnologies aim at even finer degrees of integration, and it is reasonable to assume that at the hardware level new computational paradigms may emerge because of that. However, as in the pioneering stage of any technology, the span from theory and implementation may not be short. Lin et al. proposed a hybrid FPGA architecture based on nano-RAM [48] with run-time configuration abilities and high logic density, to revert later to a CMOS SRAM implementation for immaturity of the nano-RAM fabrication processes [49].

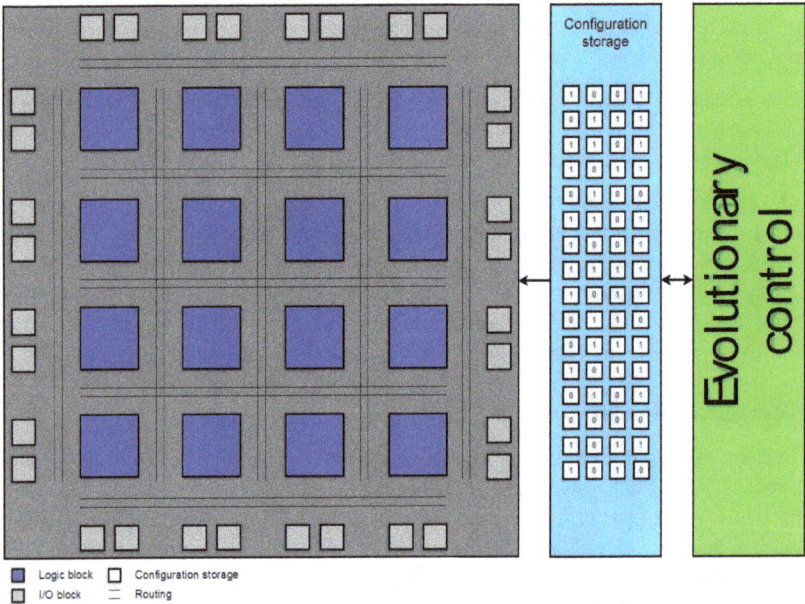

Figure 5. A schematic depiction of Evolvable Hardware. A reconfigurable device (i.e. FPGA), in gray, is coupled with a configuration storage (light blue). Configuration is updated by the control block (green) in real time according to some fitness function, as in genetic programming.

5.2. Evolvable hardware

Reconfigurable hardware is turned into a specific implementation by loading bitstreams compiled from "soft cores" coded in some hardware description language. Being a totally software matter, a host processor can perform real-time reconfiguration when needed. New paradigms blending traditional machine language and reconfigurable hardware bitstreams become thus possible.

The idea that hardware could change autonomously its own configuration seeking the best one according to some fitness function was called evolvable hardware, recurring –once again– to a biological metaphor [50, 51]. Continuing with the metaphor, the bitstream takes the role of the digital DNA (Figure 5).

Approaches from genetic and evolutionary programming are attempted on the hardware configuration bitstream. Interesting applications of EHWs to pattern recognition are those presented by Glette et al. [52, 53].

Even though most work on EHW concerns digital implementations, some evolution-oriented analog implementation are reported in literature, such as the evolvable hardware architecture based on field programmable transistor arrays [54], and quantum-inspired paradigm to be implemented in evolutionary analog hardware [55].

The enthusiasm of the early 2000s notwithstanding, EHW has not yet delivered what promised. Cancare et al. [56] investigating the reasons of this apparently missed success, and citing scalability issues as the most prominent, propose to abandon generic genetic algorithms and look at hierarchical evolution and linkage learning, encouraging support from the Evolutionary Computation community.

6. Networks

Computer networks provide many examples of global behaviors emerging from interactions of elements without centralized control. At different levels of abstraction and implementation, from medium access control and routing, to the application level protocols, algorithms drive each independent network node so that some global goal, be it communication, co- ordination, or distributed processing, is achieved. Thus, non-surprisingly, "self-" prefixed and akin terms abound in related literature.

While computer networks in general are a rather natural field to study self-organization, and many analogies with biological systems may be detected, without broadening too much our discussion, we restrict our discussion considering only one example of network of very simple nodes in which distributed processing of locally collected data is the main goal: Wireless Sensor Networks (WSNs).

These systems are composed by a number of nodes, consisting in miniaturized, battery-operated, computational elements fitted with sensors to monitor the surrounding environment, that are connected through short distance radio links. Depending on the applications, the number of nodes may vary sensibly from a few to thousands and more units. In many scenarios sensor nodes are dispersed in the environment thus their expendability becomes another important requisite.

As a consequence of these constraints on physical size, energy supply and cost per unit, processing resources are limited, and in most designs they only consist in simple microcontrollers. Comprehensive surveys on WSNs including sensor technologies, network protocols, and hardware and software architectures, are provided by Akyldiz et al. [57], and Yick et al. [58].

Even though WSNs were conceived as distributed sensing architectures, several examples are provided in literature about nodes also performing in-network pre-processing of raw sensed data [59]. The need for a trade-off between the limited available energy source, and the manifold application scenarios [60], typically calls for the application of self-organization techniques, breaking the boundaries between the traditional architectural layers in order to optimize the behavior of such nodes. Sohrabi et al. presented a number of algorithms and protocols for self-organization of wireless sensor networks [61]. Self-organization techniques to reduce energy consumption in ad-hoc networks of wireless devices were described by Olascuaga-Cabrera et al. [62]. With even more technological constraints than WSNs, Wireless Sensor Body Networks (WSBNs) consist of wearable and implantable devices. Health-monitoring usage of WSBNs is discussed by Hao and Foster [63].

Indeed, due to their ultra-low energy consumption requirements, WSBNs represent a very challenging scenario for sensor devices based on current, and even near future, general purpose processing elements, and implementing signal processing algorithms on nodes may prove unfeasible.

Alternative approaches based on application specific integrated circuits have been investigated [64]. Departing from the network oriented vision, and calling for the establishment of self-managing systems engineering, Beal et al. proposed the "amorphous medium" abstraction in which the deployed sensor network represents the physical space of the application to be engineered [65].

From an engineering perspective, the application goal is reached by programming the medium instead of the network. The former abstracts the computational model, turning sensor nodes into points of the physical space. A global behavior is described in a specifically crafted language, as it were to be executed by the abstract medium. Actually, the abstract description is compiled into code to be executed identically on each node. Besides executing the same code, nodes interact only with neighboring devices.

Beal et al. called this programming paradigm amorphous computing, revealing their inspiration to come from some properties of biological systems, such as morphogenesis and regeneration. More interestingly, even though with substantial topological differences, many similarities can be detected between the "amorphous-medium" and cellular automata.

7. Conclusions

The paradoxical fascination of simplicity producing complexity has traversed decades of research in information systems. Even more now that extremely high integration is pushing millions of highly modular circuits in few square millimeters, and inter-networking is the next –or rather, current – large scale integration.

Some research directions seem to suggest that breaking some of the fixed ties in engineered system, letting systems auto-organize in response to the environmental changes, as biological systems have been doing for millions of years, is the way to go to "put some order in the chaos". In support of these indications, self-organizing systems have provided interesting results in modeling complex processes, blurring a little the line between artificial and natural systems.

Other researches seek to extend self-organization to the extreme of self-healing systems able to recognize their own faults and self-repair, while biological applications confirm that taking into account self- organization when studying natural processes, while not an easy task, can provide more comprehensive and effective models.

If all these efforts move on the path towards truly intelligent systems, or even Artificial Life –as some have been suggesting for years– is yet to be discovered, nevertheless it is a very interesting path.

Author details

Daniele Peri and Salvatore Gaglio

DICGIM - University of Palermo, Italy, ICAR - CNR, Palermo, Italy

References

[1] Widrow B, Lehr M. 30 years of adaptive neural networks: perceptron, Madaline, and backpropagation. Proceedings of the IEEE 1990;78(9) 1415 –1442. doi:10.1109/5.58323.

[2] Kohonen T. The self-organizing map. Proceedings of the IEEE 1990;78(9) 1464 –1480. doi:10.1109/5.58325.

[3] Bednar JA, Kelkar A, Miikkulainen R. Scaling Self-Organizing Maps To Model Large Cortical Networks. Neuroinformatics 2001; 275–302.

[4] Fritzke B. Let it grow – self-organizing feature maps with problem dependent cell structure. In: Kohonen T, Ma kisara K, Simula O, Kangas J, editors, Artificial Neural Networks. North-Holland, Amsterdam, 1991;403–408.

[5] Choi DI, Park SH. Self-creating and organizing neural networks. Neural Networks, IEEE Transactions on 1994;5(4) 561 –575. doi:10.1109/72.298226.

[6] Rauber A, Merkl D, Dittenbach M. The growing hierarchical self-organizing map: exploratory analysis of high-dimensional data. Neural Networks, IEEE Transactions on 2002;13(6) 1331 – 1341. doi:10.1109/TNN.2002.804221.

[7] Macq D, Verleysen M, Jespers P, Legat JD. Analog implementation of a Kohonen map with on-chip learning. Neural Networks, IEEE Transactions on 1993;4(3) 456 – 461. doi:10.1109/72.217188.

[8] Ienne P, Thiran P, Vassilas N. Modified self-organizing feature map algorithms for efficient digital hardware implementation. Neural Networks, IEEE Transactions on 1997;8(2) 315 –330. doi:10.1109/72.557669.

[9] Chacon-Murguia M, Gonzalez-Duarte S. An Adaptive Neural-Fuzzy Approach for Object Detection in Dynamic Backgrounds for Surveillance Systems. Industrial Electronics, IEEE Transactions on 2012;59(8) 3286 –3298. doi:10.1109/TIE.2011.2106093.

[10] Dingle A, Andreae J, Jones R. The chaotic self-organizing map. In: Artificial Neural Networks and Expert Systems, 1993. Proceedings., First New Zealand International Two-Stream Conference on. 15 –18. doi:10.1109/ANNES.1993.323092.

[11] da Silva L, Sandmann H, Del-Moral-Hernandez E. A self-organizing architecture of recursive elements for continuous learning. In: Neural Networks, 2008. IJCNN 2008.

[12] Kauffman SA. The Origins of Order: Self-Organization and Selection in Evolution. 1 edition. Oxford University Press, USA, 1993.

[13] Wolfram S. Statistical mechanics of cellular automata. Rev. Mod. Phys. 1983;55 601– 644. doi:10.1103/RevModPhys.55.601.

[14] Wolfram S. Universality and complexity in cellular automata. Physica D: Nonlinear Phenomena 1984;10(1-2) 1 – 35. doi:10.1016/0167-2789(84)90245-8.

[15] Mandelbrot BB. The Fractal Geometry of Nature. WH Freeman and Co., New York, 1982.

[16] Heylighen F, Gershenson C, Staab S, Flake G, Pennock D, Fain D, De Roure D, Aberer K, Shen WM, Dousse O, Thiran P. Neurons, viscose fluids, freshwater polyp hydra-and self-organizing information systems. Intelligent Systems, IEEE 2003;18(4) 72 – 86. doi:10.1109/MIS.2003.1217631.

[17] Gershenson C, Heylighen F. When Can We Call a System Self-Organizing? In: Banzhaf W, Ziegler J, Christaller T, Dittrich P, Kim J, editors, Advances in Artificial Life, volume 2801 of Lecture Notes in Computer Science. Springer Berlin / Heidelberg, 2003;606–614. doi:10.1007/978-3-540-39432-7 65.

[18] Gershenson C. Introduction to Random Boolean Networks. In: Bedau M, Husbands P, Hutton T, Kumar S, Suzuki H, editors, Workshop and Tutorial Proceedings, Ninth

International Conference on the Simulation and Synthesis of Living Systems (ALife IX). Boston, MA, 160–173.

[19] Gershenson C. Guiding the Self-organization of Random Boolean Networks. ArXiv e-prints 2010;1005.5733.

[20] Gacs P. Reliable cellular automata with self-organization. In: Foundations of Computer Science, 1997. Proceedings., 38th Annual Symposium on. 90 –99. doi: 10.1109/SFCS.1997.646097.

[21] Shumate S, El-Shenawee M. Computational Model of Ductal Carcinoma In Situ: The Effects of Contact Inhibition on Pattern Formation. Biomedical Engineering, IEEE Transactions on 2009;56(5) 1341 –1347. doi:10.1109/TBME.2008.2005638.

[22] Chaudhary S, Shin SY, Won JK, Cho KH. Multiscale Modeling of Tumorigenesis Induced by Mitochondrial Incapacitation in Cell Death. Biomedical Engineering, IEEE Transactions on 2011;58(10) 3028 –3032. doi:10.1109/TBME.2011.2159713.

[23] Shimokawa K, Muraki S. A study on spatial and temporal visual simulation of nerve excitement propagation. In: Neural Networks, 2000. IJCNN 2000, Proceedings of the IEEE-INNS-ENNS International Joint Conference on, volume 1. 217 –221 vol.1. doi: 10.1109/IJCNN.2000.857839.

[24] Sakamoto Y, Tuchiya K, Kato M. Deformation method for surgery simulation using voxel space automata. In: Systems, Man, and Cybernetics, 1999. IEEE SMC '99 Conference Proceedings. 1999 IEEE International Conference on, volume 4. 1026 – 1031 vol.4. doi:10.1109/ICSMC.1999.812551.

[25] Wei J, Wang A, Du N. Study of self-organizing control of traffic signals in an urban network based on cellular automata. Vehicular Technology, IEEE Transactions on 2005;54(2) 744 – 748. doi:10.1109/TVT.2004.841536.

[26] Rosenblueth DA, Gershenson C. A model of city traffic based on elementary cellular automata. Complex Systems 2011;19(4) 305–322.

[27] Gershenson C, Rosenblueth DA. Self-organizing traffic lights at multiple-street intersections. Complexity 2012;17(4) 23–39. doi:10.1002/cplx.20392.

[28] Wang D, Kwok N, Jia X, Fang G. A Cellular Automata approach for superpixel segmentation. In: Image and Signal Processing (CISP), 2011 4th International Congress on, volume 2. 1108 –1112. doi:10.1109/CISP.2011.6100339.

[29] Cappellari L, Milani S, Cruz-Reyes C, Calvagno G. Resolution Scalable Image Coding With Reversible Cellular Automata. Image Processing, IEEE Transactions on 2011; 20(5) 1461 –1468. doi:10.1109/TIP.2010.2090531.

[30] Debled-Rennesson I, Margenstern M. Cellular automata and discrete geometry. In: High Performance Computing and Simulation (HPCS), 2011 International Conference on. 780 –786. doi:10.1109/HPCSim.2011.5999908.

[31] OrHai M, Teuscher C. Spatial Sorting Algorithms for Parallel Computing in Networks. In: Self-Adaptive and Self-Organizing Systems Workshops (SASOW), 2011 Fifth IEEE Conference on. 73 –78. doi:10.1109/SASOW.2011.10.

[32] Maignan L, Gruau F. Convex Hulls on Cellular Spaces: Spatial Computing on Cellular Automata. In: Self-Adaptive and Self-Organizing Systems Workshops (SASOW), 2011 Fifth IEEE Conference on. 67 –72. doi:10.1109/SASOW.2011.14.

[33] Chua L, Yang L. Cellular neural networks: theory. Circuits and Systems, IEEE Transactions on 1988;35(10) 1257 –1272. doi:10.1109/31.7600.

[34] Chua L, Yang L. Cellular neural networks: applications. Circuits and Systems, IEEE Transactions on 1988;35(10) 1273 –1290. doi:10.1109/31.7601.

[35] Yang L, Chua L, Krieg K. VLSI implementation of cellular neural networks. In: Circuits and Systems, 1990., IEEE International Symposium on. 2425 –2427 vol.3. doi: 10.1109/ISCAS.1990.112500.

[36] Ruckert U. ULSI architectures for artificial neural networks. Micro, IEEE 2002;22(3) 10 –19. doi:10.1109/MM.2002.1013300.

[37] Porrmann M, Witkowski U, Ruckert U. A massively parallel architecture for self- organizing feature maps. Neural Networks, IEEE Transactions on 2003;14(5) 1110 – 1121. doi:10.1109/TNN.2003.816368.

[38] Hopfield JJ. Neural networks and physical systems with emergent collective computational abilities. Proceedings of the National Academy of Sciences 1982;79(8) 2554– 2558.

[39] Aleksander I. Self-adaptive universal logic circuits. Electronics Letters 1966;2(8) 321 – 322. doi:10.1049/el:19660270.

[40] Shoa A, Shirani S. Run-Time Reconfigurable Systems for Digital Signal Processing Applications: A Survey. The Journal of VLSI Signal Processing 2005;39 213–235. 10.1007/s11265-005-4841-x.

[41] Hartenstein R. A decade of reconfigurable computing: a visionary retrospective. In: Design, Automation and Test in Europe, 2001. Conference and Exhibition 2001. Proceedings. 642 –649. doi:10.1109/DATE.2001.915091.

[42] Lee G, Lee S, Choi K. Automatic mapping of application to coarse-grained re- configurable architecture based on high-level synthesis techniques. In: SoC De- sign Conference, 2008. ISOCC '08. International, volume 01. I–395 –I–398. doi: 10.1109/SOCDC.2008.4815655.

[43] He C, Papakonstantinou A, Chen D. A novel SoC architecture on FPGA for ultra fast face detection. In: Computer Design, 2009. ICCD 2009. IEEE International Conference on. 412 –418. doi:10.1109/ICCD.2009.5413122.

[44] Badawi M, Hemani A. A coarse-grained reconfigurable protocol processor. In: System on Chip (SoC), 2011 International Symposium on. 102 –107. doi:10.1109/ISSOC. 2011. 6089688.

[45] Jain V, Bhanja S, Chapman G, Doddannagari L. A highly reconfigurable computing array: DSP plane of a 3D heterogeneous SoC. In: SOC Conference, 2005. Proceedings. IEEE International. 243 – 246. doi:10.1109/SOCC.2005.1554503.

[46] Hendry D, Duncan A, Lightowler N. IP core implementation of a self-organizing neural network. Neural Networks, IEEE Transactions on 2003;14(5) 1085 – 1096. doi: 10.1109/TNN.2003.816353.

[47] Starzyk J, Zhu Z, Liu TH. Self-organizing learning array. Neural Networks, IEEE Transactions on 2005;16(2) 355 –363. doi:10.1109/TNN.2004.842362.

[48] Zhang W, Jha NK, Shang L. A hybrid nano/CMOS dynamically reconfigurable system–Part I: Architecture. J. Emerg. Technol. Comput. Syst. 2009;5(4) 16:1–16:30. doi: 10.1145/1629091.1629092.

[49] Lin TJ, Zhang W, Jha NK. SRAM-Based NATURE: A Dynamically Reconfigurable FPGA Based on 10T Low-Power SRAMs. Very Large Scale Integration (VLSI) Systems, IEEE Transactions on 2011;PP(99) 1 –5. doi:10.1109/TVLSI.2011.2169996.

[50] Yao X, Higuchi T. Promises and challenges of evolvable hardware. Systems, Man, and Cybernetics, Part C: Applications and Reviews, IEEE Transactions on 1999;29(1) 87 –97. doi:10.1109/5326.740672.

[51] Forbes N. Evolution on a chip: evolvable hardware aims to optimize circuit design. Computing in Science Engineering 2001;3(3) 6 –10. doi:10.1109/5992.919259.

[52] Glette K, Torresen J, Yasunaga M, Yamaguchi Y. On-Chip Evolution Using a Soft Processor Core Applied to Image Recognition. In: Adaptive Hardware and Systems, 2006. AHS 2006. First NASA/ESA Conference on. 373 –380. doi:10.1109/AHS.2006.55.

[53] Glette K, Torresen J, Hovin M. Intermediate Level FPGA Reconfiguration for an On-line EHW Pattern Recognition System. In: Adaptive Hardware and Systems, 2009. AHS 2009. NASA/ESA Conference on. 19 –26. doi:10.1109/AHS.2009.46.

[54] Stoica A, Zebulum R, Keymeulen D, Tawel R, Daud T, Thakoor A. Reconfigurable VLSI architectures for evolvable hardware: from experimental field programmable transistor arrays to evolution-oriented chips. Very Large Scale Integration (VLSI) Systems, IEEE Transactions on 2001;9(1) 227 –232. doi:10.1109/92.920839.

[55] Wang Y, Shi Y. The application of quantum-inspired evolutionary algorithm in analog evolvable hardware. In: Environmental Science and Information Application Technology (ESIAT), 2010 International Conference on, volume 2. 330 –334. doi: 10.1109/ESIAT.2010.5567359.

[56] Cancare F, Bhandari S, Bartolini D, Carminati M, Santambrogio M. A bird's eye view of FPGA-based Evolvable Hardware. In: Adaptive Hardware and Systems (AHS), 2011 NASA/ESA Conference on. 169 –175. doi:10.1109/AHS.2011.5963932.

[57] Akyildiz I, Su W, Sankarasubramaniam Y, Cayirci E. A survey on sensor networks. Communications Magazine, IEEE 2002;40(8) 102 – 114. doi:10.1109/MCOM.2002. 1024422.

[58] Yick J, Mukherjee B, Ghosal D. Wireless sensor network survey. Computer Networks 2008;52(12) 2292 – 2330. doi:10.1016/j.comnet.2008.04.002.

[59] Gatani L, Lo Re G, Ortolani M. Robust and Efficient Data Gathering for Wireless Sensor Networks. In: Proceedings of the 39th Annual Hawaii International Conference on System Sciences, HICSS'06. IEEE Computer Society, 235–242.

[60] Anastasi G, Lo Re G, Ortolani M. WSNs for Structural Health Monitoring of His- torical Buildings. In: Proceedings of HSI'09. The 2nd Conference on Human System Interactions. IEEE, 574–579.

Chapter 11

Progress of Doppler Ultrasound System Design and Architecture

Baba Tatsuro

Additional information is available at the end of the chapter

1. Introduction

Evolution of electronic technology and semiconductor technology in recent years can realize a high-speed and high-quality signal-processing with low cost, low size, and low power consumption. Various signal-processing devices were born, and their performances are continuing developing. This article introduces the technical innovations and the effects of digital signal-processing in accordance with generations of the Doppler ultrasound system architecture. The diagnostic image of the carotid artery by a Doppler ultrasound system is shown in Fig. 1. The upside image is a tomogram called color flow mapping (CFM). A Doppler range gate is set up in the center of the blood vessel in the CFM. Bloodflow information on this position is displayed as the spectrum Doppler image in the downside. The horizontal axis is time, and the vertical axis is the flow velocity corresponding to Doppler shift frequency, and it expresses the time-change of velocity distribution of the bloodflow. The embedded technology of CFM and spectrum Doppler began from the composition of analog signal-processing and primitive logical operation elements, and resulted to accumulator devices, PAL, various memories, and changed to FPGA, CPLD, ASIC, DSP, and CPU/GPU [1].

2. Progress of Doppler signal-processing architecture

2.1. The 1st generation architecture (Fixed-point processing)

Doppler signal-processing has developed selecting the most suitable realization method in all generations. Architecture of the 1980s is shown in Fig. 2. Analog signal-processing (dark-orange block in Fig. 2) occupied most in this architecture. Henceforth, this is called the 1st generation architecture. Since only fast Fourier transform (FFT) was the digital signal-processing, analog-digital converter (ADC) was arranged before FFT. In those days, the conversion speed of ADC was hundreds kHz in 12-16 bits. Since a complex butterfly-operation was required, FFT processing was realized by accumulators (TRW: 1010J) in the first stage. After a while, a fixed point DSP (Toshiba: DSP-T9508) was used from the second half of the 1980s.

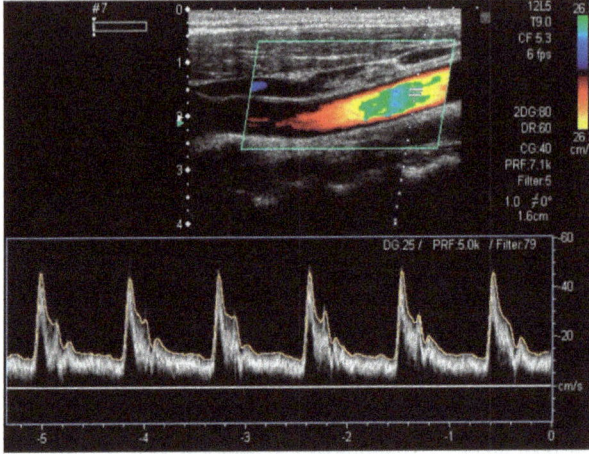

Figure 1. Diagnostic image of Doppler ultrasound system

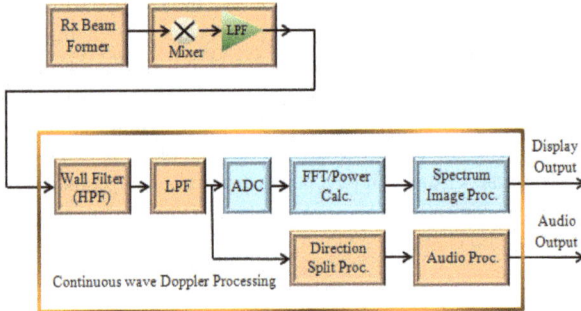

Figure 2. The 1st generation architecture

2.2. The 2nd generation architecture (Floating point DSP and ASIC)

The development of full digital system started early in the 1990s. The early digital architecture is shown in Fig. 3. Henceforth, this is called the 2nd generation architecture. An analog low-pass filter (LPF) was arranged after an analog high-pass filter (HPF) in the conventional analog signal-processing. Since it was difficult to realize a high-speed digital HPF (wall filter in Section 5) with low-cutoff, HPF and LPF were replaced. Furthermore, the LPF with high sampling frequency was divided into two subcomponents. The first stage LPF was realized by FPGAs (Altera: FPGA), and the next stage LPF and HPF were realized by floating point DSPs (NEC: μPD77240A), respectively. Furthermore, FFT processing was realized by an ASIC (Toshiba: ASIC). However, as for Doppler audio processing (direction separation of complex signal, etc.), the conventional analog-circuit was used in consideration of cost-performance. Therefore, the digital filter output was converted into analog signal again by digital-analog convertoer (DAC), and was inputted into analog-circuit.

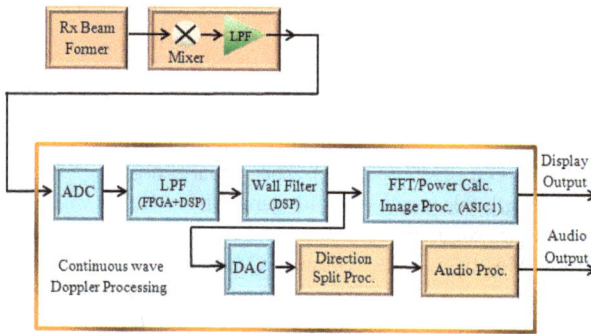

Figure 3. The 2nd generation architecture

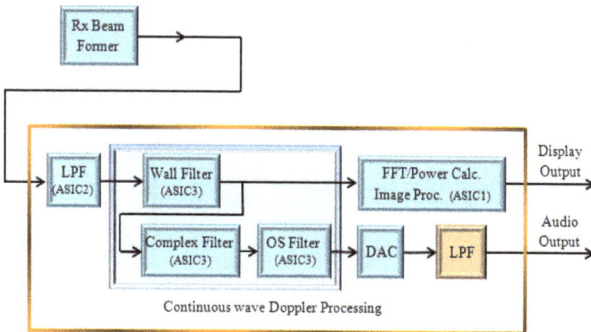

Figure 4. The 3rd generation architecture

2.3. The 3rd generation architecture (Dynamic-range expansion by ASIC)

In the second half of the 1990s, in order to merge CFM and spectrum Doppler, the development which reduces the size and cost of these systems started. Henceforth, this is called the 3rd generation architecture. In order to compare generations, only a spectrum Doppler portion is shown in Fig. 4. In the 3rd generation, since the system clock went up sharply, the floating point device was hard to use. CFM and spectrum Doppler were unified, and they were realized by five kinds of ASIC (Toshiba: fixed point ASIC). In this architecture, I adopted the newly developed digital complex IIR filter for the direction separation processing without using analog phase-shifters with heavy manual adjustments [2, 3]. Furthermore, an oversampling filter and high-speed DAC (Analog Device: DAC) were used for the Doppler audio processing. This reduced the analog-circuit, such as high-order switched capacitor filters (SCF). The scale of these large-scale ASIC reached more than twice of typical CPU (Intel: Pentium processor) respectively, but the total cost of spectrum Doppler declined in 1/3. Furthermore we were able to get the wide dynamic-range signal-processing which was difficult in analog processing. As a result, the sensitivity of bloodflow detection had improved and the diagnostic targets also spread to abdomen, surface blood vessels, and limbs etc.

2.4. The 4th generation architecture (Reduction of circuits by large scale DSP)

In the first stage of the 2000s I realized whole Doppler signal-processing using only one floating point DSP (TI: DSP TMS320C6701). The signal-processing block inside DSP is shown in Fig. 5. Henceforth this is called the 4th generation architecture. Since the clock frequency went up tens times compared with the 2nd generation floating point DSP, throughput improved sharply. Moreover, changes of the ultrasound system architecture contributed to downsizing. The interrupt cycle to DSP was changed into display frequency (Vsync: 50-75 Hz) from ultrasonic pulse repetition frequency (PRF: 1-50 kHz). Although real-time performance was spoiled a little by forming packet processing, drastic reduction of circuit scales was realized.

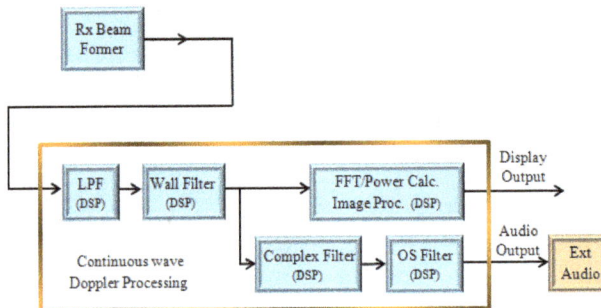

Figure 5. The 4th generation architecture

2.5. The future architecture (Real-time analysis)

Evolution of signal-processing had been influenced by the realization methods, like as from analog to digital, or from hardware to software. The size and cost of Doppler signal-processing were reduced day by day, and its performance (sensitivity or the response etc.) was also improved. It will be possible to gain more huge calculation power for signal-processing from now on. It will be also an important theme to investigate for what this power should be used. Now I am trying to apply this ability to automatic measurement and automatic diagnosis as an intellectual signal-processing [4-6]. Moreover, as another possibility, mounting the Doppler signal-processing on a Windows program is also tried. But problems, such as stability or response time, remain. In order to realize the real-time system which completes the processing within time, I think that the architecture based on DSP will remain.

3. Considerations of digital technology

As introduced in Section 2, the spectrum Doppler signal-processing architecture had changed from analog to digital in the first half of the 1990s. In this section, comparison of analog technology and digital technology and the innovations which digital technology brought about are introduced.

3.1. Comparison of analog processing and digital processing

As compared with analog processing, merits of digital processing are shown in Table 1. Digital processing realized quality improvement (reduction of variations), performance improvement (wide dynamic-range), and size reduction. Moreover, development efficiency was improved by the separation of analog power supply and digital power supply, and by the reduction of the noise in analog systems. Also the cost of digital-circuit had been improved in the 1990s more easily than analog-circuit.

Items	Analog Processing	Digital Processing
Variation caused by electrical parts	Big. (Expensive adjustment for analog-circuit)	Small.
Tolerance to noise	Noise countermeasure must be done to every sub-block. Limitation of dynamic-range	Only ADC needs adjustment. Wide dynamic-range is realizable.
Kinds of power	At least 3 kinds The isolation between digital Power supply and analog power supply is needed.	Only the digital power except for ADC and DAC
Cost performance	Before the 1990s, analog processing had high C/P ,but recently, it is low C/P compared with the digital processing.	Cost, size and power consumption are improving substantially every year.
Throughput	Real time performance is good.	Digital processing realizes the high-speed and complex processing.

Table 1. Comparison of analog processing and digital processing

3.2. Time-spatial resolutions and S/N ratio

By development of digital technology, sampling frequency and pixel size are increasing in a digital camera and a digital audio field every year. When sampling frequency and pixel size increase, finer sampling becomes possible in time and space. The spatial-resolution and the time-resolution have improved recently, and a high-definition image and a high-fidelity audio can be enjoyed now. Moreover, the product performance that exceeds human vision and hearing is also improved. However, from the viewpoint of manufacturing cost, if the target performance to demand is filled, the present performance level may be enough. But the products which exceed this performance actually appeared one after another in the market. I consider this reason as follows. The present product level does not fill the dynamic-range of human vision and hearing. Since human sensitivity perceives physical quantity by logarithm, according to the surrounding environment, a wide dynamic-range is required. That is, I think that the commercial products of digital camera or digital audio have not reached the demand dynamic-range of luminosity or sound pressure yet [7-9].

Aside from improvement in spatial resolution or time resolution, another merit of digitization is a high S/N ratio. A digital signal is sampled in a spacial axis and/or a time-axis. An ensemble mean processing can remove the noise which adjoins in space or time, so it can extract a low-frequency component with sufficient accuracy. An ensemble mean model, a signal level, a noise level, and expansion of S/N ratio are shown in Fig. 6. When the ensemble number is set to N, a signal increases to N times and a noise increases to $\sqrt{N}$ times so an S/N ratio is expanded to $\sqrt{N}$ times [10, 11]. Bandwidth restriction filters (HPF, LPF, BPF etc.) has the same effect of the ensemble mean processing. In digital processing designs, we should be cautious of an internal dynamic-range and an output noise level. In analog system, since the dynamic-range was narrow, a wide dynamic-range signal was not processed faithfully. The artifact caused by saturation or quantization occurred in the intermediate processes, so the sufficient sensitivity of bloodflow detection was not obtained. After the 3rd generation architecture, as the wide dynamic-range signal-processing was realized, the big improvement in bloodflow diagnosis was brought about.

3.3. Hardware reduction by over-sampling

The design concept of the compact disk (CD) which started in the 1980s was that the stereo digital signal (44 kHz sampling) was changed into the stereo analog signals (maintaining 20 kHz bandwidth). For this purpose, the steep analog filter (at least 7th-order) which rejects harmonics after AD conversion output (22 kHz) was required. Several years afterward, 4-times over-sampling system (about 170 kHz) appeared. It was realized by a digital moving average filter (sampling-frequency: about 170 kHz, cutoff frequency: 20 kHz, cutoff property: loose), a 4 times DA conversion, and a simple analog filter (cutoff frequency: 20 kHz, cutoff property: about 2nd-order). Since this system had many merits (reduction of cost and size, improvement of S/N ratio etc.), it became mainstream [12].

The third generation architecture of Doppler ultrasound system was designed based on this over-sampling concept. In Doppler audio processing, unlike CD, a sampling frequency changes widely (1 kHz to 50 kHz). Therefore, the cutoff frequency of digital filter before DA conversion had to be variable. The change range of sampling frequency is the same as not only Doppler audio processing but also digital filters of the 2nd generation. The effect of the over-sampling processing is shown in Fig. 7. Fig. 7(a) shows the sampling characteristic of the sampling frequency fs, and harmonic (a side lobe of -14 dB) is mixed because of a simple over-sampling (hold characteristic). In order to remove these harmonics and to keep the re-quired dynamic-range in required bandwidth, the filter with suitable bandwidth property (broken line) is required. In the case of Fig. 7(b), since fs/BW is small, a high-order filter is required. But in the case of Fig. 7(c), since fs/BW is large, a low-order filter is also fully realiz-able. The sampling frequency of high-speed digital devices is going up now. Since fs/BW is expanded, both downsizing and high-performance are realized simultaneously [13].

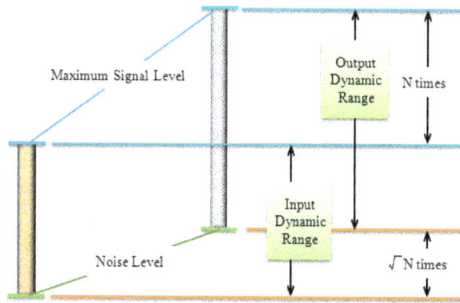

Figure 6. Principal of ensemble mean processing and S/N ratio expansion

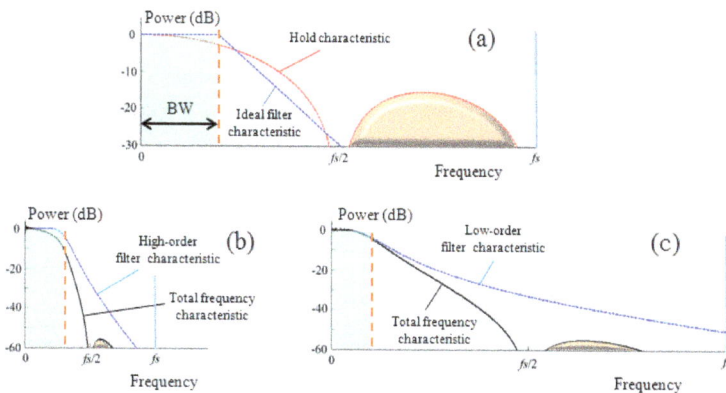

Figure 7. Over-sampling processing (a) Over-sampling model (b) Small fs/BW case (c) Large fs/BW case

4. Cascade digital filter design

The design tools for various digital filters, such as LPFs, HPFs, and BPFs, were popular and spread. However, in order to realize large-scale signal-processing combining various filters which have different sampling frequency, there are many points that should be taken into consideration, such as an aliasing and artifacts. In this section, two design examples of the cascade digital filter are introduced. One is a cascade BPF for continuous wave Doppler processing that has the down-sampling processing exceeding 10^{-4} or less. Another is a cascade LPF for Doppler audio processing that has the over-sampling processing exceeding 10^3 or more.

4.1. Continuous wave Doppler signal-processing

In the Doppler signal-processing, an HPF is effective to extract weak bloodflow. There is a clutter signal which mixes unnecessary reflective ingredients from a blood vessel wall etc. By removing it at the entrance of frequency-analysis, the dynamic-range of FFT processing and audio processing can be held down, and also the signal-processing load can be reduced. High-order HPF array (bandwidth: several kHz, cutoff-frequency: several hundred to several thousand Hz) had been used in the analog system. For digitizing this, the 2nd generation architecture was developed. The filter array for continuous-wave Doppler signal-processing is shown in Fig. 8.

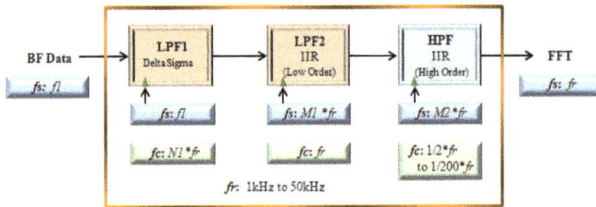

Figure 8. Digital filter design of continuous-wave Doppler signal-processing

In the conventional system, after an analog HPF, a bandwidth restriction was applied by the anti-alias LPF before ADC. To realize the digital HPF which had high-order and high sampling frequency was difficult in those days. For example, when a high-order IIR filter (several MHz and 2ch processing) was assumed, hundreds Mflps performance was required for calculation. Then, LPF was arranged before HPF in digitization. A high-speed and low-cutoff LPF was required. For example, the relative cutoff 1/1000 (that means several kHz bandwidth restrictions with tens of MHz sampling) was required. In order to prevent expansion of the tap-length (number of delay-registers) and bit-length of internal registers (inner dynamic-range), I chose the system which divided LPF into two steps and applied down-sampling. Since the output after quadrature detection was high-speed of tens MHz ($f1$ Hz in Fig. 8), in the front part of LPF the delta-sigma LPF that had cutoff frequency $N1*fr$ was adopted, and it was realized by FPGA.

The latter part of LPF had re-sample frequency $M1^*fr$, and it was realized by the single-precision floating point DSP (low-order LPF, cutoff-frequency: fr). An HPF was arranged after these two steps of LPFs. The HPF carried out scaling-processing to the LPF output by re-sampling frequency $M2^*fr$. High-order and wide-range cutoff HPF processing (cutoff-frequency: $fr/2$ to $fr/200$) was realized by the double-precision floating point DSP. By this architecture, the required bandwidth restrictions and dynamic-range could be realized even in the continuous-wave Doppler processing which had heavy mixing of clatter artifacts. The frequency characteristic of the cascade digital filter (Fig. 8) is shown in Fig. 9. Actually a chirp waveform (0 to 40 kHz) was inputted into the cascade digital filter and its performance was checked by the spectrum Doppler image of the trial product. We can check the loose bandwidth restrictions near ±6 kHz (fr) in this figure.

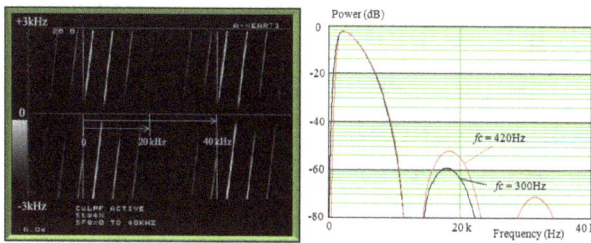

Figure 9. Frequency characteristics of continuous-wave Doppler filter of the 2nd generation architecture

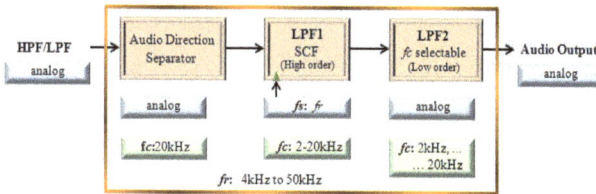

Figure 10. Analog Doppler audio system of the 2nd generation architecture

4.2. Doppler audio signal-processing

After quadrature detection, the Doppler audio system divides IQ signal into a forward component and a reverse component with a direction separation filter, and outputs them to the left and right stereo speakers. In the case of a pulse wave Doppler, since about 4 kHz sampling frequency is interlocked with fr (pulse repetition frequency), it is hard to hear blood-flow signal as it is (by the mixing of harmonics). In order to remove harmonics, it is required to realize the steep filter which has cutoff of $fr/2$ in consideration of audio bandwidth (20 Hz to 20 kHz), and to reject unnecessary harmonics. The conventional analog filter architecture is shown in Fig. 10. In the Fig. 10, LPF1 had the steep cutoff characteristic by SCF after direction separation processing. And LPF2 removed the harmonics generated in LPF1 (SCF noise). The S/N ratio of SCF was 50 dB or less in audio-range, and sound quality was quite bad compared with the present system.

The Doppler audio filter of the 3rd generation is shown in Fig. 11. The signals of direction separation (complex BPF output) were oversampled by LPF1 whose sampling frequency ($M3*fs$) was hundreds times larger than fr. And LPF1 had a loose bandwidth restriction by a moving average. In the following stage LPF2, scaling was applied by $f2$ (same as the ADC clock frequency). LPF2 had the bandwidth restriction by the IIR filter (cutoff frequency: $fr/2$). After AD conversion, in order to remove harmonics, loose bandwidth restriction was again applied by LPF3. This cascade filter processing could realize a high-quality Doppler audio (S/N ratio: more than 90 dB). Moreover, since the oversampling frequency of LPF1 and the conversion frequency of ADC were set up more highly, the simple filter (lower-order) was used and drastic hardware reduction of LPF2 and LPF3 were realized [14].

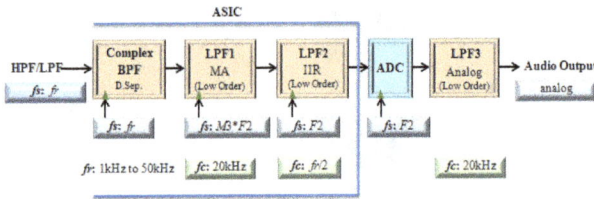

Figure 11. Digital Doppler audio system of the 3rd generation architecture

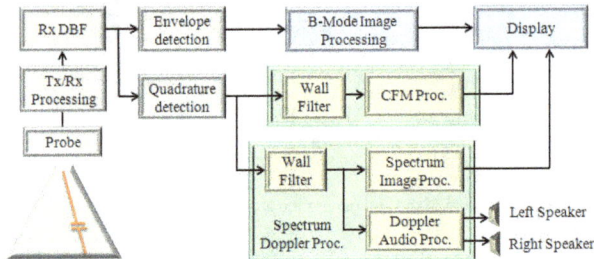

Figure 12. Wall filter arrangements of Doppler ultrasound system

5. High-precision digital filter design

Doppler ultrasound diagnostic method spread to many diagnostic fields, such as cardiac and abdomen. On the other hand, the improvement of bloodflow detection (sensitivity and velocity-range) had been desired for a long time. For this purpose, since it was required to separate weak bloodflow signal from high-power artifacts, like a blood vessel wall, the steep HPF (called a wall filter) had been arranged before frequency-analysis. With digitization, I investigated a new wall filter designs.

5.1. Purpose of wall filter

The locations of wall filters in Doppler ultrasound system are shown in Fig. 12. Quadrature detection outputs (IQ signal) are divided into CFM processing and the spectrum Doppler processing. In both CFM processing and the spectrum Doppler signal-processing, to save dynamic-range, wall filters are arranged before frequency analyses respectively [15].

Beam scanning methods and the display modes corresponding to them are shown in Fig. 13. Fig. 13(a) shows the scanning method of a B-mode echo image (tomogram), and it scans a beam from the right to the left. Fig. 13(b) shows the scanning method of spectrum Doppler, and it scans the same beam in a tomogram continuously. Fig. 13(c) shows the scanning method of CFM, and it scans a beam from the right to the left like Fig. 13(a), but the same beam is scanned twice or more. The sampling methods of Fig. 13(b) and Fig. 13(c) are shown in Fig. 14(a) and Fig. 14(b). Beam data is sampled by fr, and includes the information on the depth direction. As shown in Fig. 14(a), since the spectrum Doppler has a long time series signal, a detailed frequency analysis can be realized. On the other hand, as shown in Fig. 14(b), CFM has plurality data series (hundreds points) on the depth direction. Because CFM processing consists of a finite wall filter and a complex autocorrelation processing, the analysis data of CFM is same as sampling beam number (5 in the case of Fig. 14(b)), and is very small compared with that of the spectrum Doppler.

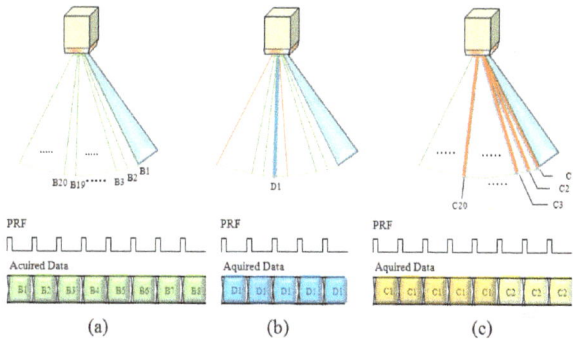

Figure 13. Ultrasound beam scan mode (a) B-mode (b) Spectrum Doppler (c) CFM

The clutter is the high-power and low-frequency component at the quadrature detection output. Compared with a bloodflow signal, it has high-power low-frequency component in abdomen (about 20 dB) and heart (more than 40-60 dB). The quadrature detection outputs and wall filter outputs collected in the heart (left ventricle outflow) are shown in Fig. 15. The horizontal axes of Fig. 15(a) and Fig. 15(b) are time, and vertical axes are amplitude. The horizontal axis of Fig. 15(c) is frequency and a vertical axis is power. The wall filter has the 4th-order Butterworth characteristic with 200 Hz cutoff frequency. The power spectra of Fig. 15(c) show that a big clatter component (20dB bigger) is removed. The wall filter is required high-order (steep) and low-cutoff characteristic [16].

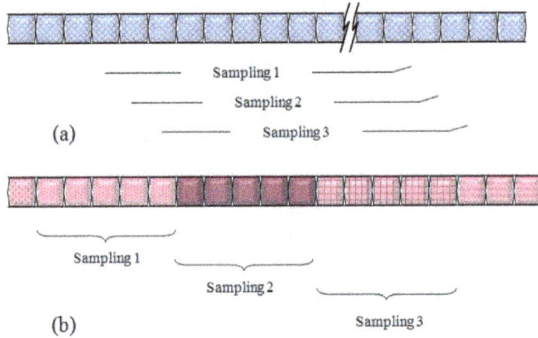

Figure 14. Sampling methods (a) Spectrum Doppler (b) CFM

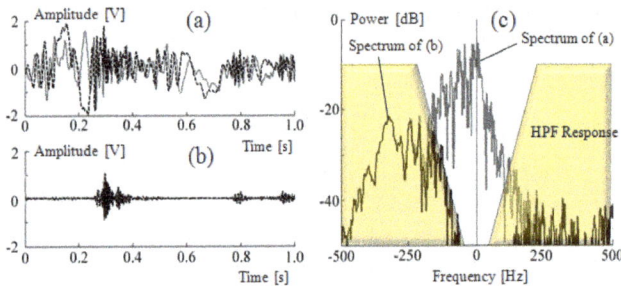

Figure 15. Effect of wall filter (a) Input signals of wall filter (b) Output signals of wall filter (c) Spectra of (a) and (b)

5.2. Wall filter of CFM

The high-order analog filter had been used for the wall filter of spectrum Doppler, and a finite digital filter had been used for the wall filter of CFM for a long time. The step responses of infinite impulse response (IIR) Butterworth filter in the case of changing cutoff and order are shown in Fig. 16. Fig. 16(a) shows the responses when order is changed at the relative cutoff frequency 1/16 (normalized by sampling frequency), and Fig. 16(b) shows the responses at the relative cutoff frequency 1/128. A transient response becomes long at the time of low-cutoff and high-order (steep). Relation between cutoff and order when the transient response is set to -20 dB (10% of step input amplitude) is shown in Fig. 16(c). Since performance of the wall filter with finite input is insufficient, the technology for reducing a transient response was required. The wall filter systems of CFM are shown in Fig. 17. In the finite impulse response (FIR) system of Fig. 17(a), if the number of delay registers N is small, sufficient performance cannot be obtained. So the IIR system of Fig. 17(b) became main-stream. However, unlike the wall filter of spectrum Doppler, the transient response of IIR filter has a serious influence to frequency-analysis. In order to solve this problem the adaptive filter which is consistuted by a time-variant FIR filter shown in Fig. 17(c) appeared recently.

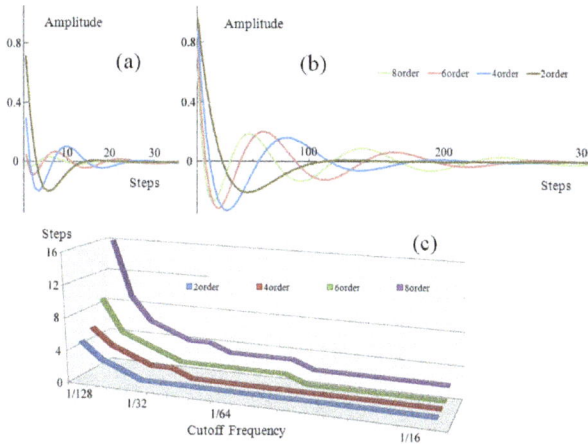

Figure 16. Step responses of HPF (a) Responses @ relative cutoff: 1/16 (b) Responses @ relative cutoff: 1/128 (c) Cutoff and transient response

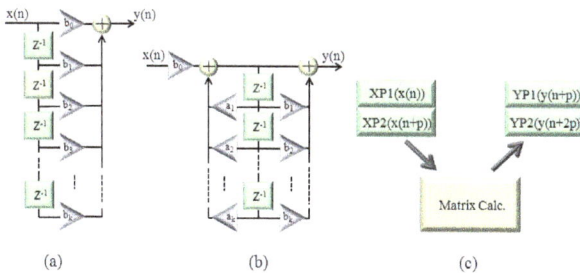

Figure 17. Wall filter systems of CFM (a) FIR filter (b) Biquad filter (c) Time valiant FIR filter

Since there is no feedback in the FIR system of Fig. 17(a), saturation does not occur easily, and the internal dynamic-range can be made small. However, since it is necessary to increase the number of taps in order to obtain low-cutoff, it is disadvantageous in a response and size. Bi-quad filter system of Fig. 17(b) has good response. However, since big internal dynamic-range is required at the time of low-cutoff, there are problems of a quantizing noise or a transient response. Compared with these, although time delay is given as a packet unit, a time-variant FIR filter system shown in Fig. 17(c) has many advantages. The filter response is calculated from packet data based on matrices. By progress of signal-processing device in recent years, the development of an adaptive filter based on the time-variant FIR system became also easy. A time-variant FIR filter with input x(n), output y(n), and the state variable $v(n)$ shown in Fig. 17(c) consists of the state equation and output equation in equation (1).

$$
\begin{aligned}
v(n+1) &= F * v(n) + q * x0(n) \\
y(n) &= g^T * v(n) + d * x(n)
\end{aligned}
\tag{1}
$$

The coefficients of an IIR filter (fig. 17(b)) are transposed into the matrix F, q, g and d in equation (2). The signal-processing equivalent to an IIR system can be realized by a time-variant FIR system.

$$
F = \begin{bmatrix}
0 & 1 & 0 & \cdots & 0 \\
0 & 0 & 1 & \cdots & 0 \\
 & & \cdots & & \\
0 & 0 & 0 & \cdots & 1 \\
-a_K & -a_{K-1} & -a_{K-2} & \cdots & -a_1
\end{bmatrix}
\qquad
q = \begin{bmatrix} 0 \\ 0 \\ \cdots \\ 0 \\ 1 \end{bmatrix}
$$

$$
g = \begin{bmatrix}
b_K - b_0 * a_n \\
b_{K-1} - b_0 * a_{K-1} \\
\cdots \\
b_1 - b_0 * a_1
\end{bmatrix}
\qquad
d = b_0
\tag{2}
$$

Thus, since the time-variant FIR system shown in Fig. 17(c) can solve both the problem of the internal dynamic-range and a transient response, it will develop as a new wall filter of CFM from now on.

5.3. Wall filter of spectrum Doppler

The step responses of 8th-order Butterworth filter (4 cascade biquads, relative cutoff 1/128) are shown in Fig. 18. Fig. 18(a) shows the step input x(n) and the output y(n). Since it is HPF, its output approaches DC with a damped oscillation. The responses of the inner registers for each stage (Z1(n), Z3(n), Z5(n) and Z7(n) in Fig. 20(b)) are shown in Fig. 18(b). Although Z3(n), Z5(n) and Z7(n) are converged on DC with about tens times amplitude of an input, Z1(n) holds about 400 times amplitude of an input. Thus, when the HPF prevents saturation or keeps internal accuracy, wide dynamic-range of internal registers is required. The relation between cutoff and the dynamic-range (the bit-length of inner register) is shown in Fig. 19. In order to realize a high-precision digital filter, accuracy of operation registers and filter co-efficients is important. I checked the minimum bit-length that was not influenced by quantizing noise. The responses of the fixed point 8th-order Butterworth filters were simulated. If quantizing noise is mixed, the unstable oscillation such as a limit cycle etc. will occur. I changed cutoff frequency and measured the limit of stability. As a result, in order to realize low cutoff, it turned out that sufficient mantissa-length of operation-registers and sufficient multiplication-coefficient length of multipliers were required. In fact, since cutoff frequency became about 1/200 in the spectrum Doppler processing, a huge internal dynamic-range (about 200dB) was required.

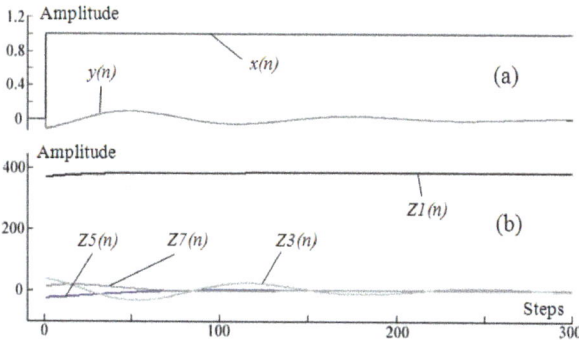

Figure 18. Step response of HPF and transit response (a) Input $x(n)$ and output $y(n)$ (b) Responses of internal resisters

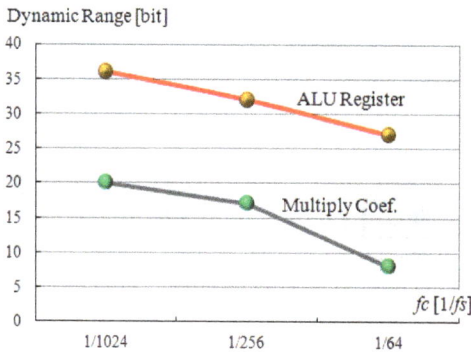

Figure 19. Dynamic-range (bit-length) and cutoff frequency of HPF

The signal-processing architecture which reduces the internal dynamic-range of a digital filter was developed in the 2nd generation architecture. Simultaneously, the algorithm which reduces calculation in a real-time system was also investigated. The systems to realize 8th-order digital filter are shown in Fig. 20. Fig. 20(a) shows the loop system which makes the internal dynamic-range small with four delay-registers. Fig. 20(b) shows the loop biquad filter system with two delay-registers. The upper Z^{-1} corresponds to $Z1(n)$, $Z3(n)$, $Z5(n)$ and $Z7(n)$ of Fig. 18(b). Fig. 20(c) shows the system with eight delay-registers in series. While the calculation cycle becomes small, the dynamic-range of internal registers becomes large.

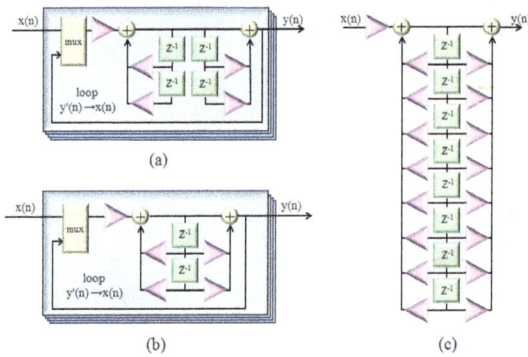

Figure 20. Wall filter systems of spectrum Doppler (a) Sysytem1: IIR+FIR loop system (b) System2: Biquad loop system (c) System3: Direct IIR system

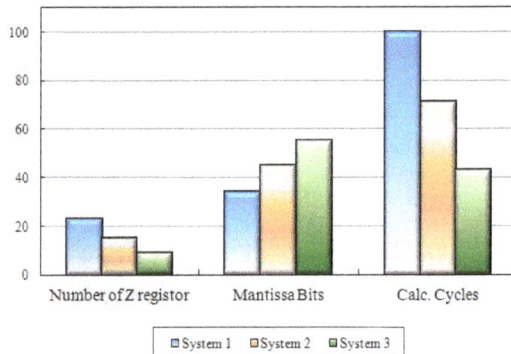

Figure 21. Benchmark based on floating point DSP (µPD77240)

Fig. 21 shows the evaluation of above systems. As benchmark condition, NECµPD77240 was used for floating point DSP. The 8th-order Butterworth HPF was chosen as benchmark processing, and its relative cutoff was set to 1/256. In the system 3 (equivalent to Fig. 20(c)), although both the number of double-precision registers and the operation cycles were small, internal bit-length became large. Since inner bit-length exceeded above 50 bits, even the double floating point arithmetic (mantissa-length: 48 bits) run short of accuracy, and was difficult to realize. In the system 1 (equivalent to Fig. 20(a)), many operation registers were required although internal bit-length was small. Since low-speed external memory access was required, its operation cycle increased. In the system 2 (equivalent to Fig. 20(b)), internal bit-length did not exceed the range of double-precision floating-point arithmetic, and an operation cycle was comparatively small. As mentioned above, the system 2 was judged the best system for mounting [17].

6. Wide dynamic-range system design

The dynamic-range of conventional system was insufficient for some clinical applications. In recent years, diagnostic ultrasound system was improved through the use of high-frequency electronics and integrated circuits. A new diagnostic method became effective by the higher dynamic-range system. However, the higher dynamic-range system also means more complicated gain control. It is possible to optimize the gain automatically through the use of ultrasound system parameters. This technology reduces the size of hardware and reduces the gain control range substantially.

6.1. Signal-processing of ultrasound system

Conventional ultrasound signal-processing is shown in Fig. 22. The transceiver processor (Tx/Rx Proc.) receives signals from the probe elements and the receive signals are amplified by the preamplifier. Gain compensation is applied to the signal to correct for range-distance attenuation (STC: sensitivity time control) and an analog gain correction for probe characteristics (frequency, sensitivity, etc) is applied. The signal is then sent to an ADC. After AD conversion the digital beam former (DBF) applies a delay pattern to the data to focus it and produce beam data. This data is processed by B-Mode Image Proc. and Doppler Image Proc., then displayed as a tomogram image and/or a spectrum Doppler image in the display processor (Display). The S/N ratio is increased in the DBF as the number of channels (corresponds to transducer elements) to which delay calculations are applied increases [18, 19]. In Doppler signal-processing, quadrature detection (Mixer) is applied to the DBF output and a BPF provides band-limitation and clutter rejection. The result is a base-band Doppler signal. At this time, the S/N ratio is sharply increased because of the band-limitation of the BPF. In the Doppler signal-processing, I apply range-gate integration (RG) across the range direction of the ROI. This also increases the S/N ratio. In the case of continuous wave Doppler the dynamic-range is even larger. And the HPF applied to this data must be more sophisticated. The dynamic-range of the signal leaving the HPF is also much larger than in the pulse wave case, in the order of 100 dB. After FFT the S/N ratio is greatly increased because of the butterfly integration. The dynamic-range of the signal is now very large and considerable gain adjustment and display compression must be done in order to display the data.

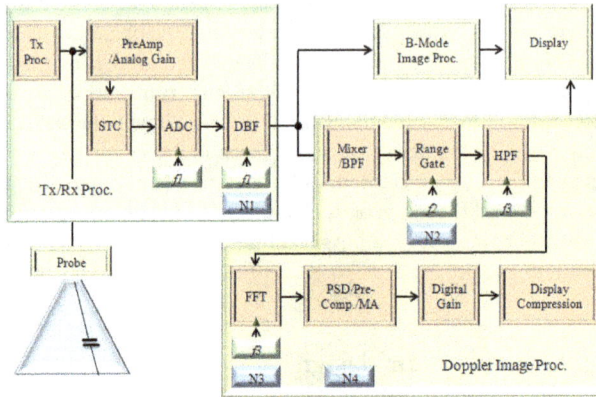

Figure 22. Ultrasound signal-processing system

It is also necessary to take into consideration that the dynamic-range is increased in Doppler signal-processing. In addition, mirror effect and/or quantization artifacts are introduced when performing automatic gain compensation. Although the beam data leaving the DBF has a frequency in the order of about 10-100 MHz, it is re-sampled at about 1-100 kHz. So the input dynamic-range of FFT increased by the band-limitation effect. Moreover the S/N ratio of FFT output is increased in a similar manner to that of ensemble mean processing.

6.2. Purpose of gain adjustment

Gain adjustment corrects for diagnostic target, bloodflow sensitivity, and the difference of user's skill. In addition to this, gain adjustment compensates for variation in other equipment parameters, such as the number of summing channel in DBF, apodization function, bandwidth of the BPF, the integration length of the range-gate (RG), FFT number, window function, and the number of the shift addition of power spectrum according to sweep speed etc. The maximum signal level and a noise level change because of change of these equipment parameters. In order to realize highly sensitive Doppler bloodflow diagnosis without saturation, a system with wide dynamic-range must perform gain compensation according to all these parameters. Table 2 illustrates the rough estimation of the dynamic-range and S/N ratio based on the virtual system [20].

The model of the signal-processing accompanied by expansion of the S/N ratio is shown in Fig. 23(a). The noise level and the maximum signal level of the incoming signal are expanded by signal-processing. But the expansion of the noise level differs from the expansion of the maximum signal level, and the overall S/N ratio is increased. Under optimal gain adjustment (range shown in light green in Fig. 23(a)) that there is no saturation of the maximum signal level and quantizing noise and signal are not mixed in the output. When gain adjustment is unsuitable (range shown in light pink in Fig. 23(a)), mirror effect or quantization artifacts occur on the spectrum image due to saturation or omission. As for signal amplitude, when quantization accuracy is inadequate, quantizing noise mixes. The gain adjustment that takes into consideration when detecting the weak Doppler signal around a system noise, and has sufficient quantizing margin is required. The influence of quantization is shown in Fig. 23(b). It is a spectrum when inputting sinusoidal (0.02*fs) including white noise. The horizontal axis is time and the vertical axis is frequency normalized by a sampling frequency fs. The quantizing level of an input range was changed every 2 seconds with 3, 5, 9, 17. It turns out that the harmonics components (-20 to -30dB) by quantization has occurred near the frequency -0.3*fs, -0.2*fs, and +0.25*fs. The mirror effect is an imaginal image symmetrically generated with a real image on both sides of a baseline. In the analog system, it is mainly caused by the phase error of quadrature detection, or the small gain difference between IQ signals. In a digital system, although these influences do not receive, a mirror effect generates them owing to saturation. As shown in Fig. 23(c), on both sides of 0 Hz, a symmetrical mirror effect occurs in the spectrum image. Fig. 23 (c) is the spectrum image which raised the gain 6 dB at a time every 2 seconds to the sinusoidal input including white noise. In this figure FFT input dynamic-range is 16 bit. The horizontal axis is time and the vertical axis is the frequency that normalized by a sampling frequency fs. The mirror effect component (-0.2*fs) has occurred by saturation to an original signal component (+0.2*fs). So in conventional ultrasound design both the mirror artifact and quantization artifact are caused by insufficient system dynamic-range in Doppler signal-processing.

Module	Cause	Effect of D.R. Increment (*1)	Conventional System Fig. 24(a) (dB)	New System Fig. 24(b) (dB)
ADC output	(Analog Gain)	50dB	DR1 @f1	DR1 @f1
DBF	BeamSum effect($N1$ channel)	+50dB	DR2=DR1 + 20log($N1$)	DR2_opt=DR1 + 20log($\sqrt{N1}$)
Mixer/BPF	BandLimitation effect ($f1/f2$)	+30dB	DR3=DR1+DR2 + 20log($f1/f2$)	DR3_opt=DR1+DR2_opt +20log($\sqrt{(f1/f2)}$)
RG	RG Integration effect($N2$ tap)	+40dB	DR4=DR1+DR2+DR3 + 20log($N2$)	DR4_opt=DR1+DR2_opt +DR3_opt + 20log($\sqrt{N2}$)
FFT	FFT number and window($N3$ sampling)	+50dB	DR5=DR1+DR2+DR3+DR4 + 20log($N3$)	DR5_opt=DR1+DR2_opt +DR3_opt +DR4_opt + 20log($\sqrt{N3}$)
PSD/Pre-Compres.	Power to Amplitude	-	DR5	DR5_opt
MA	Moving Average effect($N4$ average)	+10dB	DR6=DR1+DR2+DR3+DR4 +DR5 + 20log($N4$)	DR6_opt=DR1+DR2_opt +DR3_opt +DR4_opt+DR5_opt + 20log($\sqrt{N4}$)

Module	Cause	Effect of D.R. Increment (*1)	Conventional System Fig. 24(a) (dB)	New System Fig. 24(b) (dB)
Digital Gain	Digital Gain Input	D.R. before Input	230dB (DR6)	140dB (DR6_opt)
	Digital Gain Output	D.R. after Output	70dB (DR7)	70dB (DR7)
	Gain Control Range	-	160dB(DR6-DR7)	70dB(DR6_opt - DR7)

(*1) This estimation is based on virtual model of the Doppler ultrasound system

Table 2. Comparison of inner dynamic-range and gain control range

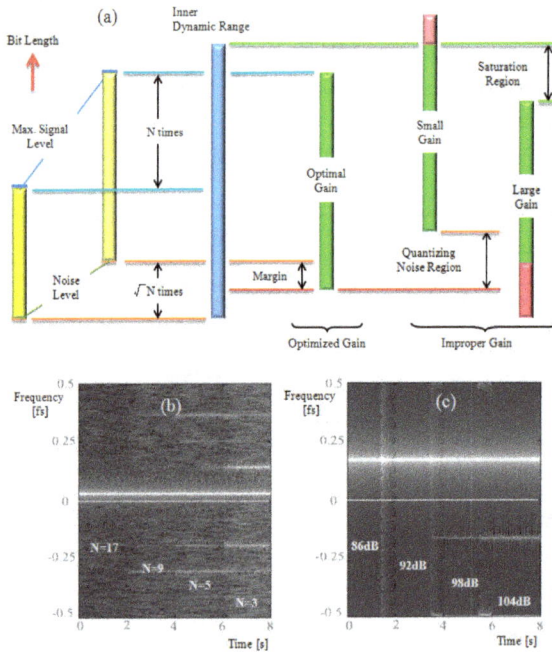

Figure 23. Artifacts caused by inadequate gain control (a) Problems of inadequate gain control (b) Artifacts caused by quantization (c) Artifacts caused by mirror effect.

6.3. Wide dynamic-range design and its optimization

Table 2 depicts the dynamic-range increment and the gain-control range increment of the conventional system and new system based on the signal-processing block diagram shown in Fig. 22. The DBF has the beam-summing effect of $N1$ channels. The Mixer/BPF has the band-limitation effect of $f1/f2$. The RG has the integration effect of $N2$ taps. The FFT has the integration effect (weighted by the window and operator) of $N3$. The PSD/Pre-Comp. just transfer the dimension (amplitude into power) using the square-root. The MA has the moving average effect of $N4$ columns according to the sweep speed of spectrum display. Fig. 24 shows the gain charts of the conventional system and new system based on Table 2. The gain chart of the conventional system which does not take the realization scale of hardware into consideration is shown in Fig. 24(a). This time, I developed the system that can reduce the gain-control range and can also reduce an internal dynamic-range. The automatic gain compensation according to the change-range of system parameters is realized for every sub-block of Doppler signal-processing accompanied by improved S/N ratio. Since the ranges of system parameters are known, the improvement of S/N ratio, the maximum signal level, and the noise level are calculable. An internal dynamic-range scale and the gain adjustment range can be optimally designed for every sub-block. By connecting the partial optimal sub-block in series and uniting it the internal dynamic-range of the system can be reduced so the system size and the total gain control range can both be sharply reduced. The internal S/N ratio is increased by $\sqrt{N}$. Then, supposing an input signal dynamic-range is DRin [dB], the range expansion equivalent to $20*\log(\sqrt{N})$ [dB] will occur. Moreover, the internal dynamic-range DRproc [dB] which added more than the margin ($20*\log(\sqrt{12})$) to quantizing noise is roughly calculable using equation (3).

$$DRproc \geq DRin + 20 \cdot \log\left(\sqrt{N}\right) + 20 \cdot \log\left(\sqrt{12}\right) \tag{3}$$

When digitizing, the Least Significant Bit (LSB) must be rounded not truncated otherwise an error of 1/2*LSB will exist. The RMS value of the quantizing noise is equivalent to 1/2*LSB/ $\sqrt{3}$. So an additional margin of $\sqrt{12}$ for dynamic-range must be maintained so rounding can be performed accurately. Although the internal dynamic-range DRproc is enough only in the automatic gain compensation with system parameters, it is necessary to consider a margin to the original gain adjustment that adjusts for diagnostic target and Doppler sensitivity variation at an internal dynamic-range. The gain chart of the sub-block signal-processing accompanied by range expansion is shown in Fig. 24(b). The gain-control range of the conventional system (corresponds to Fig. 24(a)) is DR6-DR7, and the gain-control range of the new system (corresponds to Fig. 24(b)) is DR6_opt-DR7. In the general Doppler signal-processing DR6 is above 200 dB, and DR7 (the digital gain output) is display luminance range (about 70 dB). So the gain-control range of conventional system should be more than 130 dB. This is very large. On the other hand the DR6_opt of new system is smaller than DR6 about 100 dB. I can reduce not only gain-control range but also the inner dynamic-ranges of sub-modules at the same time [21].

The effect of the automatic gain optimization using the new system was checked in a simulation of RG-integral processing. The spectrum images of the conventional system and the new system when changing RG-width are shown in Fig. 25(a) and Fig. 25(b). The horizontal axis is time and the vertical axis is the frequency normalized by a sampling frequency fs. The range-gate was adjusted from 1mm to 4 mm to 16 mm in 1.8 s intervals. A sinusoidal signal including white noise was used as an input. In the conventional system of Fig. 25(a), the signal level and noise level increase with expanding RG-width. For this reason, the user should reducing gain manually when the RG-width is expanded. In the new system in Fig. 25(b), although the signal level will rise if RG-width is expanded, it turns out that a noise level does not change. As mentioned above, by Doppler automatic gain compensation, the input bit length of each signal-processing block could be made smaller, and also the gain adjustment range could be made small to necessary minimum.

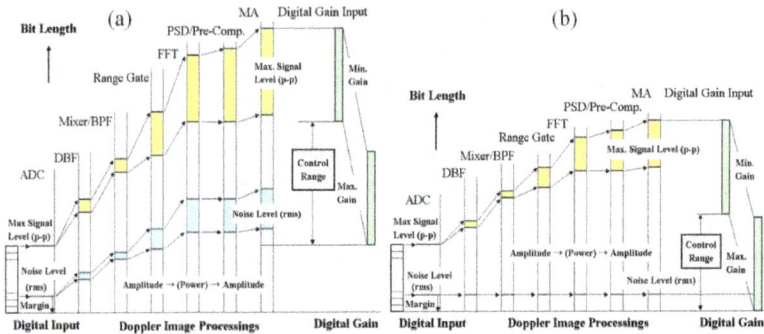

Figure 24. Comparison of gain control systems (a) Gain chart of conventional system (b) Gain chart of new system.

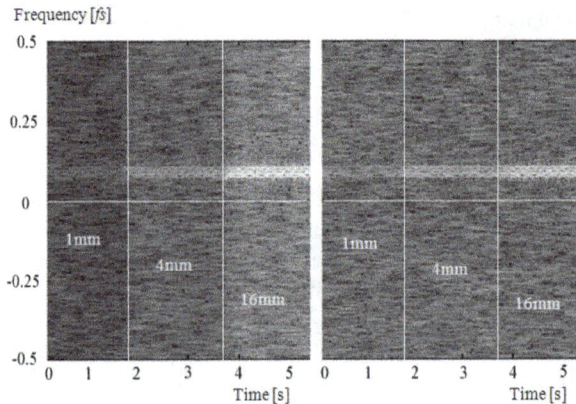

Figure 25. Effect of automatic gain compensation: example of RG integration process. (a) Conventional system (b) New system

7. Conclusion

The technical innovations which digital signal-processing brought about and their results were introduced based on some examples of the Doppler ultrasound system architectures. Not only extensive improvement of cost, size, power consumption, and adjustment, but also the improvement of sensitivity and accuracy has been realized by digital technology. Although DSP is most suitable for real-time system at present, the system architecture will be mounted as software when the calculation power of CPU/GPU improves further. In the future we will be able to acquire huge calculation ability easily, and it will be possible to apply it to real-time automatic diagnostic technology etc. besides conventional signal-processing.

Author details

Baba Tatsuro

Address all correspondence to: EZD03014@nifty.ne.jp

Toshiba Medical Systems Corporation, Japan

References

[1] Baba, T. (2009). Progress of Doppler ultrasound system architecture and considerations: The problems caused by digital system and their solutions. *Society of Signal Processing Applications and Technology of Japan*, 12(1), 2-8.

[2] Baba, T. (2005). Investigation on direction split technique of Doppler ultrasound: Comparison of six kinds of Doppler audio processing. *Society of Signal Processing Applications and Technology of Japan*, 8(2), 14-20.

[3] Baba, T. (2007). Direction separation in Doppler audio of ultrasound diagnosis equipment: Signal processing for Doppler audio dealiasing. *Acoustical Science and Technology*, 28(3), 202-210, DOI: ast.28.202.

[4] Baba, T. (2005). Research on Doppler ultrasound automatic heartbeat cycle detection: The investigation of heartbeat cycle detection from Doppler waveform using adaptive BPF. *The Journal of the Acoustical Society of Japan*, 61(11), 629-635.

[5] Baba, T. (2006). Velocity range tracking in Doppler diagnostic ultrasound systems: Range optimization using Doppler trace wave form histograms. *The Journal of the Acoustical Society of Japan*, 62(4), 327-331.

[6] Baba, T., Ohmae, N., & Osuka, K. (2008). The Optimization of Ultrasound System Doppler Velocity Range using Hybrid Control. *Transactions of the Society of Instrument and Control Engineers*, 44(9), 760-765.

[7] Ohashi, T. (2001). Recording of World Heritage on the High Definition Audio-visual Media: Documentation of History and Tradition. *ITE Transactions on Media Technology and Applications*, 55(1), 37-46.

[8] Yoshikawa, S. (2002). Present status of high definition audio. *The Journal of the Acoustical Society of Japan*, 58(4), 250-255.

[9] Misawa, T. (2004). The image sensor for digital cameras. *Japanese Journal of Optics*, 33(9), 544-549.

[10] The Physical Society of Japan. (1978). Physics experiment data processing by a computer, the 4th edition. Tokyo: SAIENSU-SHA Co., Ltd.

[11] Miyagawa, H. (1981). Digital Signal Processing The 9th edition. Tokyo: CORONA PUBLISHING Co., LTD.

[12] Nakajima, T. (1996). Compact disk reader, the 3rd edition. Tokyo: Ohmsha Ltd.

[13] Baba, T., & Toshiba, Corp. (2008). Ultrasonic diagnostic equipment. Japanese Patent 4068208; Jan. 18.

[14] Baba, T., & Toshiba, Corp. (2006). Ultrasonic diagnostic equipment. *Japanese Patent 3746113*, Feb. 15.

[15] Jensen, JA. (1996). Estimation of Blood Velocities Using Ultrasound: A Signal Processing Approach:. *Cambridge University Press*, 0-521-46484.

[16] Baba, T. (2008). Evaluation of Post Wall Filter for Doppler Ultrasound Systems. *Acoustical Imaging*, 29-133, 978-1-40208-822-3.

[17] Baba, T. (2006). Investigation of wall filters in Doppler ultrasound system. *Society of Signal Processing Applications and Technology of Japan*, 9(2), 14-19.

[18] Kozak, M., & Karaman, M. (2001). Digital Phased Array Beamforming Using Single-Bit Delta-Sigma Conversion with Non-Uniform Oversampling. *IEEE Transactions on UFFC*, 48(4), 922-931, DOI: 10.1109/58.935709.

[19] Engelberg, S. (2006). Implementing a $\Delta\Sigma$ DAC in Fixed Point Arithmetic. *IEEE Signal Processing Magazine:*, DOI:10.1109/SP-M.2006.248716, 66-69.

[20] Baba, T., & Toshiba, Corp. (2005). Ultrasonic diagnostic equipment. *Japanese Patent 3663206*.

[21] Baba, T. (2009). Investigation of gain optimization technique in Doppler ultrasound system. *Acoustical Science and Technology*, 30(2), 61-71, DOI: ast.30.67.

Chapter 12

FPGA Based Serial and Single-Clock Cycle Pipelined Fast Fourier Transforms in a Radio Detection of Cosmic Rays

Zbigniew Szadkowski

Additional information is available at the end of the chapter

1. Introduction

Results from various cosmic rays experiments located on the ground level, point to the need for very large aperture detection systems for ultra-high energy cosmic rays. With its nearly 100% duty cycle, its high angular resolution, and its sensitivity to the longitudinal air-shower evolution, the radio technique is particularly well-suited for detection of Ultra High-Energy Cosmic Rays (UHECRs) in large-scale arrays. The present challenges are to understand the emission mechanisms and the features of the radio signal, and to develop an adequate measuring instrument. Electron-positron pairs generated in the shower development are separated and deflected by the Earth's magnetic field [1], [2], hence they introduce an electromagnetic emission. During shower development, charged particles are concentrated in a shower disk of a few meters thickness. This results in a coherent radio emission up to about 100 MHz. Short but coherent radio pulses of 10 ns up to a few 100 ns duration are generated with an electric field strength increasing approximately linearly with the energy of the primary cosmic particle inducing the extended air showers (EAS), i.e. a quadratic dependence of the radio pulse energy vs. primary particle energy. In contrast to the fluorescence technique (e.g. used in the Pierre Auger Observatory [3]) with a duty cycle of about 12% (fluorescence detectors can work only during moonless nights), the radio technique allows nearly full-time measurements and long range observations due to the high transparency of the air to radio signals in the investigated frequency range.

The radio detection technique will be complementary to the water Cherenkov detectors and allows a more precise study of the electromagnetic part of air showers in the atmosphere. In addition to a strong physics motivation, many technical aspects relating to the efficiency, saturation effects and dynamic range, the precision for timing, the stability of the hardware developed, deployed and used, as well as the data collecting and system-health monitoring processes will be studied and optimized.

EAS are investigated in several experiments utilizing different detection techniques (scintilators, water Cherenkov and fluorescence detectors). Signals in the detectors depend on several parameters such as the energy, the type of the primary particle, a distance from the core, the angle of registered shower, etc. Usually the triggering conditions are chosen such as to detect as wide as possible classes of events. However, sometimes the standard trigger conditions are not optimized for the specific class of events, which are either not registered at all or for which the registration efficiency is poor. In experiments utilizing water Cherenkov detectors, signals from photo-multipliers (PMTs) are usually digitized in ADCs and next processed by often-sophisticated electronics. In order to increase the signal/noise ratio coincidence techniques are widely used. Typically signals from PMTs are analyzed on-line in both amplitude and time domains. Strong signals in all PMT channels, corresponding to energetic showers detected near the core, are registered because of many-fold coincidence single bin trigger with a fixed thresholds. Showers detected far from the core give much lower signals usually spread in time. Such events are detected by the other type of trigger investigating the structure of signal in some period (in a sliding time window).

The structure of signals detected in water Cherenkov tanks and generated by horizontal showers depend strongly on the point of the EAS initialization. "Old" showers generated by hadrons early in the atmosphere give flat muonic front; showers generated by deeply interacting neutrinos are characterized by a curved front (radius of curvature of a few km), a large electromagnetic component and with particles spread over a few microseconds interval [4]. In both cases muonic front produces a bump, which can be a starting signature of horizontal showers. The bump for the "old" showers is shorter and sharper than for the "young" ones and results in a larger contribution in higher Fourier coefficients. For "young" showers, with relatively smooth shape of a signal profile, the lower Fourier components should dominate. The on-line analysis of the Fourier components may trigger specific events.

The existing software procedures, available as commercial IP routines, can calculate Fourier coefficients effectively utilizing a FFT algorithm. However the software implementation is too slow to be able to trigger events in the real time. On-line triggering requires the hardware implementation calculating multi-point DFT with a sufficient speed. Modern powerful FPGAs can do this job, however, the resource requirement increases dramatically with the number of points. The analysis time interval should be a reasonable compromise between the time resolution and the resources occupancy in the FPGA.

2. DFT

The discrete Fourier transform (DFT), of length N, calculates the sampled Fourier transform of a discrete-time sequence at N evenly distributed points $\omega k = \frac{2\pi k}{N}$ the unit circle. The following equation shows the length-N forward DFT of a sequence x(n):

$$\bar{X}_k = \sum_{n=0}^{N-1} x_n e^{\frac{-2i\pi kn}{N}} \qquad k = 0, 1, ... N - 1 \qquad (1)$$

ne length-N inverse DFT:

$$\sum_{=0}^{V-1} \bar{X}_n e^{\frac{2i\pi kn}{N}} \qquad k = 0, 1, ...N - 1 \tag{2}$$

DF1 direct computation can be significantly reduced by using fast ...ms that use a nested decomposition of the summation in equations one and two-in addition to exploiting various symmetries inherent in the complex multiplications. Such algorithms are the Radix-r Decimation-in-Time (DiT) or Radix-r Decimation-in-Frequency (DiF) Fast Fourier Transforms (FFT), which recursively divides the input/output sequence into N/r sequences of length r and requires $log_r N$ stages of computation.

The commercially offered FFT processors for FPGA applications require several clock cycles to accomplish calculation of all complex DFT coefficients. Each stage of the decomposition typically shares the same hardware, with the data being read from memory, passed through the FFT processor and written back to memory. Each pass through the FFT processor is required to be performed $log_r N$ times. Popular choices of the Radix are r = 2, 4, and 16. Increasing the Radix of the decomposition leads to a reduction in the number of passes required through the FFT processor at the expense of device resources. Such an approach is very widely useful for many applications, where timing is not crucial. However, there are areas, where the FFT coefficients (based on a new set of samples) have to be known in each clock cycle. Commercial FFT processors, unfortunately, cannot be used. This approach requires special algorithms optimized for a particular solution.

2.1. Radix-2 : Decimation-in-Time and Decimation-in-Frequency

The Radix-2 algorithm is the simplest FFT one. The decimation-in-time (DIT) Radix-2 FFT recursively partitions a DFT into two half-length DFTs of the even-indexed and odd-indexed time samples. For the Radix-2 DiT, we get :

$$\bar{X}_k = \sum_{n=0}^{N-1} x_n e^{-2i\pi kn/N} = \sum_{n=0}^{\frac{N}{2}-1} x_{2n} e^{-i\frac{2\pi kn}{N/2}} + e^{-i\frac{2\pi k}{N}} \sum_{n=0}^{\frac{N}{2}-1} x_{2n+1} e^{-i\frac{2\pi kn}{N/2}} = \tag{3}$$

$$= DFT_{\frac{N}{2}}[x_0, x_2, ..., x_{N-2}] + W_N^k \times DFT_{\frac{N}{2}}[x_1, x_3, ..., x_{N-1}]$$

For the Radix-2 DiF, we get :

$$\bar{X}_{2k} = \sum_{n=0}^{N-1} x_n W_N^{2kn} = DFT_{\frac{N}{2}}\left[x_n + x_{n+\frac{N}{2}}\right] \tag{4}$$

$$\bar{X}_{2k+1} = \sum_{n=0}^{N-1} x_n W_N^{(2k+1)n} = DFT_{\frac{N}{2}}\left[\left(x_n - x_{n+\frac{N}{2}}\right) W_N^n\right]$$

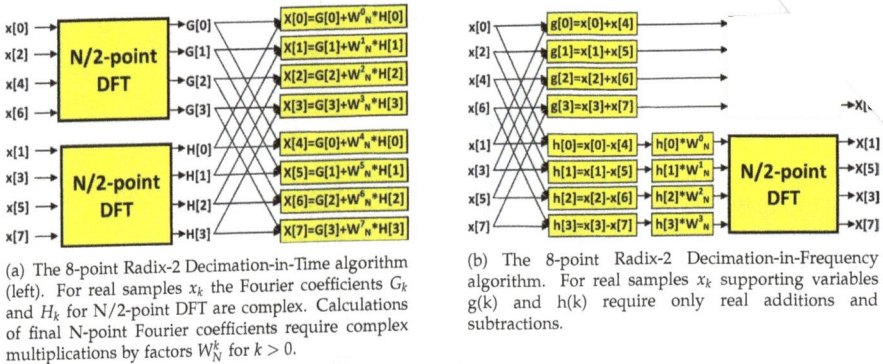

(a) The 8-point Radix-2 Decimation-in-Time algorithm (left). For real samples x_k the Fourier coefficients G_k and H_k for N/2-point DFT are complex. Calculations of final N-point Fourier coefficients require complex multiplications by factors W_N^k for $k > 0$.

(b) The 8-point Radix-2 Decimation-in-Frequency algorithm. For real samples x_k supporting variables g(k) and h(k) require only real additions and subtractions.

Figure 1. Splitting of N-point DFT on two N/2-point parallel procedures for Decimation-in-Time (left) and Decimation-in-Frequency (right), respectively, on the basis of the 8-point Radix-2 algorithms.

The N-point DFT can be easily split on two N/2-point transforms. The outputs from DFT procedures are complex. So, a calculation of final DFT coefficients by using DiT algorithm requires complex multiplication for final merging data from parallel DFT procedures with lower order (i.e. multiplication of twiddle factors W_N^k) :

$$W_N^k = e^{-i\frac{2\pi k}{N}} \tag{5}$$

by $G[k]$ and $H[k]$ in Figure 1. For the DiF algorithm the 1st stage requires additions and subtractions only. Odd indexes require additional multiplications, however, even indexes remain without modifications for the next N/2-point DFT procedure (compare Figures 1a and 1b).

2.2. Radix-4 algorithm

The Radix-4 algorithm consists of four inputs and four outputs. The FFT length is 4^p, where p is the number of stages. A stage is half of Radix-2. The Radix-4 DIF FFT divides an N-point DFT into four N/4 -point DFTs, then into 16 N/16 -point DFTs, and so on.

For Radix-4 DiF, we get :

$$\tilde{X}_k = \sum_{n=0}^{N-1} x_n e^{\frac{-2i\pi kn}{N}} = \sum_{n=0}^{N/4-1} x_n e^{\frac{-2i\pi kn}{N}} + \sum_{n=N/4}^{N/2-1} x_n e^{\frac{-2i\pi kn}{N}} + \sum_{n=N/2}^{3N/4-1} x_n e^{\frac{-2i\pi kn}{N}} + \sum_{n=3N/4}^{N-1} x_n e^{\frac{-2i\pi kn}{N}} =$$

$$= \sum_{n=0}^{N/4-1} e^{-\frac{2i\pi kn}{N}} \left[x_n + (-i)^k x_{n+N/4} + (-1)^k x_{n+N/2} + (i)^k x_{n+3N/4} \right] \tag{6}$$

This algorithm is widely used, however, as it is shown in a next section, the simple application of the DiT or DiF algorithms in all sequential steps remains still an area for further optimization.

2.3. Architectures of the Altera®'s FFT MegaCore® functions.

2.3.1. Streaming architecture

The Radix-4 decomposition, which divides the input sequence recursively to form four-point sequences, has the advantage that it requires only trivial multiplications in the 4-point DFT, it is the chosen Radix algorithm in the Altera® FFT MegaCore® function. This results in the highest throughput decomposition, while requiring non-trivial complex multiplications in the post-butterfly twiddle-factor rotations only. In cases where N is an odd power of two, the FFT MegaCore automatically implements the Radix-2 pass on the last pass to complete the transform.

To maintain a high signal-to-noise ratio throughout the transform computation, the FFT MegaCore function uses a block-floating-point architecture, which is a compromise point between fixed-point and full-floating point architectures. In the fixed-point architecture, the data precision needs to be large enough to correctly represent all intermediate values throughout the transform computation. For large FFT transform sizes, the FFT fixed-point implementation that allows for word growth can make either the data width excessive or can lead to a loss of precision.

In the floating-point architecture each number is represented as a mantissa with an individual exponent, while this leads to greatly improved precision, floating-point operations tend to demand increased device resources.

In the block-floating point architecture, all of the values have an independent mantissa but share a common exponent in each data block. Data is input to the FFT function as fixed point complex numbers (Figure 2).

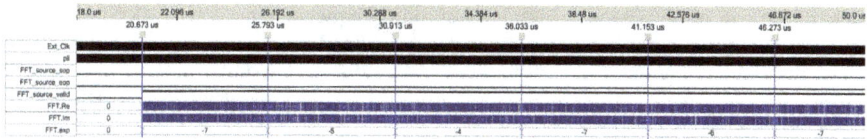

Figure 2. A simulation of the Fourier transform for the Altera® library routine of 1024 points and for streaming architecture. Each block of 1024 Fourier coefficients (Fc) is scaled by the factor FFT.exp. Fourier coefficients are provided in a serial way, each pair of real and imaginary parts of a single Fc in a single time bin. All Fc are calculated in 1024 time bins. FFT.sop (start of package) and FFT.eop (end of package) indicate begin and end of each 1024-point block.

The block-floating point architecture ensures full use of the data width within the FFT function and throughout the transform. After every pass through the Radix-4 FFT, the data width may grow up to $log_2(4\sqrt{2}) = 2.5$ bits. The data is scaled according to a measure of the block dynamic range on the output of the previous pass. The number of shifts is accumulated and then output as an exponent for the entire block. This shifting ensures that the minimum of least significant bits (LSBs) are discarded prior to the rounding of the post-multiplication output. In effect, the block-floating point representation acts as a digital automatic gain control. To yield uniform scaling across successive output blocks, you must scale the FFT function output by the final exponent [5].

2.3.2. Variable streaming architecture

The variable streaming architecture uses two different types of architecture, depending on which representation: the fixed-point or the floating-point is selected. For the fixed-point data representation, the FFT variation uses a Radix-2^2 single delay feedback architecture, which is a fully pipelined architecture. For the floating point representation, the FFT variation uses a mixed Radix-4/2 architecture. For a length N transform, log4(N) stages are concatenated together. The Radix-2^2 algorithm has the same multiplicative complexity of the fully pipelined Radix-4 architecture, but the butterfly unit retains the Radix-2 architecture. In the Radix-4/2 algorithm, a combination of Radix-4 and Radix-2 architectures are implemented to achieve the computational advantage of the Radix-4 architecture while supporting FFT computation with a wider range of transform lengths. The butterfly units use the DIF decomposition.

The fixed point representation allows for natural word growth through the pipeline. The maximum growth of each stage is 2 bits. After the complex multiplication the data is rounded down to the expanded data size using convergent rounding. The overall bit growth is less than or equal to log2(N)+1. The floating point internal data representation is the single precision floating point (32-bit). Floating point operations provide more precise computation results but are costly in hardware resources. To reduce the amount of logic required for floating point operations, the variable streaming FFT uses "fused" floating point kernels. The reduction in logic occurs by fusing together several floating point operations and reducing the number of normalizations that need to occur [5].

3. An FPGA based RFI filter for radio detection of cosmic rays

3.1. A physical background

The energy threshold of radio detection of cosmic rays is limited by the considerable radio background and noise. The very high level of radio frequency interferences (RFI) in the FM and the short wave band has to be eliminated by a band pass filter amplifier. Within the remaining receiver bandwidth of 30 to 80MHz the noise at the quiet-rural environment of cosmic-rays experiments is dominated by the frequency dependent galactic noise [6] with noise temperatures of 5000K at 60 MHz.

In addition to galactic noise, there is still a human made background. This background consists of continuous signals, as from a few radio and TV stations, and transients produced by machines. Without an effective trigger, a stable and low-level energy threshold is not guaranteed. Furthermore, the data rate for communication of the triggered data to the central DAQ would exceed the available power budget.

For self-triggered measurements, the data will be digitized and processed in real time by a powerful FPGA chip. The narrow peaks in the frequency domain due to radio frequency interferences have to be strongly suppressed before building a trigger. These peaks are removed by a median filter. The filter works in the frequency domain using the Fast Fourier Transform (FFT) routine provided by Altera®. Furthermore, the phase of the signal deformed by the steep band pass filter is reconstructed by a deconvolution in the frequency domain.

The median FPGA filter eliminates mono-frequent carriers, but broadband radio pulses from cosmic showers are not affected. After a second inverse FFT, signals are converted back to the time domain. This chain of the digital signal processing strongly enhances the signal to noise ratio, and thus improves the radio pulse detection sensitivity (Figure 3).

Due to the Nyquist theorem, the used 80 MHz band should be sampled with at least 160 MHz. An application of 16-bit ADCs with such a sampling rate would be a challenge for the price, the power consumption and PCB routing to keep a reasonable noise level. The used practical option is an 12-bit ADC with 180 MSPS, leaving sufficient space for the anti-aliasing filter and implementing a high and low gain channel to obtain the required dynamic range.

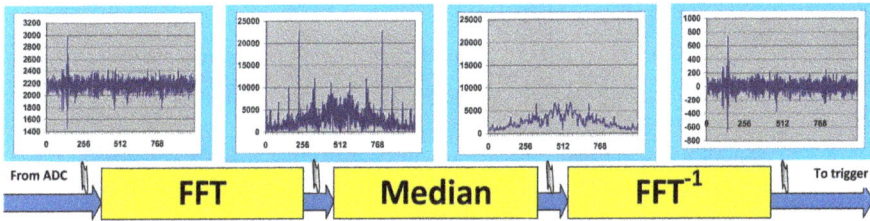

Figure 3. A diagram showing a (FFT + Median filter + iFFT) chain cleaning the signal from the RFI contamination. The 1st graph shows the ADC input as unsigned data with an offset of ca. 2300 ADC-counts, the 2nd - the absolute values of FFT coefficients in the frequency domain, the 3rd - FFT coefficients "decontaminated" by the median filter and 4th - signal converted back to the time domain. Additionally, the 0th FFT coefficient has been zeroed. Thus, the cleaned signal in the time domain is represented as signed data without the offset. The amplitude of the signal remains roughly the same and the noise is considerably reduced.

The necessary filtering accuracy requires at least 1024-point Fourier transforms. For the 180 MHz sampling, it corresponds to 360 kHz resolution in the frequency domain. Shorter transformation blocks give too rough filtering and may affect real signals from showers. For these parameters, the RFI filter has been developed and optimized [7].

3.2. Selection of the FFT architecture

The Altera® FFT MegaCore offers 4 types of FFT engines with various architectures :

- streaming
- variable streaming
- burst
- buffered burst

calculating the FFT and iFFT in real-time. All architectures can be implemented a fixed point FFT, whereas the variable streaming architecture can also be configured in a floating point data representation. A comparison of resource occupancy of different architectures is given in Table 1. Parameters are shown for 12-bit and 16-bit data processing.

architecture	TCC	BTC	LE wizard 12-bits	memory bits 12-bits	DSP 12-bits	LE wizard 16-bits	memory bits 16-bits	DSP 16-bits
streaming	1024	1024	3723	155 648	24	4952	155 648	24
variable streaming								
fixed bit reverse	1024	1024	6139	31792	48	7175	39 976	56
floating bit reverse	1024	1024	23000	73 568	128	23000	73 568	128
fixed natural order	1024	1024	6139	82 992	48	7175	100 380	56
floating natural order	1024	1024	23000	139 104	128	23000	139 104	128
burst (single engine)	1113	3162	2814	57 344	24	3804	57 344	24
burst (4 engines)	345	2394	7864	114 688	96	11 136	114 688	96
buffered burst (single engine)	1103	1291	3202	122 880	24	4197	122 880	24
buffered burst (4 engines)	335	1099	8517	245 760	96	11 885	245 760	96

Table 1. An utilization of resources for various FFT architectures at 12-bit and 16-bit data processing. The 2nd column shows Transform Calculation Cycles (TCC), required by the Altera® wizard, the 3rd - Block Throughput Cycles (TBC), the 4th - required Logic Elements (LE), the 5th - required memory bits, the 6th - required Digital Signal Processing blocks. Parameters in columns 4th to 6th correspond to the 12-bit data processing, in columns 7th to 9th to the 16-bit processing, respectively.

For the RFI filtering scheme shown on Figure 3 sampled ADC data have to be processed continuously in real-time. "Continuously" means that any dead-time is not acceptable. Data can be processed in blocks of fixed length, but no any sample can be ignored. This requirement eliminates two architectures: burst and buffered burst, because for i.e. 1024-point (and 1024 clock cycles when samples appear from the ADC output) these architectures require more than 1024 clock cycles for processing (BTC = 3162, 2394, 1103 and 1099 for burst and buffered burst and single and 4 engines, respectively). For any configuration the fundamental requirement of no dead-time is not obeyed.

The floating point representation for the variable streaming architecture requires huge amount of logic elements and DSP blocks. For two cascade FFT engines for two polarization channels almost all resources could be utilized for the FFT engines only. There would not be resources for other tasks. Additionally, the Altera®'s documentation shows that the registered performance for this architecture is much below our expectations (on the level of 110 MHz, while we need at least 180 MHz for the signal processing).

Some FFT applications require the FFT + the user operation + the iFFT chain. In this case, a careful selection of the input and output order can significantly save a memory and a latency. If the input to the first FFT is in the natural order and the output is in the bit-reversed order, the FFT engine operates in a mode with a minimal resource utilization (called Engine-only mode). Thus, if the iFFT operation is configured to accept bit-reversed inputs and produces natural order outputs (iFFT is operating again in Engine-only mode), only the minimum amount of memory is required, which provides a saving of N complex memory words, and a latency saving of N clock cycles, where N is the size of the current transform.

However, in the case of the RFI filtering by the median filter the sequence of FFT coefficients in the frequency domain has to be natural, to eliminate/suppress narrow-band peaks. The FFT routines have to be working with Engine with bit-reversal modes only. Two architectures: (a) streaming and (b) variable streaming with the natural order and the fixed-point data representation survived the selection.

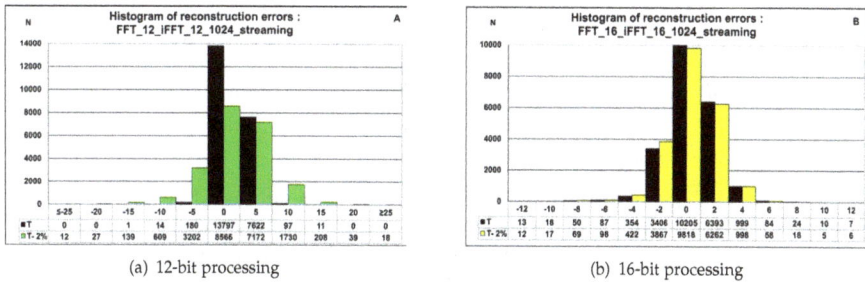

Histogram of reconstruction errors : FFT_12_iFFT_12_1024_streaming											A
N	≤-25	-20	-15	-10	-5	0	5	10	15	20	≥25
■T	0	0	1	14	180	13797	7622	97	11	0	0
□T-2%	12	27	139	809	3202	8566	7172	1730	208	39	18

Histogram of reconstruction errors : FFT_16_iFFT_16_1024_streaming													B
N	-12	-10	-8	-6	-4	-2	0	2	4	6	8	10	12
■T	13	18	50	87	354	3406	10205	6393	999	84	24	10	7
□T-2%	12	17	69	96	422	3867	9818	6262	998	58	18	5	6

(a) 12-bit processing (b) 16-bit processing

Figure 4. Histograms of reconstruction errors for the streaming architecture (differences between the original ADC data and data after application of the 12-bit (a) and 16-bit (b) wide FFT and inverse FFT). "T" denotes ideal configuration with zeroed offset. However, for a small shift of 2% only, error distributions become wider and more flat.

3.3. Streaming architecture

The streaming architecture accepts as input a two's complement format with a complex data vector of length N, where N is the desired transformation block length. The function output is given as a complex vector in the natural order. An accumulated block exponent is given to indicate any data scaling that has occurred during the transformation to maintain precision and maximize the internal numerical signal-to-noise ratio.

The signed block exponent, used for scaling of internal signal values, remains constant for a full data block. For relatively small variations of the signal samples x_n (typical for noise background), but with not negligible pedestal the Fourier component $\bar{X}_0$, may be relatively large whereas the $\bar{X}_{n \neq 0}$ components are rounded off to relatively small values. This may cause large errors of the reconstructed signals after going through the FFT/iFFT chain. Hence, the pedestal has to be subtracted carefully from the input signal. Errors of the reconstruction for the 1024-point transforms of a real event signal recorded in real cosmic rays experiment are shown in Figure 4.

The streaming architecture introduces, unfortunately, significant distortions of signals in a data processing for the FFT+iFFT cascade chain. The reconstruction errors for the 12-bit processing are on unacceptable level of 10 and more ADC-counts. The 16-bit configuration introduces smaller reconstruction errors and maybe used for real data processing, however, an influence of the data processing errors have to be carefully take into account for the final trigger and recorded data.

Figure 5 shows a possible optimization, where 12-bit data is processing in 14-bit FFT engine and 2 lower significant bits are grounded and treated as potentially fractional part.

3.4. Variable streaming architecture

The 12-bit input FFT routine with the variable streaming architecture yields 25-bit Re/Im Fourier coefficients. Processing of both buses with the full width in the iFFT procedure would be too spendthrift and slows down the speed significantly. A reasonable compromise for a selection of the input lines driving the iFFT routine is required.

Figure 6 shows that cropping the output FFT bus to 12 bits provides already a good reconstruction. The error is on the level of one ADC-count. This is achieved at the expenses of 2000 additional LEs and 24 additional DSP blocks. However, this architecture's maximum clock frequency of roughly 200 MHz (for selected FPGA from Cyclone® III family) is too low.

3.5. Aliasing and leakage removal

The incoming data stream must be chopped into blocks to be processed by the FFT routine. If signal pulses are located close to the border of a block, aliasing occurs. It manifests by a spurious contribution in the opposite border of the block and in the neighboring block as well. This effect may cause spurious pulses and has to be eliminated. The leakage effect is caused by the finite length of the blocks, acting like an applied rectangular window function. Thus, a signal amplitude leaks from one frequency bin to another. By using a suitable window function, the leakage effect can be reduced. To keep algorithmic costs low, we use a window function with a constant middle part like a trapezoidal shape or a Tukey-window.

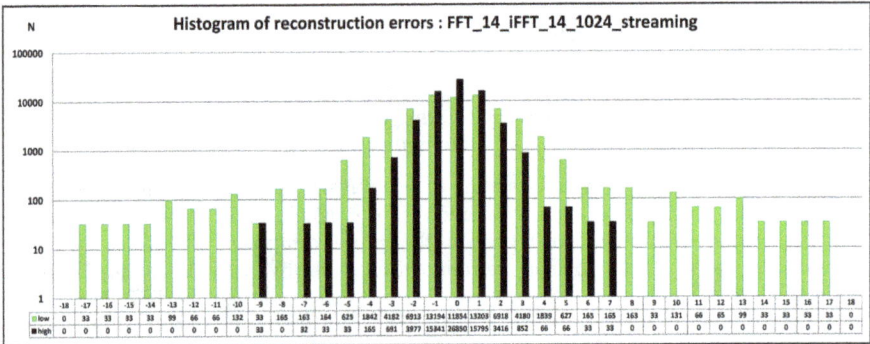

Figure 5. Histograms of reconstruction errors for the streaming architecture (differences between the original ADC data and data after application of the 14-bit wide FFT and inverse FFT). The width of input data is 12 bits connected to low 12 bits (starting from LSB) ("low") or to higher 12 bits (starting from MSB (high). For "low" configuration 13th and 12th input bits are connected to the sign (11th) bit. A distribution of the reconstruction errors is rather wide. For the "high" configuration 0th and 1st are grounded and they play role of a fractional zeroed input part. For a such modification of input connection only, the error distribution is significantly narrower.

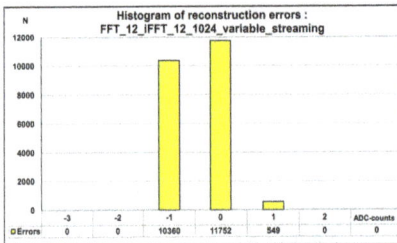

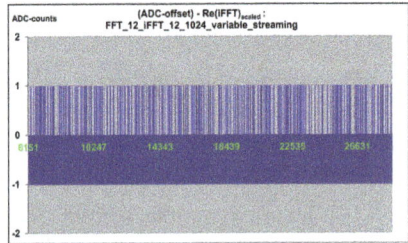

(a)   (b)

Figure 6. Histogram (a) of reconstruction errors for the variable streaming architecture (differences between the original ADC data and data after application of the 12-bit wide FFT and inverse FFT). The right plot (b) shows differences for raw data.

Both problems can only be solved, without introducing dead time between the blocks, by using an overlapping routine [7]. Therefore the filter engine must run in another clock domain with higher frequency. Preliminary estimation shows that for an overlapping of N = 32 errors due to an aliasing contribution is acceptable, however for a better safety margin N = 64 is preferred. N = 128, allows a total removal of aliasing effect, however this option requires too high over-clocking according to Table III. An odd value like N = 73 seems to be a valid compromise, although requiring some special modules to assure a seamless hand over of the data stream between the different clock domains.

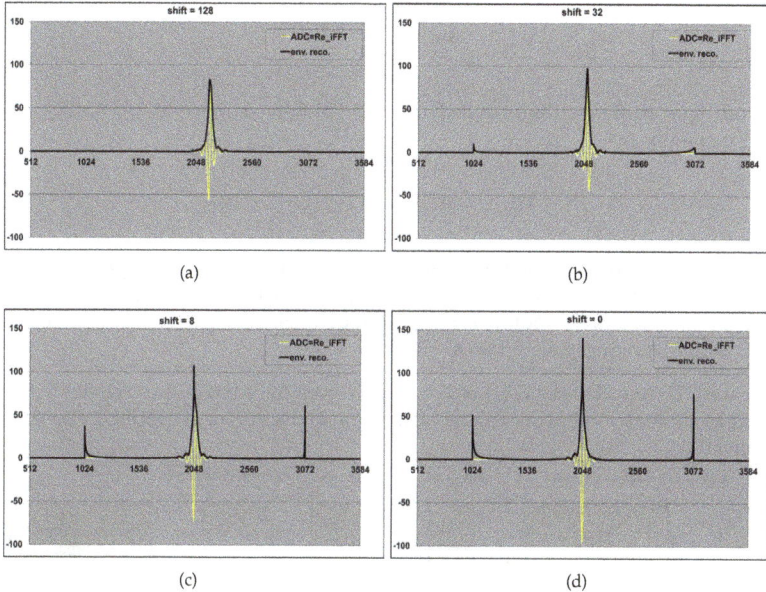

(a) (b)

(c) (d)

Figure 7. An example of spurious envelopes due to aliasing, when a signal appears close to the border of converted blocks, 128, 32, 8 and exactly on the border, respectively

Figures 7 show a potential danger if the aliasing were ignored. If the signal appears relatively far from the end of the block border (i.e. 128 time bins for 1024-point conversion) the envelope of the signal is reconstructed rather good (Figure 7a). There is no any false peaks, which could be recognized as spurious triggers. If the signal appears relatively close the end of the block border (Figure 7b) one can observe some spurious wings on the borders of neighboring blocks. However, if a relatively strong signal appears close to the block border (Figure 7c) the spurious peaks are created on both borders and there is a very high danger that these spurious peaks can be mistakenly taken as a trigger. If the signal appears exactly on the border of two blocks (Figure 7d), the spurious peaks can get an amplitude of more than 30 % of real signal. An additional procedure removing a spectral leakage has to be absolutely used to keep a high reliability of the system.

3.6. Simultaneous processing of two signals with perpendicular polarizations

Each antenna station measures radio signals in two opposite polarization channels. Thus, it would be straightforward to use two FFT engines for calculating the frequency domain signal, while setting their imaginary input to zero. A more efficient way is to exploit the symmetries of the FFT. Therefore the data streams of both antenna channels (N windowed signal samples f_j and g_j) are connected to the real respectively imaginary component input of the FFT engine. The resulting output components, H_n, are given in (7).

$$H_n = \sum_j e^{2\pi ijn/N}(f_j + ig_j) \tag{7}$$

The H_n can then easily be disentangled into the Fourier components, F_n and G_n, by the following equations (8)

$$H_n + H^*_{N-n} = 2F_n , \quad H_n - H^*_{N-n} = 2iG_n, \tag{8}$$

The (N-n) indices in (8) in a real time system correspond to a time reversed order. The H_n and H^*_{N-n} are synchronized by a routine inverting the order of the H^*_n like First In Last Out (FILO) and by using a delay routine for the H_n in parallel. Doing so, the amount of needed FFT engines can be reduced from two to one.

After the iFFT, the envelopes $f_{env}(t)$ and $g_{env}(t)$ (Figure 8) of the output signal $x(t)$ have to be created to allow the following trigger algorithms to discriminate specific pulse shapes in each channel.

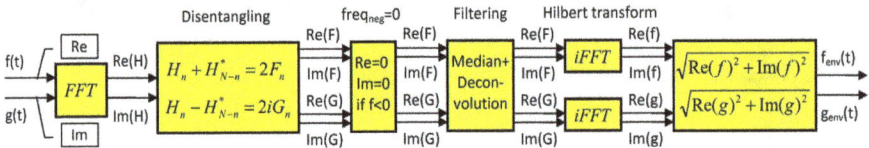

Figure 8. Schematic view of the resources-optimized implementation of the used two antenna chains with opposite polarization, each consisting of FFT, median filter, deconvolution, Hilbert transform Im(f) and Im(g) and FIR filters.

4. Wavelets

Let us investigate a time series X, with values of x_n, at time index n. Each value is separated in time by a constant time interval Δt. The wavelet transform $W_n(s)$ is just the inner product (or convolution) of the wavelet function with our original time series:

$$W_n(s) = \sum_{m=0}^{N-1} x_m \psi^* \left[\frac{(m-n)\Delta t}{s} \right] \tag{9}$$

where the asterisk (*) denotes complex conjugate. The above sum can be evaluated for various values of the scale s (usually taken to be multiples of the lowest possible frequency), as well as all values of n between the start and end dates.

It is possible to compute the wavelet transform in the time domain according to (9). However, it is much simpler to use the fact that the wavelet transform is the convolution between the two functions X and ψ, and to carry out the wavelet transform in Fourier space using the Fast Fourier Transform (FFT). In the Fourier domain, the wavelet transform is :

$$W_n(s) = \sum_{k=0}^{N-1} \tilde{X}_m \Psi^*(s\omega_k) e^{i\omega_k n\Delta t} \tag{10}$$

Unlike the convolution, the FFT method allows the computation of all n points simultaneously, and can be efficiently coded using any standard FFT package.

Wavelets coefficients allow an estimation of the signal power. The global wavelet spectrum, defined as the time average over a series of p-wavelet powers, can be expressed as [8]:

$$\bar{W}^2(p) = \frac{1}{N} \sum_{k=0}^{N-1} |W_k(p)|^2 = \frac{1}{N} \sum_{k=0}^{N-1} |\tilde{X}_k \times \Psi_k(p)|^2 \tag{11}$$

A sum of products of Fourier coefficients calculated in a FFT32 routine for ADC data (x_n) in each clock cycle with pre-calculated Fourier coefficients of a reference wavelet gives an estimation of the signal power for selected type of the wavelet. Only a single FFT32 routine for the on-line calculation of Fourier coefficients for data is needed. Fourier coefficients for various wavelets can be calculated earlier and be available for final power estimation as constants.

A fundamental limitation for the on-line wavelet analysis in the FPGA is an amount of embedded DSP multipliers. A multiplication by an utilization of logic elements is rather inefficients. The Quartus® II environment for an Altera® FPGA programming provides parametrized FFT routines with various architectures: streaming, variable streaming, burst and buffered burst. However, all routines deliver the FFT coefficients in a serial form (Figure 4). No any Altera® routine allows calculating all FFT coefficients simultaneously.

If FFT coefficients are spread in time, the wavelet transform can be also calculated in a serial way (in a single clock cycle only a single pair of $\tilde{X}_n$ is multiplied by a single pair of ψ^*), however, a product will strongly depend on a relative position of $\tilde{X}_n$ and ψ^*. If the variables are shifted between themselves, even strong signal may give a negligible final contribution. Some additional procedure is needed, which could tune a wavelet transform regarding to the Fourier transform of ADC samples.

This problem can be automatically solved if all Fourier coefficients were provided simultaneously in each clock cycle. A synchronous multiplication with Fourier coefficients of wavelets would give required power estimation independently of any relatively configurations of these variables. The Fourier coefficients of selected wavelets are fixed, a sliding window of N ADC samples gives all Fourier coefficients in each clock cycle. This assures that for some set of samples (if a signal appears) the product of both transforms may give a significant contribution and may be used as a trigger.

The radio signal is spread in a time interval of an order of couple hundred nanoseconds, most of registered samples gave a time interval below 200 ns. The frequency window in the atmosphere, where a signal suppression is on an acceptable level (the atmosphere is relatively transparent) is ca. 30-80 MHz. According to the Nyquist theorem the sampling frequency should be at least twice higher than the maximal frequency in a investigated spectrum. The anti-aliasing filter should have the cut-off frequency of ca. 85 MHz. Taking into account some width of the transition range for the filter (from pass-band to stop-band) the final sampling frequency should not be lower than 180 MHz (200 MHz in our considerations). This frequency corresponds to 5 ns between rising edges of the clock.

The interval of 160 ns (estimated as sufficient time interval for radio signals) requires 32-point Fourier transform calculated in each clock cycle.

5. General algorithm

Let us consider a DFT $\tilde{X}$ of dimension N

$$\tilde{X}_k = \sum_{n=0}^{N-1} x_n W^{nk} \quad where \quad W = e^{-2i\pi/N} \quad and \quad k = 0, \ldots N-1 \tag{12}$$

If N is the product of two factors, with $N = N_1 N_2$, the indices n and k we can redefined as follows: $n = N_1 n_2 + n_1$,

where $n_2 = 0, \ldots, N_2\text{-}1$ and $n_1 = 0, \ldots, N_1\text{-}1$, $k = N_2 k_1 + k_2$, $k_2 = 0, \ldots, N_2\text{-}1$ and $k_1 = 0, \ldots, N_1\text{-}1$

$$\tilde{X}_{N_2 k_1 + k_2} = \sum_{n_1=0}^{N_1-1} W^{N_2 n_1 k_1} W^{n_1 k_2} \times \sum_{m_2=0}^{N_2-1} x_{N_1 m_2 + m_1} W^{N_1 m_2 k_2} \tag{13}$$

For the Radix-2 algorithm: $N = 2^t$, $N_1 = 2$ and $N_2 = 2^{t-1} = N/2$. Hence,

$$\tilde{X}_k = \sum_{n=0}^{N/2-1} (x_{2n} + W^k x_{2n+1}) W^{2nk} \tag{14}$$

If we split the sum as follows

$$\tilde{X}_k = \sum_{n=0}^{N/4-1} x_{2n} W^{2nk} + \sum_{n=N/4}^{N/2-1} x_{2n} W^{2nk} + + \sum_{n=0}^{N/4-1} x_{2n+1} W^{2nk} + \sum_{n=N/4}^{N/2-1} x_{2n+1} W^{2nk} \tag{15}$$

and afterwards, if we redefine indices and group the sums, we get

$$\tilde{X}_k = \sum_{n=0}^{N/4-1} (x_{2n} + (-1)^k x_{2(n+N/4)}) W^{2nk} + + W^k \Big(\sum_{n=0}^{N/4-1} (x_{2n+1} + (-1)^k x_{2(n+N/4)+1}) W^{2nk} \tag{16}$$

We can introduce the new set of variables defined for n = 0,...,N/4-1 as follows:

$$A_{2n} = x_{2n} + x_{2n+N/2} \qquad A_{2n+1} \quad = x_{2n+1} + x_{2n+1+N/2} \qquad (17)$$

$$A_{2n+N/2} = x_{2n} - x_{2n+N/2} \qquad A_{2n+1+N/2} = x_{2n+1} - x_{2n+1+N/2}, \qquad (18)$$

we get

$$\tilde{X}_k = \sum_{n=0}^{N/4-1} (A_{2n} + W^k A_{2n+1}) W^{2nk} \qquad (19)$$

$$\tilde{X}_k = \sum_{n=0}^{N/4-1} (A_{2n+N/2} + W^k A_{2n+1+N/2}) W^{2nk} \qquad (20)$$

for k even and odd respectively.

x_n represent signals in time domain. They can be easily available from outputs of shift registers clocked synchronously with the ADC. The DFT coefficients $\tilde{X}_k$ can be expressed by new set of variables A_m, Because A_m are simple linear combination of x_m, they can be calculated by typical adders (eqs.(17) and sub-tractors (eqs.(18) in a single clock cycle. The input values x_n are real and positive, since they represent the signal in the real time.

Coefficients of DFT in the real domain additional simplify due to the following symmetry:

$$Re(\tilde{X}_k) = +Re(\tilde{X}_{N-k}) \qquad Im(\tilde{X}_k) = -Im(\tilde{X}_{N-k}) \qquad (21)$$

The Radix-2 algorithm allows regrouping of inputs elements in the DFT expression in order to utilize some symmetries of Fourier coefficients. In a single step of the Radix-2 algorithm we can redefine the "new" set of variables by some mathematical expression of the "old" ones. This step will correspond to an elementary process in the pipeline chain. The redefinition of variables in eqs.(17) corresponds to the 1st stage of the pipeline. Splitting the sum (14) reduces of coefficient W^k set from 0,...,N-1 for input x_n to 0,...,$\frac{N}{2}$-1 for (19-20). The 1st stage utilizes the feature of the twiddle factors related to the 1st stage of the pipeline.

$$W_A = W^{N/2} = e^{-i\pi} = -1 \qquad (22)$$

So, the 1st stage can be implemented in a very simple way. The implementation of the multi-points algorithm requires multiple pipeline stages and apart from adders and sub-tractors also requires multipliers, which correspond to the W^k coefficients relating to the fractional "angle" $e^{-2ik\pi/N}$. The Radix-2 algorithm used in the next stage reduces again the abundance of W^k coefficients due to the next twiddle factors' related to the 2nd stage of the pipeline.

$$W_B = W^{N/4} = e^{-i\pi/2} = -i \qquad (23)$$

The W_B suggests the similar splitting structure in the 2nd pipeline stage as in the 1st one (minus in (23) as in (22)), however the imaginary unit imposes the DFT calculation separately for their real and imaginary parts. If we split the sum in (20) similar as in (15), we get for k = 0,2,...,N-2

$$\bar{X}_k = \sum_{n=0}^{N/8-1} [(A_{2n} + (-i)^k A_{2n+N/4}) W^{2nk} + +(A_{2n+1} + (-i)^k A_{2n+1+N/4}) W^{(2n+1)k}] \quad (24)$$

Let us consider separately two subset of odd indices: k=4n and k=4n+2 (n = 0,...,N/4-1)

$$\bar{X}_{4p} = \sum_{n=0}^{N/8-1} [(A_{2n} + A_{2n+N/4}) W^{8np} + (A_{2n+1} + A_{2n+1+N/4}) W^{(2n+1)4p}] \quad (25)$$

Notice that $\bar{X}_0$ and $\bar{X}_{N/2}$ are real.

$$\bar{X}_{4p+2} = \sum_{n=0}^{N/8-1} [(A_{2n} - A_{2n+N/4}) W^{8n(4p+2)} + (A_{2n+1} - A_{2n+1+N/4}) W^{(2n+1)(4p+2)}] \quad (26)$$

If we introduce new variables

$$\begin{align}
B_{2n} &= A_{2n} + A_{2n+N/4} & B_{2n+1} &= A_{2n+1} + A_{2n+1+N/4} & (27) \\
B_{2n+N/2} &= A_{2n} - A_{2n+N/4} & B_{2n+1+N/2} &= A_{2n+1} - A_{2n+1+N/4} & (28)
\end{align}$$

we get

$$\bar{X}_{4p} = \sum_{n=0}^{N/8-1} (B_{2n} + B_{2n+1} W^{4p}) W^{8np} \quad (29)$$

$$\bar{X}_{4p+2} = \sum_{n=0}^{N/8-1} (B_{2n+N/4} + B_{2n+1+N/4} W^{4p+2}) W^{8np+4n} \quad (30)$$

However, repeating the above procedure for odd indices related to the eq.(20) gives more complicated formulas, which cannot be simplified due to complex coefficients $W^{4(n+p)}$ (eq.31).

$$\bar{X}_{4p+q} = \sum_{n=0}^{N/8-1} W^{2n(4p+q)} [(A_{2n+N/2} + A_{2n+1+N/2} W^{4p+q}) \mp j(A_{2n+3N/4} + A_{2n+1+3N/4} W^{4p+q})] \quad (31)$$

where $\mp$ corresponds to q = 1,3 respectively. Next simplification is possible due to symmetries of trigonometric functions. However, general considerations give relatively complicated formulas, which seem to be unnecessary here.

6. 16-point algorithm

For N = 16 and odd indices we get

$$\bar{X}_{4p+q} = (A_8 \mp jA_{12}) + (i)^p(A_9(-1)^p W^q - iA_{15}W^{4-q}) +$$

$$+ (-i)^{\frac{q-1}{2}}(-1)^p W^2(A_{10} \mp iA_{14}) \pm (i)^p(A_{11}W^{4-q} - iA_{13}(-1)^p W^q) \tag{32}$$

Since of $W^4 = -i$, all coefficients can be expressed as a linear combination of the complex base W^1, W^2, W^3

$$W^1 = e^{-i\frac{\pi}{8}} = \cos(\frac{\pi}{8}) - i \cdot \sin(\frac{\pi}{8}) = \alpha - i \cdot \beta \tag{33}$$

$$W^2 = e^{-i\frac{\pi}{4}} = \cos(\frac{\pi}{4}) - i \cdot \sin(\frac{\pi}{4}) = \gamma(1 - i) \tag{34}$$

$$W^3 = e^{-i\frac{\pi}{8}} = \cos(\frac{3\pi}{8}) - i \cdot \sin(\frac{3\pi}{8}) = \beta - i \cdot \alpha \tag{35}$$

Symmetries in (33-35) allow the following simplification. Notice that

$$W^2(A_{10} \mp iA_{14}) = \gamma(A_{10} \mp A_{14})(1 - i) \tag{36}$$

$$A_9(-1)^p W^q - iA_{15}W^{4-q} = \widehat{\mathcal{X}}[(-1)^p A_9 - A_{15})] - i\widehat{\mathcal{Y}}[(-1)^p A_9 + A_{15})] \tag{37}$$

$$A_{11}W^q - iA_{13}(-1)^p W^{4-q} = \widehat{\mathcal{Y}}[A_{11} - (-1)^p A_{13})] - i\widehat{\mathcal{X}}[A_{11} + (-1)^p A_{13})] \tag{38}$$

where $\widehat{\mathcal{X}} = \alpha, \beta, \widehat{\mathcal{Y}} = \beta, \alpha$ for q = 1,3.

We can extend the set of variables $(27 - 28)$ also to odd indices of $\bar{X}$

$$B_{8,12} = A_{8,12} \qquad B_{9,15} = A_9 \pm A_{15} \qquad B_{10,14} = A_{10} \pm A_{14} \qquad B_{11,13} = {}_{11} \pm A_{13} \tag{39}$$

Formulae (39) show that the entire 2nd pipeline stage can be built also from only adders and sub-tractors. Signals $A_{8,12}$ have to be delayed in parallel shift registers in order to assure synchronization with adjacent ones.

For N = 16 the DFT coefficients can be expressed by the B_n variables as follows

$$
\begin{aligned}
Re(\bar{X}_0) &= & B_0 + B_1 + B_2 + B_3 \qquad & Re(\bar{X}_8) = B_0 - B_1 + B_2 - B_3 & (40)\\
Re(\bar{X}_4) &= & B_0 - B_2 \qquad & & (41)\\
Re(\bar{X}_2) &= & B_4 + \gamma \cdot (B_5 - B_7) \qquad & Re(\bar{X}_6) = B_4 - \gamma \cdot (B_5 - B_7) & (42)\\
Im(\bar{X}_2) &= & -B_6 - \gamma \cdot (B_5 + B_7) \qquad & Im(\bar{X}_6) = B_6 - \gamma \cdot (B_5 + B_7) & (43)
\end{aligned}
$$

$$Re(\bar{X}_{1,7}) = \quad B_8 \pm \alpha B_{15} + \gamma B_{14} \pm \beta B_{13} \qquad Re(\bar{X}_{3,5}) = \quad B_8 \pm \beta B_{15} - \gamma B_{14} \mp \alpha B_{13} \quad (44)$$

$$Im(\bar{X}_{1,7}) = \mp B_{12} - \beta B_9 \mp \gamma B_{10} - \alpha B_{11} \qquad Im(\bar{X}_{3,5}) = \pm B_{12} - \alpha B_9 \mp \gamma B_{10} + \beta B_{11} \quad (45)$$

The next, 3rd pipeline stage requires implementation of 10 multipliers calculating products from (44-45), 3 adders, 3 sub-tractors and 4 shift registers : according to the following formulae

$$C_0 = B_0 + B_2 \qquad C_1 = B_1 + B_3 \qquad C_2 = B_0 - B_2 \qquad C_3 = B_1 - B_3 \qquad (46)$$
$$C_5 = B_5 + B_7 \qquad C_7 = B_5 - B_7 \qquad (47)$$
$$C_4 = B_4 \qquad C_6 = B_6 \qquad C_8 = B_8 \qquad C_{12} = B_{12} \qquad (48)$$
$$C_{9A} = \alpha \cdot B_9 \qquad C_{11A} = \alpha \cdot B_{11} \qquad C_{13A} = \alpha \cdot B_{13} \qquad C_{15A} = \alpha \cdot B_{15} \qquad (49)$$
$$C_{9B} = \beta \cdot B_9 \qquad C_{11B} = \beta \cdot B_{11} \qquad C_{13B} = \beta \cdot B_{13} \qquad C_{15B} = \beta \cdot B_{15} \qquad (50)$$
$$C_{10} = \gamma \cdot B_{10} \qquad C_{14} = \gamma \cdot B_{14} \qquad (51)$$

The 4th stage utilizes 2 multipliers, 5 adders, 5 sub-tractors and 4 shift registers

$$D_{5,7} = \gamma \cdot C_{5,7} \qquad D_{0,1} = C_0 \pm C_1, \qquad D_{8,14} = C_8 \pm C_{14}, \qquad D_{10,12} = C_{10} \pm C_{12} \qquad (52)$$
$$D_9 = C_{9A} - C_{11B} \qquad D_{11} = C_{11A} + C_{9B} \qquad D_{15} = C_{13A} - C_{15B} \qquad D_{13} = C_{13B} + C_{15A} \quad (53)$$
$$D_{2,3,4,6} = C_{2,3,4,6} \qquad (54)$$

Finally, the set of DFT $\bar{X}_k$ in the 5th stage coefficients is calculated by 6 adders and 6 sub-tractors supported by 4 shift registers.

$$Re\bar{X}_{0,4,8} = D_{0,2,1} \qquad Re\bar{X}_{1,7} = D_8 \pm D_{15} \qquad Re\bar{X}_{2,6} = D_4 \pm D_7 \qquad Re\bar{X}_{3,5} = D_{14} \mp D_{13} \quad (55)$$
$$Im\bar{X}_4 = -D_3, \qquad Im\bar{X}_{1,7} = \mp D_{10} - D_{11} \quad Im\bar{X}_{2,6} = \mp D_6 - D_5 \quad Im\bar{X}_{3,5} = \mp D_{12} - D_9 \quad (56)$$

Figure 9 shows the internal structure of the 16-point FFT algorithm. As shown later (compare Figure 15), this algorithm is higher optimized in comparison to a pure DiF approach.

The algorithm with the 16-point FFT was tested on the 3rd generation of the Auger surface detector Front-End Board (Figure 10) [9], [10]. The 1st [12] and the 2nd [13] generations of the Front-End Boards could not support the FFT algorithms due to a lack of FPGA resources. However, the FFT algorithm seems to be less efficient than the DCT approach. The DCT algorithm implemented into the 4th generation Front-End with the CycloneIII® EP3C40F324C7 (Figure 11) passed successfully tests on the field recognizing short peaks with an exponentially attenuated tails characteristically for signals generated by very inclined showers.

7. 32-point FFT algorithm

For 32-point Discrete Fourier Transform $\bar{X}$

$$\bar{X}_{k=0,\dots,31} = \sum_{n=0}^{31} x_n e^{-i\pi kn/16} \tag{57}$$

where x_n as samples from an ADC chip are real. The formula (57) can be split on two or more parts by rearranging of the sum and indices. The standard approach of a formula simplification is a Radix-2 Decimation-in-Time (DiT) (Figure 1a) or Decimation-in-Frequency algorithm (DiF) (Figure 1b) one.

For Radix-2 DiT, we get the formula 3. N-point DFT can be easily split on two N/2-point transforms. Outputs from DFT procedures are complex. So, a calculation of final DFT coefficients by using DiT algorithm requires the complex multiplication for final merging

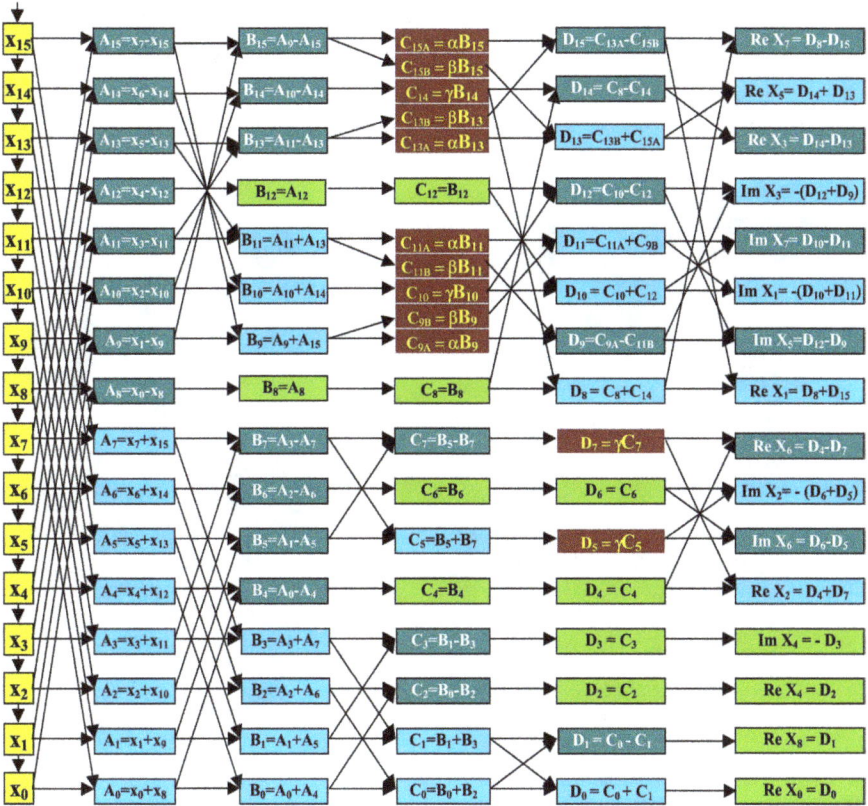

Figure 9. A global pipeline internal structure of FFT_16 [11] .

Figure 10. The 3rd generation of the Front-End Board with Cyclone® FPGA EP1C12Q240I7 used in more than 800 surface detectors in the Pierre Auger Observatory on the Argentinean pampas. The EP1C12Q240I7 does not contain DSP blocks. The multipliers had to be implemented from logic elements according to the scheme on the Figure 9.

Figure 11. The 4th generation of the Front-End Board with CycloneIII® FPGA EP3C40F324I7. The EP3C40F324I7 contains DSP blocks and it is possible to implement even a sophisticated algorithm like DCT engines for a recognition of horizontal or very inclined showers. This board has been used also for preliminary testing of the wavelet trigger and the signal filtering based on a chain: FFT+Median filter+iFFT.

data from parallel DFT procedures with a lower order i.e. multiplication of twiddle factors W_N^k :

$$W_N^k = e^{-i\frac{2\pi k}{N}} \tag{58}$$

by $G[k]$ and $H[k]$ in Figure 1. Altera® provides a library routine of the complex multiplication in the FPGA (Figure 12a), however, for i.e. 16x16 bits operation requires 6 DSP embedded 9x9 multipliers even in most economical (canonical) mode. Generally, the complex multiplication in the FPGA is rather resource-spendthrift and if possible it should be replaced by the multiplication of real variables.

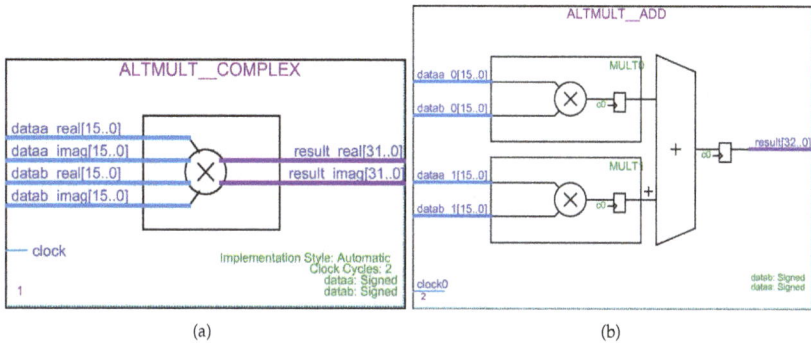

(a)

(b)

Figure 12. The ALTMULT_COMPLEX and ALTMULT_ADD procedures provided by Altera®. For a calculation of $|W_k|^2$, dataa_0 = datab_0 and dataa_1 = datab_1. The ALTMULT_ADD routine requires 4 DSP 9 × 9 multipliers. It is used in E_bin pipeline stage for odd FFT indices (Figure 17). Inputs $dataa\_0, 1$ are used for C_k, $datab\_0, 1$ for constants α, β, ξ, η, σ and ρ. The routine requires two clock cycles. Sub-products are registered in MULT0 and MULT1 DSP blocks, respectively. Thus, the sum appears in the next register stage.

For the Radix-2 DiF, we get the formula 4. The standard Radix-2 Decimation-in-Frequency algorithm (DiF) rearranges the DFT equation (57) into two parts: computation of the even-numbered discrete-frequency indices $X(k)$ for k=[0,2,4,...,30] and computation of the odd-numbered indices k=[1,3,5,...,31]. This corresponds to a splitting N-point DFT into two k = N/2-point routines. The first corresponding twiddle factor is $e^{-i\frac{2\pi}{N}\frac{N}{2}} = -1$. The first operations are simple sums and subtractions of real variables (see Figure 1b). Each operation related to the consecutive twiddle factor will be performed in a single clock cycle.

The algorithm of Decimation in Frequency used for the 32-point DFT allows splitting eq. 57 as follows:

$$\bar{X}_{k=2p} = \sum_{n=0}^{15} A_n e^{-i\pi kn/8} \quad \Rightarrow \quad FFT16_{even} \tag{59}$$

$$\bar{X}_{k=2p+1} = \sum_{n=0}^{15} A_{n+16} e^{-i\pi(2p+1)n/16} \tag{60}$$

$$A_n = x_n + x_{n+16} \qquad A_{n+16} = x_n - x_{n+16} \qquad n = 0, 1, ..., 15 \tag{61}$$

The next twiddle factors are:

$$W_B = e^{-i\pi/2} = -i \qquad W_C = e^{-i\pi/4} = \gamma(1-i) \qquad W_D = e^{-i\pi/8} = \alpha - i\beta \qquad (62)$$

$$W_E = e^{-i\pi/16} = \xi - i\eta \qquad W_F = e^{-3i\pi/16} = \sigma - i\rho \qquad (63)$$

$$\gamma = cos(\pi/4) \qquad \alpha = cos(\pi/8) \qquad \xi = cos(\pi/16) \qquad \sigma = cos(3\pi/16) \qquad (64)$$

$$\beta = sin(\pi/8) \qquad \eta = sin(\pi/16) \qquad \rho = sin(3\pi/16) \qquad (65)$$

The scheme developed on the pure Radix-2 Decimation in Frequency algorithm is presented in Figure 15. The algorithm takes into account only FFT coefficients with indices k = 0,...,15. Due to real input data ($x_{0,...31}$) the higher FFT coefficients have well known symmetry : $Re\tilde{X}_{32-n} = Re\tilde{X}_n$ and $Im\tilde{X}_{32-n} = -Im\tilde{X}_n$ ($n > 0$). The calculation of $\tilde{X}_{0,...15}$ according the pure Radix-2 DiF algorithm requires 8 pipeline stages. For $\tilde{X}_{0,4,8,12,16}$ 2 pipeline stages are necessary only for a synchronization.

According to the eq. (59) all $\tilde{X}_{0,2,4,...,14}$ with even indices could be calculated by the algorithm presented in [11]. Variables x_n in Figure 2 in [11] were be replaced by variable of A_n according to eq. (61). An application of a modified algorithm reduces an amount of 9×9 multipliers from 12 to 10 only and shorten a pipeline chain on stages (the last 2 stages are simple registers for synchronization) (see Figure 16).

Let us notice that for the odd indices stages B and C for k=16,...,19 and k = 24,...27 are pure delay lines, while for neighboring indices k=20,...,23 and k = 28,...31 mathematical operation are performed in a cascade. Let us multiply $A_{16,...19}$ and $A_{24,...27}$ by the factor $\lambda = \gamma^{-1}$. Then to adjust variables in the C stage for odd FFT coefficients (for k = 20,21,22,33 and k = 28,29,30,31)

$$C_k = \lambda \times \gamma = B_k \qquad (66)$$

Thus, by such a redefinition, The C stage for the odd FFT indices is a pure pipeline stage. It can be removed with one of pipeline stage for the even FFT indices. In order to come back to the correct values coefficients in F stage can be simple redefined

$$\alpha' = \gamma \times \alpha \quad \beta' = \gamma \times \beta \quad \xi' = \gamma \times \xi \quad \eta' = \gamma \times \eta \quad \sigma' = \gamma \times \sigma \quad \rho' = \gamma \times \rho \qquad (67)$$

but for indices k = 16, 20, 24 and 28 we have to use additional 4 multipliers. Nevertheless, at this cost we save one pipeline stage and depending on a width of buses in the final FFT coefficients we save at least of 1000 logic elements.

We can save a next pipeline stage and more ca. 1000 logic elements but again at the cost of additional utilized multipliers. The algorithm used for indices k = 2,6,10,14 is neither Decimation in Time nor Decimation in Frequency. The eq. (60) can be rewritten as follows:

$$\bar{X}_{k=2p+1} = A_{16} + A_{24} + \sum_{n=1}^{7} (cos\phi B_{n+24} - isin\phi B_{n+16}) \qquad \phi = \frac{\pi(2p+1)n}{16}$$

$$B_{16,24} = A_{16,24}$$
$$B_{n+16} = A_{n+16} + A_{32-n} \qquad n = 1, ..., 7 \tag{68}$$
$$B_{n+24} = A_{n+16} - A_{32-n} \qquad n = 1, ..., 7$$

A development of the algorithm according to eq. (22) would allow a reduction of the next pipeline stage, but unfortunately at the cost of additional 16 ALTMULT_ADD routines (64 DSP blocks) (see Figure 12b).

If the speed is not a factor, sums of products in the E_bin routine can be performed in a single clock cycle instead of two cycles as shown on Figure 17. Thus, $D_{26,20,24,28}$ shift registers are not necessary and can be removed. A shorter chain for the odd indices allows removing also the last pipeline chain for even indices and saving totally more than 1000 logic elements without the cost of additional multipliers. However, we should be aware, that a registered performance significantly decreases from ca. 220 MHz to only 158 MHz for EP3C120F780C7.

8. Wavelet power calculation

The reference wavelets are real, however, their Fourier transform are already complex. An elementary product from eq. (11) is a product of two complex numbers: Fourier coefficients of data and Fourier coefficient of a reference wavelet. The simplest way is to use the Altera® routine from Figure 12. However, due to a fact that the wavelet Fourier coefficients are predefined constant and finally we are going to calculate a module of a complex product as well as $|W \times \Psi|^2 = |W|^2 \times |\Psi|^2$, we can calculate only $|W|^2$ and next as real number multiply by a next real $|\Psi|^2$.

The FFT32 routine from Figure 17 utilizes 96 DSP 9×9 multipliers. For a calculation of $|W_k|^2$, the ALTMULT_ADD routine utilizes 4 DSP 9×9 multipliers for each index k, totally 60 ($|W_0|$ is trivial). $|W_k|^2 \times |\Psi_k|^2$ products use next 30 DSP 9×9 multipliers.

This algorithm can be implemented only in very powerful modern FPGA chips. The FPGA families ACEX® or Cyclone®, currently used in surface detectors, do not contain DSP blocks. Even CycloneIII® EP3C40F324I7 [14] used for DCT trigger tests ([15], [16]) does not consist of a sufficient amount of DSP blocks to implement the wavelet trigger.

The biggest FPGAs from the CycloneIII® EP3C120F780C7 (Figure 13) and CycloneIV® EP4CE115F29C7 (Figure 14) families with 576 and 532 DSP multipliers, respectively, allow the implementation of the FFT32 routine (96 DSP blocks) + "Module" block (60 DSP blocks) + 14 or 11 "engines" (30 DSP blocks each) simultaneously for a power estimation of 14 or 11 various reference wavelets, respectively.

Table 2 shows results calculated and measured in the Altera®'s development kit DK-DSP-3C120N for various variants for Cyclone® III EP3C120F780C7 (a heart of this development kit). Results do not fully agree with our expectations. A reduction of a single pipeline stage decreases a resource occupation on ca. 410 (not 640) logic elements. This

Figure 13. The test system based on a development kit with Altera® CycloneIII® FPGA EP3C120F780C7 supported by two daughter boards: AD/DA Data Conversion Card (left) with two ADCs (150MHz sampling) and two DACs (250 MHz), as well as the Industrial Communication Board (ICB-HSMC)(right) allowing a connection via the galvanic isolated RS485 ports.

Figure 14. Test system based on a development kit with Altera® CycloneIV® EP4CE115F29C7 supported by ICB-HSMC daughter board.

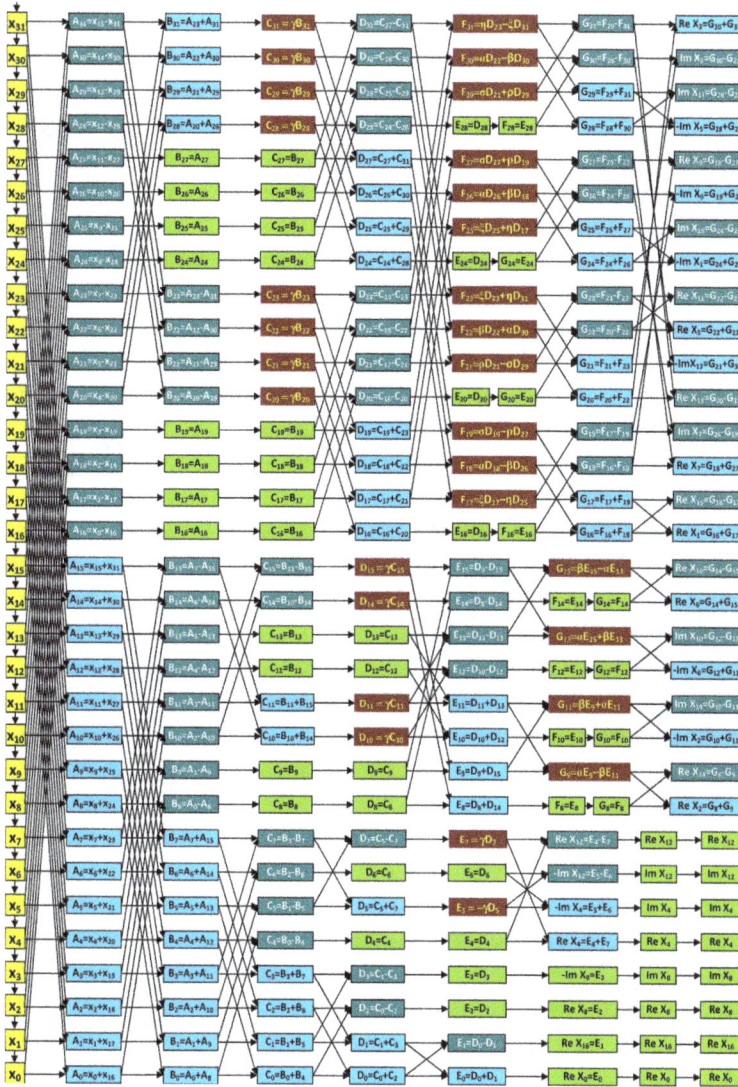

Figure 15. An internal structure of the FFT32 FPGA procedure. The algorithm uses 14 single clock-cycle multipliers (i.e. $F_7 = \gamma D_7$ - each utilizes two 9x9 DSP multipliers) and 16 two clock-cycles multipliers (i.e. $N_7 = \beta G_7 - \alpha H_7$ - each utilizes four 9x9 DSP multipliers). Totally, the algorithm needs 92 9x9 DSP multipliers.

may be due to optimization processes performed by the Quartus® II compiler to achieve the maximal registered performance. Nevertheless, for all comparisons the speed in the "optimized" design is higher than for the "pure DiF". For a development of wavelet engines the "optimized" variant has been selected as potentially faster.

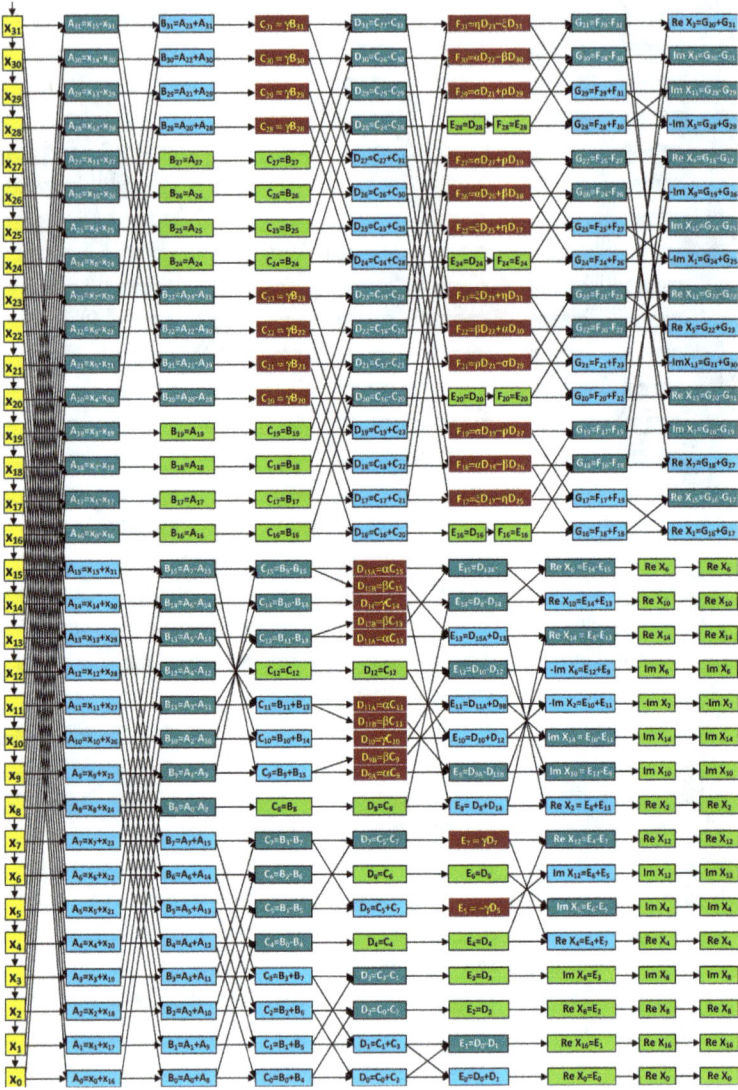

Figure 16. A modified structure for $\tilde{X}_{2,6,10,14}$ allowing a reduction of two 9×9 multipliers and shorten a pipeline chain on two stages (shift registers still used for synchronization).

The Quartus® II compiler estimated a power consumption for the core, a static mode and for the I/O sector. As possible, the output of registers were multiplexed to reduce an amount of output pins (all pins were achieved to HSMC connectors on the development board). According to expectation the power for I/O increase ca. linear with a number of used pins. The static power consumption is on a level ~100 mW. It is a reasonable level. In comparison

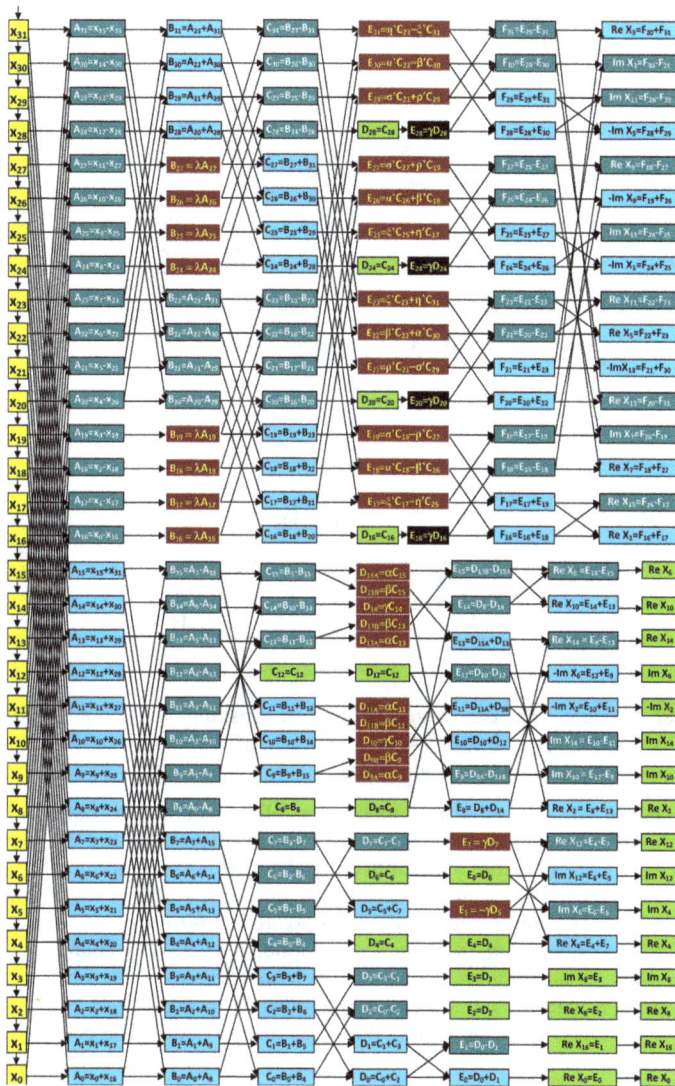

Figure 17. An optimized structure with a reduced a single pipeline stage at the cost of only 4 additional multipliers (8 DSP 9×9 blocks).

the Stratix® III chips have a huge power consumption in a static mode of ~600 mW, which significantly limited their application in systems supplied from solar panels. The power consumption for the "optimized" variant is ~35 mW higher than for the "pure DiF" solution. The additional 35 mW is not a factor, if it allows an improvement of the safety margin for the register performance. The EP3C120F780C7 allows the implementation of 14 wavelet engines.

config	logic elements	DSP	pins	fmax (MHz)	power sim. core (mW)	power sim. I/O (mW)	power mea. core (mW)	power mea. I/O (mW)
pure FFT32 pure DiF	4712 - 4%	92 - 16%	25 - 5%	236	557	65	580	170
pure FFT32 optimized	4301 - 4%	96 - 16%	25 - 5%	241	589	65	588	170
plus Module pure DiF	4990 - 4%	152 - 26%	25 - 5%	245	750	68	779	170
plus Module opt	4541 - 4%	156 - 27%	25 - 5%	246	787	68	783	170
1 wavelet - 24-bit opt	4726 - 4%	186 - 32%	29 - 5%	235	861	88	840	240
1 wavelet - 16-bit opt	4265 - 4%	186 - 32%	21 - 4%	228	814	66	790	170
4 wavelets - 16-bit opt	5478 - 5%	276 - 48%	81 - 15%	212	1134	215	1040	240
8 wavelets - 16-bit opt	5967 - 5%	396 - 69%	161 - 30%	204	1591	413	1363	360
12 wavelets - 16-bit opt	7060 - 6%	516 - 90%	241 - 45%	208	1980	612	1691	478

Table 2. Resources Occupancy and Power Consumption for the Cyclone III FPGA - EP3C120F780C7 for 200 MHz PLL Global Clock

Family	FPGA	config	logic elements	DSP	Slack Fast (ns)	Fmax Slow 0°C (MHz)	Fmax Slow 85°C (MHz)
Cyclone IV	EP4CE115F29C7	12 wavelets	7120 - 6%	516 - 97%	2.594	234	214
Cyclone V	5CGXFC7D6F31C6	12 wavelets	6933 - 6%	156 - 100%	2.111	195	196
Cyclone V	5CGXFC7D6F31C6	4 wavelets	3177 - 3%	111 - 71%	2.169	227	228

Table 3. Resources Occupancy and Timing for the Cyclone® IV and Cyclone® V FPGAs for 200 MHz PLL Global Clock

A design with 12 engines has been tested. The power consumption is on a level of ∼100-110 mW per the wavelet engine. It gives ∼2 W for 12 engines. This may be a challenge for an autonomous system supplied from solar panels.

Measurements of the power consumption for all considered variants show some discrepancies with simulations. The Measured power consumption for the core increases slower with new wavelet engines than simulations show. Almost 300 mW lower power taken by the FPGA (in comparison to simulations) for 12 engines gives optimistic predictions for the future applications. The power consumption for the core seems to be ca 15% overestimated in simulations. On the other hand, the power consumption for the I/O section is unpredictable much higher than for simulations. However, differences decrease with a higher amount of active pins. This, actually, is not a problem, I/O pins have been attached for test only. In real applications almost all variables are utilized as internal nodes. The power optimization is highly recommended.

Designs have been also implemented into EP4CE115F29C7 from the Cyclone® IV family of Altera® used in a development kit DE2-115 (Terasic). According to the Altera®'s specification, the power consumption for the Cyclone® IV family is 30% less than for the Cyclone® III one. However, the Terasic's development kit does not contain any system allowing a measurement of the power consumption on the board.

For the Cyclone®IV EP4CE115F29C7 timing shows a pretty good safety margin.

9. Spectral leakage

For the serial FFT processing the input data have to be chopped into blocks to be processed by the FFT routine. If signal pulses are located close to the border of a block, aliasing occurs. It manifests by a spurious contribution in the opposite border of the block and in the neighboring block as well. This effect may cause spurious pulses and has to be eliminated. The problem can only be solved, without introducing dead time between the blocks, by using an overlapping routine. Therefore the FFT engines have to be over-clocked. Practically for 1024-length blocks aliasing is reduced to a negligible level, when two blocks are overlapped during 64 time bins [7]. For parallel data processing, when all set of Fourier coefficients is available for each clock cycle FFT engines aliasing can be eliminated by a selection of a set of these coefficients not significantly affected. If a reduced set of Fourier coefficients is taken for data analysis, there is a possibility to increase an amount of wavelet engines for simultaneously analysis of more reference wavelets.

10. Design improvement

The new Altera®'s FPGA family - Cyclone® V provides the industry's lowest system cost and power, along with performance levels that make the device family ideal for high-volume applications. A total power consumption compared with the previous generation (Cyclone® IV) is reduced up to 40%.

The biggest FPGA from the Cyclone® V E family 5CEA9 (with logic only without ARM-based hard processor system (HPS) contains 684 DSP 18×18 multipliers + 342 variable-precision DSP blocks (DSP blocks include three 9×9, two 18×19, and one 27×27 multiplier). Assuming roughly a single 18×18 multiplier is equivalent to two 9×9 ones, 5CEA9 could implement FFT32 + 18 engines for various 18 reference wavelets. However, the 5CEA9 FPGA is not yet available even for compilation (latest Quartus® II version 12.0). An estimation for 12 wavelet engines for 5CGXFC7 FPGA shows the scarcity of DSP blocks. Fast multipliers are replaced by logic elements, which significantly reduced the register performance for slow models, below our requirements. Nevertheless, if all multiplication all implemented in the fast DSP blocks (see Table 3 Cyclone® V for 4 wavelet engines only), timing is perfect. This allows anticipating also a perfect timing for the 5CEA9 chip. Expected total 58% less power consumption (30% and next 40% of reduction of power consumption from Cyclone® III to Cyclone® V) gives an estimation of 840 mW for 12 and 1260 mW for 18 wavelet engines, respectively. It is acceptable level of the power consumption for currently used supply systems in cosmic rays experiments.

11. Conclusions

The FFT32 routine has been successfully and cost-effectively implemented into the powerful FPGA EP3C120F780C7 from the Cyclone® III family used in a development kit DK-DSP-3C120N (Altera®) and EP4CE115F29C7 from the Cyclone® IV family of Altera® used in a development kit DE2-115 (Terasic).

Nevertheless, both FPGAs from Cyclone® III and IV families were treated as an engineering test platform for a development of the algorithm and a timing verification. The prototype targeted for real detection of radio signals coming from air showers developing in the atmosphere will be built on a basis of Cyclone® V family.

The Pierre Auger Observatory is worldwide the largest cosmic ray experiment and operates its southern observatory since 2004. Results from Auger South have shown that the spectrum of cosmic rays has a characteristic cut-off at ca. 50 EeV; that events with higher energy arrive anisotropic; and that cosmic rays at highest energies are probably built from heavy nuclei. These results define the requirements for the next generation experiment: it needs to be considerably increased in size, it needs a better sensitivity to composition, and it should cover the full sky. Such a facility, AugerNext, will be specified within the next 3-5 years.

The innovative research studies are needed in order to prepare an AugerNext proposal fulfilling the demands. Requested resources are primarily focused in the areas: consolidation of the detection of cosmic rays using MHz radio antennas, proof-of-principle of cosmic rays microwave detection, testing the large-scale application of new generation photo sensors, generalization of data communication techniques, and developing a new technique of muon detection with surface arrays. Studies for such a next generation cosmic ray experiment and the utilization of detection methods are principle elements of the ASPERA /ApPEC roadmaps.

ASPERA-2 [18] supporting these efforts is the project of "The Innovative Research Studies for the Next Generation Ground-Based Ultra-High Energy Cosmic-Ray Experiment: AugerNext".

Acknowledgements

This chapter has been supported by the National Center of Researches and Development (Poland) under the Grant No. ERA-NET-ASPERA/02/11.

Author details

Zbigniew Szadkowski

* Address all correspondence to: zszadkow@kfd2.phys.uni.lodz.pl

University of Łódź, Department of Physics and Applied Informatics, Faculty of High-Energy Astrophysics, Łódź, Poland
The author is the member of the Pierre Auger Collaboration since 1999

References

[1] H. R. Allan, "Radio emission from extensive air showers", Prog. in Elem. Part. and Cos. Ray Phys., vol. 10, pp. 171, 1971.

[2] H. Falcke, P.W. Gorham, "Detecting radio emission from cosmic ray air showers and neutrinos with a digital radio telescope", Astropart. Phys. vol. 19, pp. 477-494, July 2003, ISSN: 0927-6505

[3] J. Abraham et al., [Pierre Auger Collaboration], "Properties and Performance of the Prototype Instrument for the Pierre Auger Observatory", Nucl. Instr. Meth., ser. A, vol. 523, pp. 50-95, May 2004, ISSN: 0168-9002

[4] X. Bertou, P. Billoir, O. Deligny, C. Lachaud, A. Letessier-Selvon, "Tau Neutrinos in the Auger Observatory : A New Window to UHECR Sources ", astro-ph/ 0104452.

[5] http://www.altera.com/products/ip/dsp/transforms/m-ham-fft.html

[6] G. A. Dulk, W. C. Erickson, R. Manning, and J.-L. Bougeret, "Calibration of low-frequency radio telescopes using the galactic background radiation", *A&A*, vol. 365, pp. 294-300, Jan. 2001, ISSN (Print Edition): 0004-6361, ISSN (Electronic Edition): 1432-0746

[7] A. Schmidt, H. Gemmeke, A. Haungs, K-H Kampert, C. Rühle, Z. Szadkowski, "An FPGA Based Trigger and RFI Filter for Radio Detection of Cosmic Rays", *IEEE Trans. Nucl. Science.* vol. 58, no 4, pp. 1621-1627, Aug. 2011, ISSN: 0018-9499

[8] Z.Ge, Significance tests for the wavelet power and the wavelet power spectrum, *Ann. Geophys.*, 25, pp. 2259-2269, 2007, ISSN: 1593-5213

[9] Z. Szadkowski, K-H Becker, K-H Kampert, "Development of a new first level trigger for the surface array in the Pierre Auger Observatory based on the Cyclone™ Altera® FPGA", *Nucl. Instr. Meth*, vol. A545, pp. 793-802, June 2005, ISSN: 0168-9002

[10] Z. Szadkowski, et. al, "The 3rd Generation Front-End Cards of the Pierre Auger Surface Detectors: Test Results and Performance in the Field", *Nucl. Instr. Meth*, vol. A606, pp. 439-445, July 2009, ISSN: 0168-9002

[11] Z. Szadkowski, "16-point Discrete Fourier Transform based on the Radix-2 FFT algorithm implemented into Cyclone™ FPGA as the UHECR trigger for horizontal air showers", *Nucl. Instr. Meth*, vol. A560, pp. 309-316, May 2006, ISSN: 0168-9002

[12] Z. Szadkowski, D. Nitz, "Implementation of the first level surface detector trigger for the Pierre Auger Observatory Engineering Array", *Nucl. Instr. Meth*, vol. A545, pp. 624-631, June 2005 ISSN: 0168-9002

[13] Z. Szadkowski, "The concept of an ACEX® cost-effective first level surface detector trigger in the Pierre Auger Observatory", *Nucl. Instr. Meth*, vol. A551, pp. 477-486, Oct. 2005, ISSN: 0168-9002

[14] Z. Szadkowski. "Trigger Board for the Auger Surface Detector with 100 MHz Sampling and Discrete Cosine Transform". *IEEE Trans. Nucl. Science*, vol. 58, no 4, pp. 1692-1700, Aug. 2011, ISSN: 0018-9499

[15] Z. Szadkowski, "A spectral 1st level FPGA trigger for detection of very inclined showers based on a 16-point Discrete Cosine Transform for the Pierre Auger Experiments", *Nucl. Instr. Meth*, vol. A606, pp. 330-343, July 2009, ISSN: 0168-9002

[16] Z. Szadkowski "An optimization of 16-point Discrete Cosine Transform Implemented into a FPGA as a Design for a Spectral First Level Surface Detector Trigger in Extensive Air Shower Experiments", *Applications of Digital Signal Processing*, InTech, ISBN 978-953-307-406-1, Croatia, 2011.

[17] Z. Szadkowski, "FPGA implementation of the 32-point DFT for a wavelet trigger in cosmic rays experiments", Real Time Conference, Berkeley, CA, June 2012

[18] http://www.aspera-eu.org/

Permissions

The contributors of this book come from diverse backgrounds, making this book a truly international effort. This book will bring forth new frontiers with its revolutionizing research information and detailed analysis of the nascent developments around the world.

We would like to thank Gustavo A. Ruiz and Juan A. Michell, for lending their expertise to make the book truly unique. They have played a crucial role in the development of this book. Without their invaluable contribution this book wouldn't have been possible. They have made vital efforts to compile up to date information on the varied aspects of this subject to make this book a valuable addition to the collection of many professionals and students.

This book was conceptualized with the vision of imparting up-to-date information and advanced data in this field. To ensure the same, a matchless editorial board was set up. Every individual on the board went through rigorous rounds of assessment to prove their worth. After which they invested a large part of their time researching and compiling the most relevant data for our readers. Conferences and sessions were held from time to time between the editorial board and the contributing authors to present the data in the most comprehensible form. The editorial team has worked tirelessly to provide valuable and valid information to help people across the globe.

Every chapter published in this book has been scrutinized by our experts. Their significance has been extensively debated. The topics covered herein carry significant findings which will fuel the growth of the discipline. They may even be implemented as practical applications or may be referred to as a beginning point for another development. Chapters in this book were first published by InTech; hereby published with permission under the Creative Commons Attribution License or equivalent.

The editorial board has been involved in producing this book since its inception. They have spent rigorous hours researching and exploring the diverse topics which have resulted in the successful publishing of this book. They have passed on their knowledge of decades through this book. To expedite this challenging task, the publisher supported the team at every step. A small team of assistant editors was also appointed to further simplify the editing procedure and attain best results for the readers.

Our editorial team has been hand-picked from every corner of the world. Their multi-ethnicity adds dynamic inputs to the discussions which result in innovative outcomes. These outcomes are then further discussed with the researchers and contributors who give their valuable feedback and opinion regarding the same. The feedback is then collaborated with the researches and they are edited in a comprehensive manner to aid the understanding of the subject.

Apart from the editorial board, the designing team has also invested a significant amount of their time in understanding the subject and creating the most relevant covers. They scrutinized every image to scout for the most suitable representation of the subject and create an appropriate cover for the book.

The publishing team has been involved in this book since its early stages. They were actively engaged in every process, be it collecting the data, connecting with the contributors or procuring relevant information. The team has been an ardent support to the editorial, designing and production team. Their endless efforts to recruit the best for this project, has resulted in the accomplishment of this book. They are a veteran in the field of academics and their pool of knowledge is as vast as their experience in printing. Their expertise and guidance has proved useful at every step. Their uncompromising quality standards have made this book an exceptional effort. Their encouragement from time to time has been an inspiration for everyone.

The publisher and the editorial board hope that this book will prove to be a valuable piece of knowledge for researchers, students, practitioners and scholars across the globe.

List of Contributors

Keiichi Funaki
Computing & Networking Center, University of the Ryukyus, Okinawa, Japan

Takehito Higa
Graduate School of Engineering and Science, University of the Ryukyus, Okinawa, Japan

Alexey Petrovsky, Maxim Rodionov and Alexander Petrovsky
Department of Computer Engineering, Belarusian State University of Informatics and Radioelectronics,
Minsk, Belarus

Jian Wang
Wuhan National Laboratory for Optoelectronics, College of Optoelectronic Science and Engineering, Huazhong University of Science and Technology, Wuhan, Hubei, China

Alan E. Willner
Department of Electrical Engineering, University of Southern California, Los Angeles, California, USA

Jitendra Nath Roy
Department of Physics, National Institute of Technology, Agartala, Jirania, Tripura, India

Tanay Chattopadhyay
Mechanical operation (stage-II), Kolaghat Thermal Power station, WBPDCL, West Bengal, India

Jitendra Nath Roy
Department of Physics, National Institute of Technology, Agartala, Jirania, Tripura, India

Tanay Chattopadhyay
Mechanical operation (stage-II), Kolaghat Thermal Power station, WBPDCL, West Bengal, India

José Parera-Bermúdez, Javier Casajús-Quirós and Igor Arambasic
Department of Signals, Systems and Radiocommunications, Polytechnic University of Madrid, Spain

Stamos Katsigiannis and Dimitris Maroulis
Department of Informatics and Telecommunications, National and Kapodistrian University of Athens, Athens, Greece

Georgios Papaioannou
Department of Informatics, Athens University of Economics and Business, Athens, Greece

Georgios Georgis, George Lentaris and Dionysios Reisis
Electronics Laboratory, Physics Deparment, National and Kapodistrian University of Athens (NKUA), Greece

Pavel Karas and David Svoboda
Centre for Biomedical Image Analysis, Faculty of Informatics, Masaryk University, Brno, Czech Republic

Alan T. Murray and Steven R. Weller
School of Electrical Engineering and Computer Science, University of Newcastle, Callaghan, NSW 2308, Australia

Daniele Peri and Salvatore Gaglio
DICGIM - University of Palermo, Italy, ICAR - CNR, Palermo, Italy

Baba Tatsuro
Toshiba Medical Systems Corporation, Japan

Zbigniew Szadkowski
University of Łód´z, Department of Physics and Applied Informatics, Faculty of High-Energy Astrophysics, Łód´z, Poland